Saline Soil-based Agriculture by Halotolerant Microorganisms

Manoj Kumar • Hassan Etesami
Vivek Kumar
Editors

Saline Soil-based Agriculture by Halotolerant Microorganisms

Springer

Editors
Manoj Kumar
Department of Life Sciences
Central University of Jharkhand
Ranchi, Jharkhand, India

Hassan Etesami
Department of Soil Science
University of Tehran
Tehran, Iran

Vivek Kumar
Himalayan School of Biosciences
Swami Rama Himalayan University,
Jolly Grant,
Dehradun, Uttarakhand, India

ISBN 978-981-13-8334-2 ISBN 978-981-13-8335-9 (eBook)
https://doi.org/10.1007/978-981-13-8335-9

This Springer imprint is published by the registered company Springer Nature Singapore Pte Ltd.
The registered company address is: 152 Beach Road, #21-01/04 Gateway East, Singapore 189721, Singapore

Preface

Food security is one of the fundamental needs that cannot be compromised by any healthy society. The widespread increases in environmental damage and human populations have an adverse effect on global food production, which may soon be mixed with insufficient food before the global swelling population. To meet the additional food demand, a significant increase (an estimated 50%) in yields of major food crops is required for the projected population (around ten billion people) by the next 50 years. The world's population is constantly increasing, while agricultural soils are constantly decreasing due to soil salinity. The costs associated with this stress are potentially enormous (US$12 billion per annum globally) and rising. Due to the decrease in the level of fertile land, saline soil-based agriculture has been developing rapidly in recent years. A restriction to this agricultural approach is low salt tolerance of agricultural crops and trees. In such saline soils, the plants that are salt-resistant can produce significant yields. To sustain future agricultural production in salt-affected soils, we must be able to develop salt-tolerant food and fiber plants, which can successfully grow in salt-affected areas. Developing salt-tolerant crops by traditional breeding and genetic engineering has been a much desired scientific goal but with little success. An alternative strategy to improve crop salt tolerance may be to introduce salt-tolerant microbes that enhance crop growth. As soil salinity-resistant microorganisms have increased the growth of many different crops grown in salt-affected soils, this approach may succeed where it has proved difficult to develop salt-tolerant germplasm. Identifying and making use of salinity-resistant microorganisms not only meet our requirement of salt-tolerant crops but also reduce pressure on arable lands. Salinity-tolerant microorganisms also offer excellent models for understanding stress tolerance and mechanisms of adaptation and response that can subsequently be engineered to crop plants to get over stress-induced climate change. Recent advance in plant–bacteria interaction research revealed that plants are able to shape their rhizosphere and endorhiza microbiome. Under stress conditions, plants can require the presence of associated bacteria (stress tolerance) for their growth and establishment in different ecosystems. Symbiotic bacteria exist in all plants living in the natural ecosystems. This relationship may be the key factor involved in plants' stress tolerance ability. Indeed, local adaptation of plants to their environment is driven by genetic differentiation in closely associated bacteria.

Halophytes are extremely salt-tolerant plants which usually grow and survive under salt concentrations as high as 5 g l^{-1}, by maintaining negative water potential, where no cultivation occurs. The salt-tolerant microbes inhabiting the rhizosphere of halophytes may contribute to salt tolerance. The rhizosphere and endorhiza of halophytic plants represent ideal source for isolation of various groups of salt-tolerant microbes that could enhance the growth of different crops under salinity stress. These salt-tolerant microbes are positive for multiple stress-related traits that may facilitate plants to survive under growth inhibitory levels of salt. The use of halotolerant microorganisms in agriculture as biofertilizers has extremely increased as an alternative to replace agrochemicals. Halotolerant microorganisms have begun to emerge as an important alternative to recuperate abandoned farmland affected by salt in arid and semiarid areas. The microorganisms isolated from halophytes were also able to increase the growth of glycophytes (salt-sensitive crops) under salinity stress.

In this book, we offered a view of salt-tolerant microbes' ability to increase plant tolerance to salt to facilitate plant growth, the potential of halophytes' rhizosphere as reservoir of beneficial salt-tolerant microbes, the future application as bio-inoculants as alternative in agriculture, and a valuable resource for the alternative way of improving crop tolerance to salinity and promoting saline soil-based agriculture in the future. In general, this special collection of reviews will highlight some of the recent progresses made in applied aspects of plant (halophytes)-microbe interactions and its contribution toward eco-friendly approach in agri-ecosystem (saline soil-based agriculture).

We hope you like reading this book.

Ranchi, India — Manoj Kumar
Tehran, Iran — Hassan Etesami
Dehradun, India — Vivek Kumar

Contents

About the Editors

Manoj Kumar is an Associate Professor at the Department of Life Sciences, Central University of Jharkhand, India. He is committed to high-quality science that contributes to both improving our understanding of plant development and increasing the ecological niche. He has extensive professional experience and has had affiliations with Jawaharlal Nehru University and the Universities of Melbourne, Pretoria, and Ton Duc Thang. His research focus includes tree molecular genetics, plant-microbe interaction, and bioremediation of soil pollutants.

He has authored 45 research papers and 5 review articles in leading international journals as well as 13 books. He has also led several research projects and R&D facilities at industry level and collaborated internationally.

Hassan Etesami is a Research Scientist with 10 years of experience in the field of soil biology and biotechnology. He obtained his doctor's degree from the Department of Soil Science, College of Agriculture & Natural Resources, University of Tehran, Iran, where he is currently a Member of the faculty. He has a special interest in developing biofertilizers and biocontrol agents that meet farmers' demands. He has coauthored over 50 publications (research papers, review papers, and book chapters) in various areas including biofertilizers and biocontrol. He is also a Reviewer for 27 international journals.

His research areas include microbial ecology, biofertilizers, soil pollution, integrated management of abiotic (salinity, drought, heavy metals, and nutritional imbalance) and biotic (fungal pathogens) stresses, plant-microbe interactions, environmental microbiology, and bioremediation.

Vivek Kumar is an Associate Professor, involved in teaching and research and guidance. He works at the Himalayan School of Biosciences, Swami Rama Himalayan University, Jolly Grant, Dehradun, India. He currently serves on the editorial boards of numerous respected international journals. He is also a Reviewer for several prestigious international journals. In addition, he is an Evaluator of project proposals for "The Netherlands Organisation for Scientific Research (NWO)," Netherlands. He has published over 100 research papers, review articles, and book chapters and has edited several Springer books. He also served as a Microbiologist at the Department of Soil and Water Research, Public Authority of Agricultural Affairs and Fish Resources, Kuwait, for 8 years. He has been credited with the first reporting and identification of pink rot inflorescence disease of date palm in Kuwait caused by *Serratia marcescens*. He was awarded "Young Scientist Award" for the year 2002 in "Agricultural Microbiology" by the Association of Microbiologists of India (AMI). His research areas include plant-microbe interactions, sustainable agriculture, environmental microbiology, and bioremediation. He has also organized various outreach activities.

Soil Salinity as a Challenge for Sustainable Agriculture and Bacterial-Mediated Alleviation of Salinity Stress in Crop Plants

1

Hassan Etesami and Fatemeh Noori

Abstract

Nowadays, the agricultural productivity is declining due to the negative effects of climate change, global temperature rise, and increased environmental stresses. Therefore, in order to achieve sustainable development in agriculture and to increase agricultural products for feeding all of the world's people, it seems necessary to use the appropriate solutions and ecologically compatible and environmentally friendly techniques to decrease the adverse effects of these stresses on plant. Soil salinity is a problem for agricultural productivity around the world. According to available reports, crops that grow in salt-affected soils are subject to osmotic stress, poor physical soil conditions, nutritional disorders and toxicity, and reduced crop yields. Limiting crop losses due to salinity stress is a major area of concern to cope with the background of increasing food requirements. Novel agricultural technologies are needed to improve food production in salt-affected soils. Beneficial halotolerant rhizospheric bacteria associated with plant have been known to increase plant tolerance to salinity through mechanisms such as root system development, improved soil structure, increased water and nutrient uptake, reduced sodium absorption, reduced negative effects of stress ethylene, and increased expression of genes involved in resistance to salinity stress. Microbial inoculation to alleviate stresses in plants could be a more cost-effective environment-friendly option which could be

H. Etesami (✉)
Agriculture & Natural resources Campus, Faculty of Agricultural Engineering & Technology, Department of Soil Science, University of Tehran, Tehran, Iran
e-mail: hassanetesami@ut.ac.ir

F. Noori
Department of Biotechnology and Plant Breeding, Sari Agricultural Sciences and Natural Resources University, Sari, Iran

M. Kumar et al. (eds.), *Saline Soil-based Agriculture by Halotolerant Microorganisms*, https://doi.org/10.1007/978-981-13-8335-9_1

available in a shorter time frame. Such inoculants contribute to the development of sustainable agriculture under salinity-stressed conditions.

Keywords
Halotolerant PGPR · Salinity stresses · Agriculture crops · Plant · Microbe interactions · Saline soil-based agriculture

1.1 Introduction

Food security is one of the basic needs of a human society that cannot be ignored by any society. According to FAO (Food and Agriculture Organization), the total demands for agricultural products will be 60% higher in 2030 than present time. The world has depended on augmenting crop yields to supply an ever-enhancing demand for food for over half a century (Ladha et al. 1998). Due to increasing population growth, which is projected to reach 8.9 billion by 2050 according to United Nations estimates (Wood 2001), and decreasing the fertility level of agricultural soils, the ratio of acceptable agricultural land to population is decreasing rapidly. Therefore, one of the main needs is to increase yield of crop plants per unit area of land cultivation, especially in conditions of saline soils. The soil with an electrical conductivity (EC) of the saturation extract (ECe) of 4 dS/m (approximately 40 mM NaCl at 25 °C) and exchangeable sodium of 15% is a saline soil. The yield of most crop plants in such a soil is reduced, although many crop plants showed yield reduction at lower ECes (Munns 2005). Different definitions have been made for salinity. According to Shannon and Grieve (1998), salinity is the excessive concentration of soluble salts and mineral elements in water and soil solution that results in salt accumulation in the root zone, and the plant encounters difficulty in absorbing sufficient water from the soil solution. In another definition, salinity is excessive accumulation of soluble anions and cations in a way that affects the growth of plants (Çavusoglu and Kabar 2010).

Soil salinization is a process in which the accumulation of soluble salts in the surface layer of the soil increases, and as a result, the surface layer loses its potential as the growth medium of the plants. In general, the soil salinization has a primary or secondary origin. The primary origin of salinity is long-term natural processes of accumulation of salts in the soil or in surface water. Secondary salinity, which is also the result of human activity, is the result of irrigation, which is widely responsible for increasing the concentration of dissolved salts in the soil profile, which reduces plant growth and makes agricultural land unusable (Egamberdieva et al. 2007; Manchanda and Garg 2008; Munns 2005). Irrigation of agricultural land with saline water, inadequate irrigation management, recent drought and low precipitation, high surface evaporation, and poor management practices (poor cultural practices) are some of the main causes of increasing salinity at a rate of 10% annually (Jamil et al. 2011). It is estimated that about 300 million hectares of land are irrigated in the world, of which about 90% of the total water is spent generating 36% of global food (Rengasamy 2006). In the regions that are not irrigated with irrigation water and receive less rainfall and even in most agricultural land with higher temperatures, there is a more increase in salinity levels (Othman et al. 2006). According to forecasts, it has been determined that worldwide 33% of irrigated agricultural lands and 20% of total cultivated are afflicted by high salinity. Climate change may lead to even more saline landscapes in many nonirrigated regions (Othman et al. 2006). It has been estimated that more than 50% of the arable agricultural land would be salinized by the year 2050 (Jamil et al. 2011).

Among environmental stresses (i.e., extreme temperatures, high winds, drought, soil salinity, and flood that have affected the production and cultivation of agricultural crops), salinity stress is known as one of the most devastating environmental stresses (i.e., low agricultural productivity, low economic returns, and soil erosions) (Hu and Schmidhalter 2004) causing major reductions in cultivated land area (Shahbaz and Ashraf 2013; Yamaguchi and Blumwald 2005) and limiting the productivity and quality of crop plants (a 20–50% reduction in plant yield), because most of the crop plants are sensitive to salinity due to high concentrations of salts in the soil (excessive accumulation of nutrients in the form of soluble salts) (Shrivastava and Kumar 2015). In addition to diminishing the agricultural production of most crops especially salinity-sensitive crops, salinity adversely affects soil physical and chemical properties and ecological balance of the area (effect on the growth and diversity of organisms that live in or on soil such as microbes, protozoa, and nematodes) (Hu and Schmidhalter 2004; Parida and Das 2005). Due to the ever-increasing demand for food (production of 70% more food crop for an additional 2.3 billion people by 2050 worldwide) and the lack of salt-non-affected soils, we should inevitably use the potential of salinity-stressed soils to produce important crop plants such as rice (*Oryza sativa* L.), wheat (*Triticum aestivum* L.), and maize (*Zea mays* L.). It is predicted that a significant increase (an estimated 50%) in grain yield of these plants is required to fulfill the food supply requirements for the projected population by 2050 (Godfray et al. 2010). As previously mentioned, salinity significantly reduces the growth and yield of plants. In these areas, farmers are trying to compensate for losses caused by salinity through the use of more inputs such as chemical fertilizers. However, the use of chemical fertilizers, despite the high cost, low efficiency, and the risk of environmental pollution such as surface water and groundwater contamination, results in more soil salinization. Therefore, it is necessary to better manage the saline soils in order to better utilize them. Many strategies have been proposed to manage salinity soils (saline soil remediation) and increase plant growth in these soils (Bai et al. 2017; Bauder et al. 2004; Etesami and Maheshwari 2018; Qadir et al. 2000; Tejada et al. 2006; Wang et al. 2014). Although the use of strategies like leaching of salt from root zone, changed farm management practices, the use of organic amendments, resource management practices, drip or micro-jet irrigation to optimize use of water, shifting the crop calendars, the use of salt-tolerant plants, and development of salinity-tolerant varieties through breeding and genetic engineering to sustainable management can ameliorate yield reduction under salinity stress, implementation is often limited due to cost and availability of good water quality or water resource. In other words, such strategies are long drawn and expensive processes (Araus et al. 2008; Dwivedi et al. 2010; Flowers 2004; Manchanda and Garg 2008; Shrivastava and Kumar 2015; Venkateswarlu and Shanker 2009). Therefore, there is a need to develop low-cost and simple biological methods for managing salt-affected soils, which can be used on short-term basis. It is known that beneficial halotolerant microorganisms associated with crop plants could play a significant role in this respect (Etesami 2018; Etesami and Beattie 2017; Etesami and Maheshwari 2018; Shrivastava and Kumar 2015). In this chapter, the effects of salinity stress on plants and the role of halotolerant plant growth-promoting bacteria (PGPB) associated with plants in increasing the plant's resistance to salinity in saline soils are discussed.

1.2 Salinity Effect on Growth of Plants and Mechanisms of Salinity Tolerance in Plants

Salinity as a major abiotic stress hampers crop productivity worldwide and limits the increase in demand for food crops. More than 45 ha (about 20%) of the cultivated land around the world is affected by salinity stress, and the amount is enhancing day by day. Based on their adaptive evolution, some plant can withstand salinity (halophytes), and some others cannot withstand salinity and eventually die (glycophytes). Majority of major crop species that belong to glycophytes (Flowers 2004; Munns and Tester 2008) are susceptible to salinity stress (EC = 3–5 dSm^{-1}) (Horneck et al. 2007; Ondrasek et al. 2009), and their productivity is considerably reduced due to improper nutrition of the plant (Chinnusamy et al. 2005; Mantri et al. 2012).

Salinity stress affects almost all aspects of plant development (i.e., germination, vegetative growth, and reproductive development) (Bano and Fatima 2009). Depending on the severity and duration of stress, salinity stress results in changes in various metabolic and physiological processes and morphological traits of salinity-stressed crop plant such as diminished stomatal aperture (decrease in stomatal conductance), interruption of membranes, decrease in the expansion of the leaves, decrease in chlorophyll content and photosystem II efficiency, disruption of photosynthetic functions through inhibition of nitrate reductase (NR) activity, and premature aging of leaves (which affects process such as photosynthesis and results in stunted growth); causes cell swelling and reducing of energy production; adversely affects various enzymes activities (e.g., inhibition of the activity of enzymes involved in nucleic acid metabolism); adversely affects germination, plant vigor, crop yield, and the inhibition of root growth; impairs the ability to detoxify reactive oxygen species (ROS) and accelerated defoliation and senescence; adversely effects on all the major processes such as photosynthesis, protein synthesis, energy and lipid metabolism, nutritional imbalance, the inhibition of phytohormone synthesis and maturation of cell walls, hormonal imbalance like elevated levels of ethylene (C_2H_4) (which at higher concentrations imparts drastic effects on plant health, including defoliation and other unbalanced cellular processes), differences in the antioxidant enzymes, physiological disorders (like abscission, epinasty, senescence, and susceptibility to diseases), and decrease in photosynthetic activity (Arbona et al. 2005; Barnawal et al. 2014; Cramer and Nowak 1992; Dantas et al. 2005; Dolatabadian et al. 2011; Flowers 2004; Glick 2014; Hashem et al. 2015; Kang et al. 2014a; Munns 2002; Munns and Tester 2008; Nadeem et al. 2014; Netondo et al. 2004; Nunkaew et al. 2015; Parida and Das 2005; Paul and Lade 2014; Prakash and Prathapasenan 1990; Rahnama et al. 2010; Shirokova et al. 2000; Tavakkoli et al. 2011); and ultimately inhibits the production of the crop (James et al. 2011; Rozema and Flowers 2008).

In general, effects of salinity on salinity-stressed crop plants are the results of complex interactions among biochemical, physiological, and morphological processes such as plant growth, seed germination, and water and nutrient absorption (Akbarimoghaddam et al. 2011). Among them, seed germination (the stages of the development of roots) and early seedling growth are the most salt-sensitive plant growth stages, because the seedling root is in direct contact with soil and is affected by salt stress (Bae et al. 2006; Rahman et al. 2000). Due to increased osmotic pressure, higher saline concentrations could reduce the germination percentage (Atak et al. 2006; Neamatollahi et al. 2009) in various crops such as soybean (Essa 2002), wheat (Egamberdieva and Kucharova 2009), rice (Xu et al. 2011), maize (Khodarahmpour et al. 2012), and faba bean (Rabie and Almadini 2005).

Initially, salinity stress, which is also considered as hyperosmotic stress, in the form of osmotic stress suppresses plant growth, so that in the early stages of salinity stress, root water absorption capacity diminishes and osmotic stress causes loss of water from the leaves, which increases the accumulation of salt in the salinity-stressed plant (Munns 2005). In addition to creating water-deficit conditions in early phase, salinity stress, due to having higher ionic concentration, can also cause ionic toxicity in plants inducing leaf senescence and change the basic texture of the soil causing decreased soil porosity and subsequently reducing water uptake (a form of a physiological drought) (Munns and Tester 2008).

By creating a low osmotic potential of soil solution (osmotic stress), nutritional imbalances (deficiency of nutrients of N, Ca, K, P, Fe, Zn), specific ion effects (salt stress), oxidative stress, or a combination of these factors, salinity stress causes many adverse effects on plant growth and development (Ashraf 2004; Bano and Fatima 2009; Grover et al. 2011) at physiological and biochemical levels and at the molecular level (Tester and Davenport 2003) resulting in substantial loss to crop productivity.

Soil salinity significantly reduces uptake of plant nutrients such as P (due to precipitation of phosphate ions with Ca ions) (Bano and Fatima 2009), N, K, and Mg (Heidari and Jamshid 2010; James et al. 2011). Salinity also reduces nitrogen (N) incorporation to agricultural systems, which lead to a reduction in the application of chemical N fertilizers, by adversely affecting leguminous plants, which are sensitive to salinity. Only a few agronomical legumes can grow in saline soils (Ashraf and McNeilly 2004). Soil salinity particularly disturbs the symbiotic interaction between legumes and *Rhizobium* bacteria. Rhizobial bacteria fix atmospheric nitrogen through the nitrogenase complex in root nodules and make it available to the leguminous plants (Quispel 1988). For example, salinity stress has decreased nitrogen fixation by a failure of the infection and nodulation process and decrease of activity of nitrogenase enzyme in several legumes such as *Phaseolus vulgaris* (common bean), *Glycine max* (soybean), and *Vicia faba* (faba bean) (Rabie et al. 2005; Singleton and Bohlool 1984). Salinity reduces symbiotic legume growth by diminishing the growth of roots, root tips, and root hairs, thereby decreasing sites for potential rhizobial infection and further nodule development. Salinity-mediated decrease of root growth has been attributed to reduced calcium absorption by the plant in saline conditions (Bouhmouch et al. 2005).

For survival in saline soils, plants have to adapt and acclimatize to their surrounding environment (Paul and Lade 2014). Plants develop different physiological and biochemical mechanisms to survive in salt-affected soils. Many mechanisms are known in plants through which plants increase their resistance to salinity. The most important of these mechanisms are the following: (i) hormone modulation, (ii) biosynthesis of osmoprotectants and compatible solutes, (iii) ion compartmentalization and homeostasis, (iv) ion transport and uptake, (v) activation of antioxidant enzyme and synthesis of antioxidant compounds, (vi) synthesis of polyamines, and (vii) generation of nitric oxide (NO) (Gupta and Huang 2014). In addition to these mechanisms, various studies have shown that microbial (bacterial) communities associated with plant can also play a major role in increasing plant growth and tolerance of plant to environmental stresses including salinity stress in salt-affected soils (Etesami and Beattie 2018; Etesami and Maheshwari 2018).

1.3 Bacterial-Mediated Alleviation of Salinity Stress in Crop Plants

Plant-associated beneficial microorganisms including rhizoplane, rhizosphere and endophytic bacteria, and mycorrhizal fungi can play an important role in conferring plant resistance to salinity stress (Dodd and Pérez-Alfocea 2012; Etesami and Beattie 2017, 2018; Etesami and Maheshwari 2018; Gill et al. 2016; Hamilton et al. 2016; Singh et al. 2011; Vimal et al. 2016). PGPR (plant growth-promoting rhizobacteria) colonize the rhizosphere of many plant species and confer beneficial effects (Glick et al. 2007; Van Loon et al. 1998). Plants inoculated with these bacteria induce biochemical and morphological modifications leading to augmented tolerance to abiotic stresses including salinity stress defined as IST (induced systemic tolerance) (Glick 2014; Kaushal and Wani 2016; Yang et al. 2009).

Compared to previous research, which had been focused on the microorganisms that improve soil quality and fertility, current research is focused mostly on microorganisms that can alleviate abiotic stress including soil salinity (Mapelli et al. 2013). It is known that, by high osmotic strength and toxic effects, salinity negatively affects activity of soil microorganisms including soil bacteria. However, the bacteria inhabiting the sites exposed to frequent salinity stress conditions or associated with halophytic plants are likely to be more adaptive or tolerant and can survive and proliferate in the soil and in the rhizosphere in a harsh environment (Etesami and Beattie 2018; Garcia and Hernandez 1996; Paul et al. 2005). It has been proven that salinity-tolerant bacteria may serve as better plant growth promoters under salinity conditions (Etesami and Beattie 2018; Shrivastava and Kumar 2015). Among the bacteria whose habitats are the soil, rhizobacteria (root-associated bacteria) are more tolerant to salinity stress because salinity stress is higher in the area around the root (rhizosphere) because of depletion of water by the plant root, leading to an increment in both osmolality and ionic strength (Tripathi et al. 1998). These PGPR are salinity tolerant up to at least 3% NaCl (Egamberdieva et al. 2011) and are able to survive in the rhizosphere of plants due to their persistence and competitiveness under saline arid soil conditions (Mayak et al. 2004b; Yasmin et al. 2007). The bacteria isolated from these areas also had growth-promoting properties like indole-3-acetic acid (IAA) production, phosphate solubilization, 1-aminocyclopropane-1-carboxylate (ACC) deaminase activity, ammonia production, and nitrogen fixation (Etesami and Beattie 2018; Mapelli et al. 2013). The halotolerant and halophilic microorganisms could accumulate osmolytes in stress conditions (Zhou et al. 2015). Extensive research has been conducted to show the beneficial effects of halotolerant PGPR on plant growth (Table 1.1) (Etesami and Beattie 2017, 2018; Etesami and Maheshwari 2018). Salt-tolerant PGPR significantly increased the growth and yield of various plants such as wheat (Barra et al. 2016; Egamberdieva and Kucharova 2009; Nabti et al. 2010; Nia et al. 2012; Orhan 2016; Rajput et al. 2013; Ramadoss et al. 2013; Upadhyay and Singh 2015; Upadhyay et al. 2012), cucumber (Egamberdieva et al. 2011; Kang et al. 2014a; Nadeem et al. 2016), sweet potato (Yasmin et al. 2007), *Ocimum basilicum* (Heidari et al. 2011), tomato (Albacete et al. 2008; Essghaier et al. 2014; Mayak et al. 2004a; Tank and Saraf 2010), pepper, canola, bean (Egamberdieva 2011), lettuce (Barassi et al. 2006; Yildirim and Taylor 2005; Yildirim et al. 2011), cotton (Yao et al. 2010), potato (Shaterian et al. 2005b), *Glycine max* L. (Kang et al. 2014b; Naz et al. 2009), *Oryza sativa* L. (Jha et al. 2011), white clover (Han et al. 2014), *Arachis hypogaea* (Shukla et al. 2012), strawberry (Esitken et al. 2010), *Medicago truncatula* (Bianco and Defez 2009), groundnut (Saravanakumar and Samiyappan 2007), and *Galega officinalis* (Egamberdieva et al. 2013) in salt-affected soil.

1.3.1 Mechanisms of Action by Which PGPR Alleviate Salt Stress

Halotolerant PGPR including *Azospirillum*, *Rhizobium*, *Bacillus*, *Pseudomonas*, *Pantoea*, *Paenibacillus*, *Enterobacter*, *Burkholderia*, *Achromobacter*, *Microbacterium*, *Methylobacterium*, *Flavobacterium*, *Arthrobacter*, *Serratia*, *Mycobacterium*, *Oceanobacillus* sp., *Halomonas* sp., *Exiguobacterium* sp., *Zhihengliuella*, *Variovorax*, *Microbacterium*, *Brachybacterium*, *Brevibacterium*, *Haererohalobacter*, *Staphylococcus*, *Kocuria*, etc. (Adesemoye et al. 2008; Egamberdieva and Kucharova 2009; Egamberdieva et al. 2011; Egamberdiyeva 2007; Egamberdiyeva and Islam 2008; Etesami and Beattie 2018; Rajput et al. 2013; Shrivastava and Kumar 2013; Shukla et al. 2012; Siddikee et al. 2010) can use various mechanisms to increase plant tolerance to salinity stress and stimulate plant growth and development (Dimkpa et al. 2009; Etesami and Maheshwari 2018;

Table 1.1 Some PGPR that have caused induced systemic tolerance (IST) in plants under salinity stress, with their mechanism of action

PGPR	Mechanism(s) of action	Crop plant	References
Pseudomonas sp.	Increase in IAA and protein contents	*Ocimum basilicum*	Heidari et al. (2011)
Achromobacter piechaudii	ACC deaminase	Tomato	Zhang et al. (2008)
Azospirillum lipoferum	Increase in abscisic acid content	Maize	Cohen et al. (2009)
Achromobacter piechaudii	ACC deaminase	Pepper and tomato	Mayak et al. (2004b)
B. subtilis and *Arthrobacter* sp.	Increase in total soluble sugars and proline content	Wheat	Upadhyay et al. (2012)
P. pseudoalcaligenes and *B. pumilus*	Increase in osmoprotectant and antioxidant enzymes	Rice	Jha et al. (2011)
Azospirillum	Increase in N concentration	Wheat	Nia et al. (2012)
Planococcus rifietoensis	IAA, ACC deaminase, and phosphate solubilization	Wheat	Rajput et al. (2013)
S. haemolyticus and *B. subtilis*	Accumulated osmolytes	Chickpea	Essghaier et al. (2014)
Bacillus subtilis	Increase in chlorophyll content and decrease in malondialdehyde content	White clover	Han et al. (2014)
Pseudomonas fluorescens	Enhanced Fv/Fm and electron transport rate	*Pinus halepensis*	Rincón et al. (2008)
A. brasilense	Increase in leaf water content decrease in MDA content	*Arabidopsis*	Cohen et al. (2015)
P. aurantiaca	IAA production	Wheat	Egamberdieva and Kucharova (2009)
Phyllobacterium brassicacearum	Improved plant nitrogen status	*Arabidopsis*	Kechid et al. (2013)
B. cepacia and *Acinetobacter calcoaceticus*	Gibberellins production	Cucumber	Kang et al. (2014a)
Pseudomonas putida	Gibberellins production	Soybean	Kang et al. (2014b)
Bacillus subtilis	Cytokines production	Lettuce	Arkhipova et al. (2007)
P. trivialis	ACC deaminase	*Galega officinalis*	Egamberdieva and Jabborova (2013)
P. fluorescens	ACC deaminase	*Arachis hypogaea*	Saravanakumar and Samiyappan (2007)
Azospirillum lipoferum	ACC deaminase	Wheat	Zaki et al. (2004)
Azospirillum brasilense	ACC deaminase	Maize	Hamdia et al. (2004)

(continued)

Table 1.1 (continued)

PGPR	Mechanism(s) of action	Crop plant	References
Enterobacter spp.	Antioxidant activities	Tomato	Kim et al. (2014)
Azospirillum	Increased proline	Maize	Kandowangko et al. (2009)
Azospirillum brasilense	Polyamine	Rice	Cassan et al. (2009)
Bacillus spp.	Higher uptake of K^+ in plant as compared to Na^+	*Gladiolus*	Damodaran et al. (2014)
V. paradoxus	Enhanced nodulation and increased nitrogen content	Chickpea	Belimov et al. (2009)

Glick 2014; Glick et al. 2007; Grover et al. 2011; Lugtenberg and Kamilova 2009; Lugtenberg et al. 2013; Mayak et al. 2004b; Shrivastava and Kumar 2015; Yang et al. 2009). These mechanisms include:

(i) The production of phytohormones such as auxin (i.e., IAA), cytokinin, and gibberellins. It is known that phytohormones are involved in plant growth development and in stress responses including salinity stress (Shaterian et al. 2005a) and augment different cellular defense systems for the protection of plants from stressful conditions (Shaterian et al. 2005b). Under high salinity conditions, the production of hormones, auxins, gibberellins, and zeatin in the roots and leaves of salinity-stressed plants decreases (Pérez-Alfocea et al. 2010; Sakhabutdinova et al. 2003), and thereby the germination percentage and plant growth and development also diminish (Sakhabutdinova et al. 2003; Werner and Finkelstein 1995). In addition to the decrease in production of hormones, salinity also reduces the supply of hormones (i.e., cytokinin) from root to shoot (Naqvi and Ansari 1974). It has been shown that salinity has no effect on the production of hormones (i.e., auxin) of salt-tolerant PGPR (Albacete et al. 2008; Egamberdieva and Kucharova 2009). For example, in a previous study, it was found that salt-tolerant PGPR such as *A. brasilense* could produce IAA at a concentration of 200 mM NaCl (Nabti et al. 2010). In another study, salt-tolerant PGPR like *S. plymuthica* RR2-5-10, *S. rhizophila* e-p10, *P. chlororaphis* TSAU13, and *P. fluorescens* SPB2145 were shown to produce IAA at 1.5% NaCl (Egamberdieva 2011, 2012). Bacterial hormones result in increased root growth, root length (modulation of root system architecture), root surface area, and number of root tips, leading to enhanced uptake of nutrients thereby improving plant growth under salinity stress conditions (Dodd et al. 2010; Egamberdieva and Kucharova 2009; Etesami and Alikhani 2016; Etesami et al. 2015a, b; Etesami and Beattie 2017; Kurepin et al. 2015; Postma and Lynch 2011). In other words, increased rooting system (a root system with large surface area and increased number of root hairs) constitutes a greater root surface area that enables the plant to get more nutrients from the soil (Boicro et al. 2007; Egamberdieva and

Kucharova 2009; Mantelin and Touraine 2004). For example, IAA-producing PGPR (i.e., *P. aurantiaca* TSAU22, *P. extremorientalis* TSAU6, and *P. extremorientalis* TSAU20) significantly augmented wheat root growth up to 40% and shoot growth up to 52% at 100 mM NaCl compared to non-inoculated wheat plants (Egamberdieva and Kucharova 2009). Plants in response to salinity stress produce abscisic acid (ABA) causing stomatal closure to diminish transpirational loss of water and mediate root branching to augment water uptake in salinity-stressed plants hence inducing leaf growth (Tardieu et al. 2010). It has been reported that some PGPR (i.e., *Azospirillum lipoferum*) can increase ABA content in stressed plant (Cohen et al. 2009). In general, by producing phytohormones, PGPR strains can modulate hormonal balance of plant hormones thus enhancing plant growth under salinity stress conditions (Postma and Lynch 2011).

(ii) Ion homeostasis and increasing the availability of plant nutrients by production of siderophores (increased iron nutrition), enhancing transformation and acquisition of nitrogen (N), solubilization of inorganic phosphates, and solubilization of K-bearing minerals (Bell et al. 2015; Etesami et al. 2017; Etesami and Maheshwari 2018). It is known that high Na^+ concentration inhibits uptake of nutrients of K, P, and N which are essential elements for growth and development (Etesami and Maheshwari 2018). In other words, salinity creates nutritional imbalances due to disturbance in conserved intracellular ionic milieu (Nadeem et al. 2009). Salinity-stressed plants can protect themselves by restricting Na^+ entry, retrieving Na^+ from xylem stream for recirculation to roots and expelling Na^+ from roots during saline stress (Chinnusamy et al. 2006). According to previous studies, many PGPR could reduce the overaccumulation of Na^+ in salinity-stressed plants and maintain the ion homeostasis (higher N, P, and K uptake resulting in increased K^+/Na^+ ratio in plant) under salinity stress conditions (Belimov et al. 2009; Bharti et al. 2014; Damodaran et al. 2014; Etesami and Maheshwari 2018; Kang et al. 2014b; Mayak et al. 2004b; Nadeem et al. 2009; Tewari and Arora 2014).

(iii) reduction in stress-induced ethylene production. Depending upon its concentration, the ethylene is involved in the plant developmental cycle, and it may be inhibitory (an inhibitor of plant root elongation) or stimulatory (i.e., seed germination, root hair development, stem elongation, fruit ripening, etc.) (Glick 2005; Penrose et al. 2001; Siddikee et al. 2011). Under salinity stress conditions, excessive production of ethylene hormone results in reduced root and shoot growth (Glick et al. 2007). ACC deaminase-producing bacteria convert ACC (a prerequisite for ethylene production) into α-ketobutyrate and ammonia (supply of nitrogen and energy). Furthermore, by removing ACC, PGPR diminish the deleterious effect of ethylene, ameliorating stress and promoting salinity-stressed plant growth. In general, by decreasing stress ethylene, ACC deaminase-containing PGPR increase the number of lateral roots, lateral root length, and root dry weight of salinity-stressed plants (Shahzad et al. 2010). It was reported a direct correlation between bacterial ACC deaminase activity and root growth (Shaharoona et al. 2006). Bacterial ACC deaminase mediated longer roots, in turn, take up relatively more water from deep soil under salinity stress conditions, thus increasing the water use efficiency (WUE) of salinity-stressed plants (Zahir et al. 2008). The enzyme ACC deaminase is present in many PGPR (Glick 2010) including salt-tolerant PGPR (Etesami and Beattie 2018). Many studies have shown that PGPR could increase plant growth and resistance to salinity through the production of this enzyme (Ahmed et al. 2004; Etesami and Beattie 2018; Glick 2014; Hamdia et al. 2004; Mayak et al. 2004a, b; Nadeem et al. 2007, 2009, 2013; Paul and Sarma 2006; Penrose and Glick 2003; Saravanakumar and Samiyappan 2007; Zaki et al. 2004).

(iv) Synthesis of compatible solutes, also known as compatible osmolytes, such as proline, sugars, glycine betaine, polyols, and choline (Essghaier et al. 2014). Compatible osmolytes are a group of chemically diverse organic compounds that are polar, uncharged, and soluble in nature and do not interfere with the cellular metabolism even at high concentration (Ford 1984; Saxena et al. 2013). By accumulating compatible solutes to maintain the cell turgidity within boundaries acceptable for normal cellular physiology and osmotically adjust themselves, salinity-stressed plants may protect themselves from salt stress (Evelin et al. 2009; Gill and Tuteja 2010; Serraj and Sinclair 2002). Previous studies also reported that PGPR such as *Azospirillum*, *P. pseudoalcaligenes*, *Burkholderia*, *Arthrobacter*, and *Bacillus* could alleviate salinity stress in various plants through increasing the concentration of compatible solutes such as proline, glycine betaine-like quaternary compounds, free amino acids, and soluble sugars in salinity-stressed plants (Bano et al. 2013; Barka et al. 2006; Bharti et al. 2014; Cassan et al. 2009; Damodaran et al. 2014; Jha et al. 2011; Kandowangko et al. 2009; Paul and Nair 2008; Sarma and Saikia 2014; Shintu and Jayaram 2015; Sziderics et al. 2007; Vardharajula et al. 2011).

(v) Synthesis of antioxidant enzymes. Generation of reactive oxygen species (ROS), as by-products of cellular metabolism, is generally low in various plant organelles during normal growth conditions (Gill and Tuteja 2010). It is known that low levels of ROS may act as signaling molecule thus activating stress response and defense pathways (Pitzschke et al. 2006). However, response of plants to salinity stress is the increase in generation (overproduction) of ROS (i.e., hydroxyl radical, superoxide radical, singlet oxygen, hydrogen peroxide, and superoxide) (Ahmad and Umar 2011; Andersen et al. 2004; Apel and Hirt 2004; Chaves et al. 2002; Groß et al. 2013; Zhang et al. 2016), which results in oxidative damages in different cellular components, like DNA, lipids (lipid peroxidation), and proteins (disruption of membrane-bound proteins and inhibition of protein synthesis), decreased membrane fluidity, and loss of enzymatic activities, interrupting vital cellular functions of salinity-stressed plants and ultimately cell death (Dietz et al. 2016; Gill and Tuteja 2010; Gupta and Huang 2014). In other words, salinity stress causes a secondary stress that is called oxidative stress, which is production of ROS above threshold levels leading to imbalance in cell homeostasis (Gill and Tuteja 2010). Plants, which are equipped with antioxidant defense system constituting enzymatic components such as CAT (catalase), GPX (glutathione peroxidase), SOD (superoxide dismutase), APX (ascorbate peroxidase), monodehydroascorbate reductase (MDHAR), and GR (glutathione reductase) and nonenzymatic components like glutathione, cysteine, tocopherols, and ascorbate, can eliminate or neutralize ROS to cope with the oxidative stress by the antioxidant defense system and protect themselves against detrimental effects of ROS (Miller et al. 2010). Various studies have shown that there is a strong correlation between tolerance to oxidative stress and increasing concentration of antioxidant enzymes in plants (Asada 1999; Gupta et al. 2005; Sairam and Srivastava 2002). By producing enzymes of antioxidants, some PGPR can also degrade ROS. High activities of antioxidant enzymes are also linked with oxidative stress tolerance (Štajner et al. 1997). There is substantial evidence reported that PGPR could alleviate salinity-induced oxidative stress (ROS formation) in various plants through manipulation of antioxidant enzymes (Damodaran et al. 2014; Han and Lee 2005; Kim et al. 2014; Sandhya et al. 2010; Wang et al. 2012). It is noteworthy that the mechanisms underlying behind variations found in the antioxidant enzyme levels as affected by bacteria in plants imparting stress are not clear yet (Kaushal and Wani 2016), which requires further research in the future.

(vi) Production of volatile compounds (Kaushal and Wani 2016). These compounds affect the plant signaling pathways (Bhattacharyya et al. 2015).

(vii) Altering plant-water relationships and enhancing osmotic adjustment (Creus et al. 2004).

(viii) Controlling various plant diseases (a broad-spectrum of root and foliar pathogens) by antibiosis, competition for space and nutrients, and induction of systemic resistance in plants (Compant et al. 2005; Singh et al. 2011).

1.4 Conclusions

The microbial communities inhabiting the rhizosphere of halophytic plants, also known as the rhizosphere microbiome, and the microbial communities inhabiting salt-affected soils can play an important role in conferring resistance to plants against salinity stresses in a sustainable and environment-friendly manner thus ensuring increased crop yield in salt- affected soils and opening a new advanced technology for combating salinity. PGPR can also increase nutrient uptake from salinity-affected soils, thus reducing the need for using chemical fertilizers in saline soils. Such inoculants help the development of sustainable agriculture (helpful in reducing the use of agrochemicals and in restoring soil health) under salinity-stressed conditions. Despite many studies done in this area, the understanding of cross talks between halotolerant rhizobacteria and plants that mediate these responses is still one major hindrance. In addition, the application of halotolerant rhizobacteria as a bio-fertilizer requires more research in field conditions.

Acknowledgment We wish to thank the University of Tehran for providing the necessary facilities and funds for this study.

References

Adesemoye AO, Obini M, Ugoji EO (2008) Comparison of plant growth-promotion with Pseudomonas aeruginosa and Bacillus subtilis in three vegetables. Braz J Microbiol 39:423–426

Ahmad P, Umar S (2011) Oxidative stress: role of antioxidants in plants. Studium Press, New Delhi

Ahmed W, Shahroona B, Zahir ZA, Arshad M (2004) Inoculation with ACC-deaminase containing rhizobacteria for improving growth and yield of wheat. Pak J Agric Sci 41:119

Akbarimoghaddam H, Galavi M, Ghanbari A, Panjehkeh N (2011) Salinity effects on seed germination and seedling growth of bread wheat cultivars. Trakia J Sci 9:43–50

Albacete A et al (2008) Hormonal changes in relation to biomass partitioning and shoot growth impairment in salinized tomato (*Solanum lycopersicum* L.) plants. J Exp Bot 59:4119–4131

Andersen L, Williams M, Serek M (2004) Reduced water availability improves drought tolerance of potted miniature roses: is the ethylene pathway involved? J Hortic Sci Biotechnol 79:1–13

Apel K, Hirt H (2004) Reactive oxygen species: metabolism, oxidative stress, and signal transduction. Annu Rev Plant Biol 55:373–399

Araus JL, Slafer GA, Royo C, Serret MD (2008) Breeding for yield potential and stress adaptation in cereals. Crit Rev Plant Sci 27:377–412

Arbona V, Marco AJ, Iglesias DJ, López-Climent MF, Talon M, Gomez-Cadenas A (2005) Carbohydrate depletion in roots and leaves of salt-stressed potted Citrus clementina L. Plant Growth Regul 46:153–160

Arkhipova TN, Prinsen E, Veselov SU, Martinenko EV, Melentiev AI, Kudoyarova GR (2007) Cytokinin producing bacteria enhance plant growth in drying soil. Plant Soil 292:305–315

Asada K (1999) The water-water cycle in chloroplasts: scavenging of active oxygens and dissipation of excess photons. Annu Rev Plant Biol 50:601–639

Ashraf M (2004) Some important physiological selection criteria for salt tolerance in plants. Flora Morphol Distrib Funct Ecol Plants 199:361–376

Ashraf M, McNeilly T (2004) Salinity tolerance in Brassica oilseeds. Crit Rev Plant Sci 23:157–174

Atak M, Kaya MD, Kaya G, Çikili Y, Çiftçi CY (2006) Effects of NaCl on the germination, seedling growth and water uptake of triticale. Turk J Agric For 30:39–47

Bae D, Yong K, Chun S (2006) Effect of salt (NaCl) stress on germination and early seedling growth of four vegetables species. J Cent Eur Agric 7:273–282

Bai Y et al (2017) Coastal mudflat saline soil amendment by dairy manure and green manuring. Int J 478 Agron 2017:1–9

Bano A, Fatima M (2009) Salt tolerance in *Zea mays* (L). following inoculation with Rhizobium and Pseudomonas. Biol Fertil Soils 45:405–413

Bano Q, Ilyas N, Bano A, Zafar N, Akram A, Hassan F (2013) Effect of Azospirillum inoculation on maize (*Zea mays* L.) under drought stress. Pak J Bot 45:13–20

Barassi CA, Ayrault G, Creus CM, Sueldo RJ, Sobrero MT (2006) Seed inoculation with Azospirillum mitigates NaCl effects on lettuce. Sci Horticult 109:8–14

Barka EA, Nowak J, Clément C (2006) Enhancement of chilling resistance of inoculated grapevine plantlets with a plant growth-promoting rhizobacterium, Burkholderia phytofirmans strain PsJN. Appl Environ Microbiol 72:7246–7252

Barnawal D, Bharti N, Maji D, Chanotiya CS, Kalra A (2014) ACC deaminase-containing Arthrobacter protophormiae induces NaCl stress tolerance through reduced ACC oxidase activity and ethylene production resulting in improved nodulation and mycorrhization in *Pisum sativum*. J Plant Physiol 171:884–894

Barra PJ, Inostroza NG, Acuña JJ, Mora ML, Crowley DE, Jorquera MA (2016) Formulation of bacterial consortia from avocado (*Persea americana* Mill.) and their effect on growth, biomass and superoxide dismutase activity of wheat seedlings under salt stress. Appl Soil Ecol 102:80–91

Bauder TA, Davis JG, Waskom RM, Cardon GE, Follett RH, Franklin WT, Heil RD (2004) Managing saline soils. Service in action; no 0503

Belimov AA, Dodd IC, Hontzeas N, Theobald JC, Safronova VI, Davies WJ (2009) Rhizosphere bacteria containing 1-aminocyclopropane-1-carboxylate deaminase increase yield of plants grown in drying soil via both local and systemic hormone signalling. New Phytol 181:413–423

Bell CW, Asao S, Calderon F, Wolk B, Wallenstein MD (2015) Plant nitrogen uptake drives rhizosphere bacterial community assembly during plant growth. Soil Biol Biochem 85:170–182

Bharti N, Barnawal D, Awasthi A, Yadav A, Kalra A (2014) Plant growth promoting rhizobacteria alleviate salinity induced negative effects on growth, oil content and physiological status in *Mentha arvensis*. Acta Physiol Plant 36:45–60

Bhattacharyya D, Yu S-M, Lee YH (2015) Volatile compounds from *Alcaligenes faecalis* JBCS1294 confer salt tolerance in *Arabidopsis thaliana* through the auxin and gibberellin pathways and differential modulation of gene expression in root and shoot tissues. Plant Growth Regul 75:297–306

Bianco C, Defez R (2009) *Medicago truncatula* improves salt tolerance when nodulated by an indole-3-acetic acid-overproducing Sinorhizobium meliloti strain. J Exp Bot 60:3097–3107

Boiero L, Perrig D, Masciarelli O, Penna C, Cassán F, Luna V (2007) Phytohormone production by three strains of Bradyrhizobium japonicum and possible physiological and technological implications. Appl Microbiol Biotechnol 74:874–880

Bouhmouch I, Souad-Mouhsine B, Brhada F, Aurag J (2005) Influence of host cultivars and Rhizobium species on the growth and symbiotic performance of *Phaseolus vulgaris* under salt stress. J Plant Physiol 162:1103–1113

Cassan F, Maiale S, Masciarelli O, Vidal A, Luna V, Ruiz O (2009) Cadaverine production by Azospirillum brasilense and its possible role in plant growth promotion and osmotic stress mitigation. Eur J Soil Biol 45:12–19

Çavusoglu K, Kabar K (2010) Effects of hydrogen peroxide on the germination and early seedling growth of barley under NaCl and high temperature stresses. EurAsian J Biosci 4:70–79

Chaves MM et al (2002) How plants cope with water stress in the field? Photosynthesis and growth. Ann Bot 89:907–916

Chinnusamy V, Jagendorf A, Zhu J-K (2005) Understanding and improving salt tolerance in plants. Crop Sci 45:437–448

Chinnusamy V, Zhu J, Zhu J-K (2006) Salt stress signaling and mechanisms of plant salt tolerance. In: Genetic engineering. Springer, p 141–177

Cohen AC, Travaglia CN, Bottini R, Piccoli PN (2009) Participation of abscisic acid and gibberellins produced by endophytic Azospirillum in the alleviation of drought effects in maize. Botany 87:455–462

Cohen AC et al (2015) Azospirillum brasilense ameliorates the response of Arabidopsis thaliana to drought mainly via enhancement of ABA levels. Physiol Plant 153:79–90

Compant S, Duffy B, Nowak J, Clément C, Barka EA (2005) Use of plant growth-promoting bacteria for biocontrol of plant diseases: principles, mechanisms of action, and future prospects. Appl Environ Microbiol 71:4951–4959

Cramer GR, Nowak RS (1992) Supplemental manganese improves the relative growth, net assimilation and photosynthetic rates of salt-stressed barley. Physiol Plant 84:600–605

Creus CM, Sueldo RJ, Barassi CA (2004) Water relations and yield in Azospirillum-inoculated wheat exposed to drought in the field. Can J Bot 82:273–281

Damodaran T et al (2014) Rhizosphere and endophytic bacteria for induction of salt tolerance in gladiolus grown in sodic soils. J Plant Interact 9:577–584

Dantas BF, Ribeiro LS, Aragão CA (2005) Physiological response of cowpea seeds to salinity stress. Rev Bras Sementes 27:144–148

Dietz K-J, Mittler R, Noctor G (2016) Recent progress in understanding the role of reactive oxygen species in plant cell signaling. Plant Physiol 171:1535–1539

Dimkpa C, Weinand T, Asch F (2009) Plant–rhizobacteria interactions alleviate abiotic stress conditions. Plant Cell Environ 32:1682–1694

Dodd IC, Pérez-Alfocea F (2012) Microbial amelioration of crop salinity stress. J Exp Bot 63:3415–3428

Dodd IC, Zinovkina NY, Safronova VI, Belimov AA (2010) Rhizobacterial mediation of plant hormone status. Ann Appl Biol 157:361–379

Dolatabadian A, Sanavy SAMM, Ghanati F (2011) Effect of salinity on growth, xylem structure and anatomical characteristics of soybean. Notulae Sci Biol 3:41

Dwivedi S, Upadhyaya H, Subudhi P, Gehring C, Bajic V, Ortiz R (2010) Enhancing abiotic stress tolerance in cereals through breeding and transgenic interventions. Plant Breed Rev 33:31–114

Egamberdieva D (2011) Survival of Pseudomonas extremorientalis TSAU20 and P. chlororaphis TSAU13 in the rhizosphere of common bean (*Phaseolus vulgaris*) under saline conditions. Plant Soil Environ 57:122–127

Egamberdieva D (2012) Pseudomonas chlororaphis: a salt-tolerant bacterial inoculant for plant growth stimulation under saline soil conditions. Acta Physiol Plant 34:751–756

Egamberdieva D, Jabborova D (2013) Biocontrol of cotton damping-off caused by Rhizoctonia solani in salinated soil with rhizosphere bacteria. Asian Australas J Plant Sci Biotechnol 7:31–38

Egamberdieva D, Kucharova Z (2009) Selection for root colonising bacteria stimulating wheat growth in saline soils. Biol Fertil Soils 45:563–571

Egamberdieva D, Gafurova L, Islam R (2007) Salinity effects on irrigated soil chemical and biological properties in the Syr Darya Basin of Uzbekistan

Egamberdieva D et al (2011) Bacteria able to control foot and root rot and to promote growth of cucumber in salinated soils. Biol Fertil Soils 47:197–205

Egamberdieva D, Berg G, Lindström K, Räsänen LA (2013) Alleviation of salt stress of symbiotic *Galega officinalis* L. (goat's rue) by co-inoculation of Rhizobium with root-colonizing Pseudomonas. Plant Soil 369:453–465

Egamberdiyeva D (2007) The effect of plant growth promoting bacteria on growth and nutrient uptake of maize in two different soils. Appl Soil Ecol 36:184–189

Egamberdiyeva D, Islam KR (2008) Salt-tolerant rhizobacteria: plant growth promoting traits and physiological characterization within ecologically stressed environments. In: Plant-bacteria interactions: strategies and techniques to promote plant growth. p 257–281

Esitken A, Yildiz HE, Ercisli S, Donmez MF, Turan M, Gunes A (2010) Effects of plant growth promoting bacteria (PGPB) on yield, growth and nutrient contents of organically grown strawberry. Sci Horticult 124:62–66

Essa TA (2002) Effect of salinity stress on growth and nutrient composition of three soybean (*Glycine max* L. Merrill) cultivars. J Agron Crop Sci 188:86–93

Essghaier B, Dhieb C, Rebib H, Ayari S, Boudabous ARA, Sadfi-Zouaoui N (2014) Antimicrobial behavior of intracellular proteins from two moderately halophilic bacteria: strain J31 of Terribacillus halophilus and strain M3-23 of Virgibacillus marismortui. J Plant Pathol Microbiol 5:1

Etesami H (2018) Can interaction between silicon and plant growth promoting rhizobacteria benefit in alleviating abiotic and biotic stresses in crop plants? Agric Ecosyst Environ 253:98–112. https://doi.org/10.1016/j.agee.2017.11.007

Etesami H, Alikhani HA (2016) Co-inoculation with endophytic and rhizosphere bacteria allows reduced application rates of N-fertilizer for rice plant. Rhizosphere 2:5–12

Etesami H, Beattie GA (2017) Plant-microbe interactions in adaptation of agricultural crops to abiotic stress conditions. In: Probiotics and plant health. Springer, p 163–200

Etesami H, Beattie GA (2018) Mining halophytes for plant growth-promoting halotolerant bacteria to enhance the salinity tolerance of non-halophytic crops. Front Microbiol 9:148. 1–20. https://doi.org/10.3389/fmicb.2018.00148

Etesami H, Maheshwari DK (2018) Use of plant growth promoting rhizobacteria (PGPRs) with multiple plant growth promoting traits in stress agriculture: action mechanisms and future prospects. Ecotoxicol Environ Saf 156:225–246. https://doi.org/10.1016/j.ecoenv.2018.03.013

Etesami H, Alikhani HA, Hosseini HM (2015a) Indole-3-acetic acid (IAA) production trait, a useful screening to select endophytic and rhizosphere competent bacteria for rice growth promoting agents. MethodsX 2:72–78

Etesami H, Alikhani HA, Hosseini HM (2015b) Indole-3-acetic acid and 1-aminocyclopropane-1-carboxylate deaminase: bacterial traits required in rhizosphere, rhizoplane and/or endophytic

competence by beneficial bacteria. In: Bacterial metabolites in sustainable agroecosystem. Springer, p 183–258
Etesami H, Emami S, Alikhani HA (2017) Potassium solubilizing bacteria (KSB): mechanisms, promotion of plant growth, and future prospects – a review. J Soil Sci Plant Nutr 17:897–911
Evelin H, Kapoor R, Giri B (2009) Arbuscular mycorrhizal fungi in alleviation of salt stress: a review. Ann Bot 104:1263–1280
Flowers TJ (2004) Improving crop salt tolerance. J Exp Bot 55:307–319
Ford CW (1984) Accumulation of low molecular weight solutes in water-stressed tropical legumes. Phytochemistry 23:1007–1015
Garcia C, Hernandez T (1996) Influence of salinity on the biological and biochemical activity of a calciorthird soil. Plant Soil 178:255–263
Gill SS, Tuteja N (2010) Reactive oxygen species and antioxidant machinery in abiotic stress tolerance in crop plants. Plant Physiol Biochem 48:909–930
Gill SS et al (2016) Piriformospora indica: potential and significance in plant stress tolerance. Front Microbiol 7:332
Glick BR (2005) Modulation of plant ethylene levels by the bacterial enzyme ACC deaminase. FEMS Microbiol Lett 251:1–7
Glick BR (2010) Using soil bacteria to facilitate phytoremediation. Biotechnol Adv 28:367–374
Glick BR (2014) Bacteria with ACC deaminase can promote plant growth and help to feed the world. Microbiol Res 169:30–39
Glick BR, Todorovic B, Czarny J, Cheng Z, Duan J, McConkey B (2007) Promotion of plant growth by bacterial ACC deaminase. Crit Rev Plant Sci 26:227–242
Godfray HCJ et al (2010) Food security: the challenge of feeding 9 billion people. Science 327:812–818
Groß F, Durner J, Gaupels F (2013) Nitric oxide, antioxidants and prooxidants in plant defence responses. Front Plant Sci 4:419
Grover M, Ali SZ, Sandhya V, Rasul A, Venkateswarlu B (2011) Role of microorganisms in adaptation of agriculture crops to abiotic stresses. World J Microbiol Biotechnol 27:1231–1240
Gupta B, Huang B (2014) Mechanism of salinity tolerance in plants: physiological, biochemical, and molecular characterization. Int J Genomics 2014:701596
Gupta KJ, Stoimenova M, Kaiser WM (2005) In higher plants, only root mitochondria, but not leaf mitochondria reduce nitrite to NO, in vitro and in situ. J Exp Bot 56:2601–2609
Hamdia MAE-S, Shaddad MAK, Doaa MM (2004) Mechanisms of salt tolerance and interactive effects of Azospirillum brasilense inoculation on maize cultivars grown under salt stress conditions. Plant Growth Regul 44:165–174
Hamilton CE, Bever JD, Labbé J, Yang X, Yin H (2016) Mitigating climate change through managing constructed-microbial communities in agriculture. Agric Ecosyst Environ 216:304–308
Han HS, Lee KD (2005) Physiological responses of soybean-inoculation of Bradyrhizobium japonicum with PGPR in saline soil conditions. Res J Agric Biol Sci 1:216–221
Han Q-Q et al (2014) Beneficial soil bacterium Bacillus subtilis (GB03) augments salt tolerance of white clover. Front Plant Sci 5:525
Hashem A, Abd_Allah EF, Alqarawi AA, Aldubise A, Egamberdieva D (2015) Arbuscular mycorrhizal fungi enhances salinity tolerance of Panicum turgidum Forssk by altering photosynthetic and antioxidant pathways. J Plant Interact 10:230–242
Heidari M, Jamshid P (2010) Interaction between salinity and potassium on grain yield, carbohydrate content and nutrient uptake in pearl millet. ARPN J Agric Biol Sci 5:39–46
Heidari M, Mousavinik SM, Golpayegani A (2011) Plant growth promoting rhizobacteria (PGPR) effect on physiological parameters and mineral uptake in basil (Ocimum basilicum L.) under water stress. ARPN J Agric Biol Sci 6:6–11
Horneck DA, Ellsworth JW, Hopkins BG, Sullivan DM, Stevens RG (2007) Managing salt-affected soils for crop production. Oregon State University Extension Service, Covallis

Hu Y, Schmidhalter U (2004) Limitation of salt stress to plant growth. In: Hock E (ed) Plant toxicology, vol 4. Marcel Dekker, New York, pp 191–224

James RA, Blake C, Byrt CS, Munns R (2011) Major genes for Na+ exclusion, Nax1 and Nax2 (wheat HKT1; 4 and HKT1; 5), decrease Na+ accumulation in bread wheat leaves under saline and waterlogged conditions. J Exp Bot 62:2939–2947

Jamil A, Riaz S, Ashraf M, Foolad MR (2011) Gene expression profiling of plants under salt stress. Crit Rev Plant Sci 30:435–458

Jha Y, Subramanian RB, Patel S (2011) Combination of endophytic and rhizospheric plant growth promoting rhizobacteria in *Oryza sativa* shows higher accumulation of osmoprotectant against saline stress. Acta Physiol Plant 33:797–802

Kandowangko NY, Suryatmana G, Nurlaeny N, Simanungkalit RDM (2009) Proline and abscisic acid content in droughted corn plant inoculated with Azospirillum sp. and Arbuscular mycorrhizae fungi. Hayati J Biosci 16:15–20

Kang S-M et al (2014a) Plant growth-promoting rhizobacteria reduce adverse effects of salinity and osmotic stress by regulating phytohormones and antioxidants in *Cucumis sativus*. J Plant Interact 9:673–682

Kang S-M et al (2014b) Gibberellin secreting rhizobacterium, Pseudomonas putida H-2-3 modulates the hormonal and stress physiology of soybean to improve the plant growth under saline and drought conditions. Plant Physiol Biochem 84:115–124

Kaushal M, Wani SP (2016) Rhizobacterial-plant interactions: strategies ensuring plant growth promotion under drought and salinity stress. Agric Ecosyst Environ 231:68–78

Kechid M, Desbrosses G, Rokhsi W, Varoquaux F, Djekoun A, Touraine B (2013) The NRT2. 5 and NRT2. 6 genes are involved in growth promotion of Arabidopsis by the plant growth-promoting rhizobacterium (PGPR) strain Phyllobacterium brassicacearum STM196. New Phytol 198:514–524

Khodarahmpour Z, Ifar M, Motamedi M (2012) Effects of NaCl salinity on maize (*Zea mays* L.) at germination and early seedling stage. Afr J Biotechnol 11:298–304

Kim K, Jang Y-J, Lee S-M, Oh B-T, Chae J-C, Lee K-J (2014) Alleviation of salt stress by Enterobacter sp. EJ01 in tomato and Arabidopsis is accompanied by up-regulation of conserved salinity responsive factors in plants. Mol Cell 37:109

Kurepin LV, Park JM, Lazarovits G, Bernards MA (2015) Burkholderia phytofirmans-induced shoot and root growth promotion is associated with endogenous changes in plant growth hormone levels. Plant Growth Regul 75:199–207

Ladha JK, Tirol-Padre A, Punzalan GC, Castillo E, Singh U, Reddy CK (1998) Nondestructive estimation of shoot nitrogen in different rice genotypes. Agron J 90:33–40

Lugtenberg B, Kamilova F (2009) Plant-growth-promoting rhizobacteria. Annu Rev Microbiol 63:541–556

Lugtenberg BJJ, Malfanova N, Kamilova F, Berg G (2013) Plant growth promotion by microbes. Mol Microb Ecol Rhizosphere 1 & 2:559–573

Manchanda G, Garg N (2008) Salinity and its effects on the functional biology of legumes. Acta Physiol Plant 30:595–618

Mantelin S, Touraine B (2004) Plant growth-promoting bacteria and nitrate availability: impacts on root development and nitrate uptake. J Exp Bot 55:27–34

Mantri N, Patade V, Penna S, Ford R, Pang E (2012) Abiotic stress responses in plants: present and future. In: Abiotic stress responses in plants. Springer, p 1–19

Mapelli F et al (2013) Potential for plant growth promotion of rhizobacteria associated with Salicornia growing in Tunisian hypersaline soils. BioMed Res Int 2013:248078

Mayak S, Tirosh T, Glick BR (2004a) Plant growth-promoting bacteria confer resistance in tomato plants to salt stress. Plant Physiol Biochem 42:565–572

Mayak S, Tirosh T, Glick BR (2004b) Plant growth-promoting bacteria that confer resistance to water stress in tomatoes and peppers. Plant Sci 166:525–530

Miller GAD, Suzuki N, Ciftci-Yilmaz S, Mittler RON (2010) Reactive oxygen species homeostasis and signalling during drought and salinity stresses. Plant Cell Environ 33:453–467

Munns R (2002) Comparative physiology of salt and water stress. Plant Cell Environ 25:239–250
Munns R (2005) Genes and salt tolerance: bringing them together. New Phytol 167:645–663
Munns R, Tester M (2008) Mechanisms of salinity tolerance. Annu Rev Plant Biol 59:651–681
Nabti E et al (2010) Restoration of growth of durum wheat (*Triticum durum* var. waha) under saline conditions due to inoculation with the rhizosphere bacterium Azospirillum brasilense NH and extracts of the marine alga *Ulva lactuca*. J Plant Growth Regul 29:6–22
Nadeem SM, Zahir ZA, Naveed M, Arshad M (2007) Preliminary investigations on inducing salt tolerance in maize through inoculation with rhizobacteria containing ACC deaminase activity. Can J Microbiol 53:1141–1149
Nadeem SM, Zahir ZA, Naveed M, Arshad M (2009) Rhizobacteria containing ACC-deaminase confer salt tolerance in maize grown on salt-affected fields. Can J Microbiol 55:1302–1309
Nadeem SM, Zahir ZA, Naveed M, Nawaz S (2013) Mitigation of salinity-induced negative impact on the growth and yield of wheat by plant growth-promoting rhizobacteria in naturally saline conditions. Ann Microbiol 63:225–232
Nadeem SM, Ahmad M, Zahir ZA, Javaid A, Ashraf M (2014) The role of mycorrhizae and plant growth promoting rhizobacteria (PGPR) in improving crop productivity under stressful environments. Biotechnol Adv 32:429–448
Nadeem SM, Ahmad M, Naveed M, Imran M, Zahir ZA, Crowley DE (2016) Relationship between in vitro characterization and comparative efficacy of plant growth-promoting rhizobacteria for improving cucumber salt tolerance. Arch Microbiol 198:379–387. https://doi.org/10.1007/s00203-016-1197-5
Naqvi SM, Ansari R (1974) Estimation of diffusible auxin under saline growth conditions. Cell Mol Life Sci 30:350–350
Naz I, Bano A, Ul-Hassan T (2009) Isolation of phytohormones producing plant growth promoting rhizobacteria from weeds growing in Khewra salt range, Pakistan and their implication in providing salt tolerance to *Glycine max* L. Afr J Biotechnol 8:5762–5768
Neamatollahi E, Bannayan M, Darban AS, Ghanbari A (2009) Hydropriming and osmopriming effects on cumin (*Cuminum Cyminum* L.) seeds germination. World Acad Sci Eng Technol 57:526–529
Netondo GW, Onyango JC, Beck E (2004) Sorghum and salinity. Crop Sci 44:797–805
Nia SH, Zarea MJ, Rejali F, Varma A (2012) Yield and yield components of wheat as affected by salinity and inoculation with Azospirillum strains from saline or non-saline soil. J Saudi Soc Agric Sci 11:113–121
Nunkaew T, Kantachote D, Nitoda T, Kanzaki H, Ritchie RJ (2015) Characterization of exopolymeric substances from selected Rhodopseudomonas palustris strains and their ability to adsorb sodium ions. Carbohydr Polym 115:334–341
Ondrasek G, Rengel Z, Romic D, Poljak M, Romic M (2009) Accumulation of non/essential elements in radish plants grown in salt-affected and cadmium-contaminated environment. Cereal Res Commun 37:9–12
Orhan F (2016) Alleviation of salt stress by halotolerant and halophilic plant growth-promoting bacteria in wheat (*Triticum aestivum*). Braz J Microbiol 47:621–627. https://doi.org/10.1016/j.bjm.2016.04.001
Othman Y, Al-Karaki G, Al-Tawaha AR, Al-Horani A (2006) Variation in germination and ion uptake in barley genotypes under salinity conditions. World J Agric Sci 2:11–15
Parida AK, Das AB (2005) Salt tolerance and salinity effects on plants: a review. Ecotoxicol Environ Saf 60:324–349
Paul D, Lade H (2014) Plant-growth-promoting rhizobacteria to improve crop growth in saline soils: a review. Agron Sustain Dev 34:737–752
Paul D, Nair S (2008) Stress adaptations in a plant growth promoting rhizobacterium (PGPR) with increasing salinity in the coastal agricultural soils. J Basic Microbiol 48:378–384
Paul D, Sarma YR (2006) Plant growth promoting rhizhobacteria (PGPR)-mediated root proliferation in black pepper (*Piper nigrum* L.) as evidenced through GS root software. Arch Phytopathol Plant Protect 39:311–314

Paul D, Anandaraj M, Kumar A, Sarma YR (2005) Antagonistic mechanisms of fluorescent pseudomonads against Phytophthora capsici in black pepper (*Piper nigrum* L.). J Spices Aromat Crop 14:122–129

Penrose DM, Glick BR (2003) Methods for isolating and characterizing ACC deaminase-containing plant growth-promoting rhizobacteria. Physiol Plant 118:10–15

Penrose DM, Moffatt BA, Glick BR (2001) Determination of 1-aminocycopropane-1-carboxylic acid (ACC) to assess the effects of ACC deaminase-containing bacteria on roots of canola seedlings. Can J Microbiol 47:77–80

Pérez-Alfocea F, Albacete A, Ghanem ME, Dodd IC (2010) Hormonal regulation of source–sink relations to maintain crop productivity under salinity: a case study of root-to-shoot signalling in tomato. Funct Plant Biol 37:592–603

Pitzschke A, Forzani C, Hirt H (2006) Reactive oxygen species signaling in plants. Antioxid Redox Signal 8:1757–1764

Postma JA, Lynch JP (2011) Root cortical aerenchyma enhances the growth of maize on soils with suboptimal availability of nitrogen, phosphorus, and potassium. Plant Physiol 156:1190–1201

Prakash L, Prathapasenan G (1990) Interactive effect of NaCl salinity and gibberellic acid on shoot growth, content of abscisic acid and gibberellin-like substances and yield of rice (*Oryza sativa* L. var GR-3). Proc Plant Sci 100:173–181

Qadir M, Ghafoor A, Murtaza G (2000) Amelioration strategies for saline soils: a review. Land Degrad Dev 11:501–521

Quispel A (1988) Bacteria-plant interactions in symbiotic nitrogen fixation. Physiol Plant 74:783–790

Rabie GH, Almadini AM (2005) Role of bioinoculants in development of salt-tolerance of *Vicia faba* plants under salinity stress. Afr J Biotechnol 4:210

Rabie GH, Aboul-Nasr MB, Al-Humiany A (2005) Increased salinity tolerance of cowpea plants by dual inoculation of an arbuscular mycorrhizal fungus Glomus clarum and a nitrogen-fixer Azospirillum brasilense. Mycobiology 33:51–60

Rahman S, Matsumuro T, Miyake H, Takeoka Y (2000) Salinity-induced ultrastructural alterations in leaf cells of rice (*Oryza sativa* L.). Plant Prod Sci 3:422–429

Rahnama A, James RA, Poustini K, Munns R (2010) Stomatal conductance as a screen for osmotic stress tolerance in durum wheat growing in saline soil. Funct Plant Biol 37:255–263

Rajput L, Imran A, Mubeen F, Hafeez FY (2013) Salt-tolerant PGPR strain Planococcus rifietoensis promotes the growth and yield of wheat (*Triticum aestivum* L.) cultivated in saline soil. Pak J Bot 45:1955–1962

Ramadoss D, Lakkineni VK, Bose P, Ali S, Annapurna K (2013) Mitigation of salt stress in wheat seedlings by halotolerant bacteria isolated from saline habitats. SpringerPlus 2:6

Rengasamy P (2006) World salinization with emphasis on Australia. J Exp Bot 57:1017–1023. https://doi.org/10.1093/jxb/erj108

Rincón A, Valladares F, Gimeno TE, Pueyo JJ (2008) Water stress responses of two Mediterranean tree species influenced by native soil microorganisms and inoculation with a plant growth promoting rhizobacterium. Tree Physiol 28:1693–1701

Rozema J, Flowers T (2008) Crops for a salinized world. Science 322:1478–1480

Sairam RK, Srivastava GC (2002) Changes in antioxidant activity in sub-cellular fractions of tolerant and susceptible wheat genotypes in response to long term salt stress. Plant Sci 162:897–904

Sakhabutdinova AR, Fatkhutdinova DR, Bezrukova MV, Shakirova FM (2003) Salicylic acid prevents the damaging action of stress factors on wheat plants. Bulg J Plant Physiol 29:314–319

Sandhya V, Ali SZ, Grover M, Reddy G, Venkateswarlu B (2010) Effect of plant growth promoting Pseudomonas spp. on compatible solutes, antioxidant status and plant growth of maize under drought stress. Plant Growth Regul 62:21–30

Saravanakumar D, Samiyappan R (2007) ACC deaminase from *Pseudomonas fluorescens* mediated saline resistance in groundnut (Arachis hypogea) plants. J Appl Microbiol 102:1283–1292

Sarma RK, Saikia R (2014) Alleviation of drought stress in mung bean by strain *Pseudomonas aeruginosa* GGRJ21. Plant Soil 377:111–126

Saxena SC, Kaur H, Verma P, Petla BP, Andugula VR, Majee M (2013) Osmoprotectants: potential for crop improvement under adverse conditions. In: Plant acclimation to environmental stress. Springer, p 197–232

Serraj R, Sinclair TR (2002) Osmolyte accumulation: can it really help increase crop yield under drought conditions? Plant Cell Environ 25:333–341

Shaharoona B, Arshad M, Zahir ZA (2006) Effect of plant growth promoting rhizobacteria containing ACC-deaminase on maize (*Zea mays* L.) growth under axenic conditions and on nodulation in mung bean (*Vigna radiata* L.). Lett Appl Microbiol 42:155–159

Shahbaz M, Ashraf M (2013) Improving salinity tolerance in cereals. Crit Rev Plant Sci 32:237–249

Shahzad SM, Khalid A, Arshad M (2010) Screening rhizobacteria containing ACC-deaminase for growth promotion of chickpea seedlings under axenic conditions. Soil Environ 29:38–46

Shannon MC, Grieve CM (1998) Tolerance of vegetable crops to salinity. Sci Horticult 78:5–38

Shaterian J, Waterer D, De Jong H, Tanino KK (2005a) Differential stress responses to NaCl salt application in early-and late-maturing diploid potato (Solanum sp.) clones. Environ Exp Bot 54:202–212

Shaterian J, Waterer D, Jong HD, Tanino KK (2005b) Differential stress responses to NaCl salt application in early- and late-maturing diploid potato (Solanum sp.) clones. Environ Exp Bot 54:202–212. https://doi.org/10.1016/j.envexpbot.2004.07.005

Shintu PV, Jayaram KM (2015) Phosphate solubilising bacteria (Bacillus polymyxa)-an effective approach to mitigate drought in tomato (*Lycopersicon esculentum* Mill.). Trop Plant Res 2:17–22

Shirokova Y, Forkutsa I, Sharafutdinova N (2000) Use of electrical conductivity instead of soluble salts for soil salinity monitoring in Central Asia. Irrig Drain Syst 14:199–206

Shrivastava UP, Kumar A (2013) Characterization and optimization of 1-aminocyclopropane-1-carboxylate deaminase (ACCD) activity in different rhizospheric PGPR along with Microbacterium sp. strain ECI-12A. Int J Appl Sci Biotechnol 1:11–15

Shrivastava P, Kumar R (2015) Soil salinity: a serious environmental issue and plant growth promoting bacteria as one of the tools for its alleviation. Saudi J Biol Sci 22:123–131

Shukla PS, Agarwal PK, Jha B (2012) Improved salinity tolerance of Arachishypogaea (L.) by the interaction of halotolerant plant-growth-promoting rhizobacteria. J Plant Growth Regul 31:195–206

Siddikee MA, Chauhan PS, Anandham R, Han G-H, Sa T (2010) Isolation, characterization, and use for plant growth promotion under salt stress, of ACC deaminase-producing halotolerant bacteria derived from coastal soil. J Microbiol Biotechnol 20:1577–1584

Siddikee MA, Glick BR, Chauhan PS, Jong Yim W, Sa T (2011) Enhancement of growth and salt tolerance of red pepper seedlings (*Capsicum annuum* L.) by regulating stress ethylene synthesis with halotolerant bacteria containing 1-aminocyclopropane-1-carboxylic acid deaminase activity. Plant Physiol Biochem 49:427–434

Singh JS, Pandey VC, Singh DP (2011) Efficient soil microorganisms: a new dimension for sustainable agriculture and environmental development. Agric Ecosyst Environ 140:339–353

Singleton PW, Bohlool BB (1984) Effect of salinity on nodule formation by soybean. Plant Physiol 74:72–76

Štajner D, Kevrešan S, Gašić O, Mimica-Dukić N, Zongli H (1997) Nitrogen and *Azotobacter chroococcum* enhance oxidative stress tolerance in sugar beet. Biol Plant 39:441

Sziderics AH, Rasche F, Trognitz F, Sessitsch A, Wilhelm E (2007) Bacterial endophytes contribute to abiotic stress adaptation in pepper plants (*Capsicum annuum* L.). Can J Microbiol 53:1195–1202

Tank N, Saraf M (2010) Salinity-resistant plant growth promoting rhizobacteria ameliorates sodium chloride stress on tomato plants. J Plant Interact 5:51–58

Tardieu F, Parent B, Simonneau T (2010) Control of leaf growth by abscisic acid: hydraulic or non-hydraulic processes? Plant Cell Environ 33:636–647
Tavakkoli E, Fatehi F, Coventry S, Rengasamy P, McDonald GK (2011) Additive effects of Na+ and Cl–ions on barley growth under salinity stress. J Exp Bot 62:2189–2203
Tejada M, Garcia C, Gonzalez JL, Hernandez MT (2006) Use of organic amendment as a strategy for saline soil remediation: influence on the physical, chemical and biological properties of soil. Soil Biol Biochem 38:1413–1421. https://doi.org/10.1016/j.soilbio.2005.10.017
Tester M, Davenport R (2003) Na+ tolerance and Na+ transport in higher plants. Ann Bot 91:503–527
Tewari S, Arora NK (2014) Multifunctional exopolysaccharides from *Pseudomonas aeruginosa* PF23 involved in plant growth stimulation, biocontrol and stress amelioration in sunflower under saline conditions. Curr Microbiol 69:484–494
Tripathi AK, Mishra BM, Tripathi P (1998) Salinity stress responses in the plant growth promoting rhizobacteria, Azospirillum spp. J Biosci 23:463–471
Upadhyay SK, Singh DP (2015) Effect of salt-tolerant plant growth-promoting rhizobacteria on wheat plants and soil health in a saline environment. Plant Biol 17:288–293
Upadhyay SK, Singh JS, Saxena AK, Singh DP (2012) Impact of PGPR inoculation on growth and antioxidant status of wheat under saline conditions. Plant Biol 14:605–611
Van Loon LC, Bakker P, Pieterse CMJ (1998) Systemic resistance induced by rhizosphere bacteria. Annu Rev Phytopathol 36:453–483
Vardharajula S, Zulfikar Ali S, Grover M, Reddy G, Bandi V (2011) Drought-tolerant plant growth promoting Bacillus spp.: effect on growth, osmolytes, and antioxidant status of maize under drought stress. J Plant Interact 6:1–14
Venkateswarlu B, Shanker AK (2009) Climate change and agriculture: adaptation and mitigation strategies. Indian J Agron 54:226
Vimal SR, Singh JS, Karora N, Singh DP (2016) PGPR: an effective bio-agent in stress agricultural management
Wang C-J et al (2012) Induction of drought tolerance in cucumber plants by a consortium of three plant growth-promoting rhizobacterium strains. PLoS One 7:e52565
Wang L, Sun X, Li S, Zhang T, Zhang W, Zhai P (2014) Application of organic amendments to a coastal saline soil in north China: effects on soil physical and chemical properties and tree growth. PLoS One 9:e89185
Werner JE, Finkelstein RR (1995) Arabidopsis mutants with reduced response to NaCl and osmotic stress. Physiol Plant 93:659–666
Wood NT (2001) Nodulation by numbers: the role of ethylene in symbiotic nitrogen fixation. Trends Plant Sci 6:501–502
Xu G-Y et al (2011) A novel rice calmodulin-like gene, OsMSR2, enhances drought and salt tolerance and increases ABA sensitivity in Arabidopsis. Planta 234:47–59
Yamaguchi T, Blumwald E (2005) Developing salt-tolerant crop plants: challenges and opportunities. Trends Plant Sci 10:615–620
Yang J, Kloepper JW, Ryu C-M (2009) Rhizosphere bacteria help plants tolerate abiotic stress. Trends Plant Sci 14:1–4
Yao L, Wu Z, Zheng Y, Kaleem I, Li C (2010) Growth promotion and protection against salt stress by Pseudomonas putida Rs-198 on cotton. Eur J Soil Biol 46:49–54
Yasmin F, Othman R, Saad MS, Sijam K (2007) Screening for beneficial properties of Rhizobacteria isolated from sweet potato rhizosphere. Biotechnology 6:49–52
Yildirim E, Taylor AG (2005) Effect of biological treatments on growth of bean plants under salt stress. Science 123:1
Yildirim E, Turan M, Ekinci M, Dursun A, Cakmakci R (2011) Plant growth promoting rhizobacteria ameliorate deleterious effect of salt stress on lettuce. Sci Res Essays 6:4389–4396
Zahir ZA, Munir A, Asghar HN, Shaharoona B, Arshad M (2008) Effectiveness of rhizobacteria containing ACC deaminase for growth promotion of peas (*Pisum sativum*) under drought conditions. J Microbiol Biotechnol 18:958–963

Zaki NM, Ahmed MA, Hassanein MS (2004) Growth and yield of some wheat cultivars irrigated with saline water in newly cultivated land as affected by nitrogen fertilization. Ann Agric Sci Moshtohor 42:515–525

Zhang H, Kim M-S, Sun Y, Dowd SE, Shi H, Paré PW (2008) Soil bacteria confer plant salt tolerance by tissue-specific regulation of the sodium transporter HKT1. Mol Plant-Microbe Interact 21:737–744

Zhang M, Smith JAC, Harberd NP, Jiang C (2016) The regulatory roles of ethylene and reactive oxygen species (ROS) in plant salt stress responses. Plant Mol Biol 91:651–659

Zhou W, Qin S, Lyu D, Zhang P (2015) Soil sterilisation and plant growth-promoting rhizobacteria promote root respiration and growth of sweet cherry rootstocks. Arch Agron Soil Sci 61:361–370

Restoration of Plant Growth Under Saline Soil by Halotolerant Plant Growth-Promoting Rhizobacteria (PGPR)

2

Leila Bensidhoum, Sylia Ait Bessai, and Elhafid Nabti

Abstract

Saline soils are among the big obstacles in agricultural activities in arid and semiarid regions. Numerous cultivated areas in the world's irrigated soils are severely affected by both of salinity and salinization. Therefore, the development and sustainability of profitable agricultural systems are seriously threatened. Sustainable and cost-effective plant growth becomes even more important due to the need of energy-efficient plant growth for biomass and bioenergy production, especially in soils of lower quality, in addition to food production.

The utilization of halotolerant PGPRs (plant growth-promoting rhizobacteria) could help plant growth to adapt to increased salinity by the potential osmoregulatory mechanism implicated in the cell function regulation. Consequently, PGPRs provide plants with their activities to challenge osmotic stress by supporting them in the restoration of essential activities. These bacteria stimulate plant growth under saline conditions by diverse mechanisms such as phytohormone production and ACC deaminase synthesis to decrease inhibitory effect of ethylene occurring in plants under salt stress conditions. Halotolerant PGPRs can also use the natural osmoprotectants realized from marine algae to challenge salt stress. It is important to highlight that halotolerant PGPRs stimulate plant growth under elevated salinity by using similar mechanisms like other halosensitive bacteria.

Keywords

Halotolerant rhizobacteria · Plant growth · Salts stress · Inoculation · Restoration

L. Bensidhoum · S. Ait Bessai · E. Nabti (✉)
Laboratoire de Maitrise des Energies Renouvelables, Faculté des Sciences de la Nature et de la Vie, Université de Bejaia, Bejaia, Algeria

FSNV, University de Bejaia, Bejaia, Algeria
e-mail: nabtielhafid1977@yahoo.com

M. Kumar et al. (eds.), *Saline Soil-based Agriculture by Halotolerant Microorganisms*, https://doi.org/10.1007/978-981-13-8335-9_2

2.1 Introduction

The rapidly increasing human population is causing a number of challenges to sustain life on earth (Ashraf et al. 2012). In 2004, the United Nations estimated that at the end of 2025 the human population will cross 8 billion (6.8 billion will be in Asia and Africa), while at the end of 2050 it could be over 10 billion. Therefore, to satisfy the nutrient demands of this growing population, the global food production will need to increase by 38% by 2025 and by 57% by 2050 if food supply to the growing world population is to be maintained at current levels (Wild 2003).

This growing population requires increasing food quantity and quality, especially in countries where resources are damaged due to uncontrolled human activities and environmental degradation. Agriculture is the main source of food for humans, and vegetables play an important role in humans' diet by providing and assisting the body with a variety of important constituents such as minerals, vitamins, complex carbohydrate, high dietary fiber, low levels of fat, and high amount of water. Agricultural production needs to jump by 70–100% to satisfy global increased demands in the next few years (Lafe 2013). However, agriculture practices are threatened by various biotic and abiotic stresses, which cause large losses in crop production (Singh et al. 2014).

It is well known that the crop losses due to abiotic stresses are more severe than those caused by the biotic ones (Ashraf et al. 2012). Nearly 82% of potential yield of crops is lost due to abiotic stresses every year, and the available productive areas of arable land are decreasing worldwide (Mona et al. 2016). The agricultural lands are threatened by various environmental stresses such as salinity, drought, extreme temperatures, pH, mineral deficiencies or toxicities, heavy metals, nutrient depletion, soil erosion, etc. that seriously reduce plant yields (Chedlly et al. 2008; Anonymous 2015).

Among these constraints, the salinization of agricultural lands imposes enormous challenges, both for scientists and for farmers, demanding a growing exploitation of natural resources in search of promising solutions (Ashraf et al. 2012).

To alleviate salt stress in crop plant, several strategies have been applied; these approaches include chemical amendment, improved irrigation practices, plant breeding techniques, and plant genetic modification which are time-consuming, difficult, and costly (Saghafi et al. 2013; Selvakumar et al. 2014; Shrivastava and Kumar 2015). Using biological product, particularly plant growth-promoting rhizobacteria (PGPRs), may prove to be a beneficial, cost-effective, and eco-friendly strategy that can improve crop productivity and soil health under normal and stressed environment. It presents a great scientific and agronomic interest since they play an important role in biological functioning of the rhizosphere and therefore improve growth, yield, and health of agricultural crops.

PGPRs, mainly bacteria belonging to the genera *Arthrobacter*, *Azotobacter*, *Azospirillum*, *Bacillus*, *Enterobacter*, *Pseudomonas*, *Serratia* (Gray and Smith 2005), and *Streptomyces* spp. (Tokala et al. 2002), influence positively plant health by improving its growth and/or protecting it against abiotic and biotic stress. These microorganisms are able to grow and compete with other microorganisms to occupy rhizosphere zone. Plant growth-promoting rhizobacteria colonize the rhizosphere using root exudates as nutrient substrates but unlike other rhizosphere bacteria improve plant growth and health via a variety of mechanisms (Vacheron et al. 2013).

Recent focus of research involves the use of PGPRs to alleviate salt stress in plant. Kasim et al. (2016) showed that the deleterious effect of salt stress on some plant growth parameters (seedling length, fresh and dry masses, as well as relative water content) was alleviated by bacterial inoculation.

2.2 Salinity and Salinization

Salinity is the second major limiting factor after drought, which is putting down crop production and adversely impacts plant growth and crop production worldwide, mainly in arid and semiarid regions (Pessarakli 1991; Saranga et al. 2001; Yadav et al. 2011).

Soil salinization is a term used for the accumulation of soluble salts in soils at a level that negatively affects agricultural productivity, environmental health, and economic welfare (Rengasamy 2006) that results in the formation of a salt-affected soil (Ciseau 2006). It costs the world 12 billion US$ a year in reduced income (Ghassemi et al. 1995).

In 2005, the FAO has estimated that over 6% of the total continent area equal to an area of about 800 million hectares is salt affected in the world, by either sodicity (434 million h) or salinity (397 million h). When there is a high concentration of Na^+ (sodium) in soil, it is referred to as "sodicity," and when Cl^- (chloride) or other salts are involved, it is referred as "salinity." It has been estimated that worldwide 20% of total cultivated and 33% of irrigated agricultural lands are afflicted by high salinity (Shrivastava and Kumar 2015).

2.2.1 Salinity

Salinity is considered as one of the most brutal environmental factors limiting the crop plant productivity because most of the crop plants are sensitive to high concentrations of salts in soil, and the area of land affected by it is increasing day by day (Shrivastava and Kumar 2015). Of the global threats that collectively compromise about 10 ha arable land per minute (Griggs et al. 2013), salinization contributes about 30% (Buringh 1978), and due to this problem, the area that can be used for agriculture is reducing by 1–2% every year, hitting hardest in arid and semiarid regions (FAO 2002; Shilpi et al. 2008). It has been expected that 30% of arable land is likely to become saline by 2025 (Munns 2002) and more than 50% of all agricultural soils will be affected by salinity increase by 2050 (Jamil et al. 2011; Dikilitas and Karakas 2012; Stanković et al. 2015).

This alarming increase of salt-affected areas is due to various reasons such as low precipitation, high surface evaporation (reduction of soil water content by removing pure water as vapor), weathering of native rocks, irrigation with saline water, exaggerated use of fertilizers, low rainfall, high temperature, and over-irrigation causing capillary movement of hidden salts from lower layers in soil and other poor cultural practices (Yao et al. 2010; Shrivastava and Kumar 2015). According to Patel et al. (2011), this continuing increase in salt-affected areas is due to the introduction of irrigation in new areas, especially in arid and semiarid regions, because such environment received less rainfall and then more water is transpired to air and ultimately net water movement is upward leaving salts on soil surface.

2.2.2 Salinization

Soil salinization consists of water-soluble salt accumulation in soil that includes cations such as sodium (Na^+), potassium (K^+), calcium (Ca^{2+}), and magnesium (mg^{2+}) along with anions chloride (Cl^-), sulfate (SO_4^{2-}), nitrate (NO_3^-), bicarbonate (HCO_3^-), and carbonate (CO_3^{2-}) (Diby and Harshad 2014).

Salts occur as ions, which may be released from weathering minerals in soil, may be applied through irrigation water or as fertilizers, or sometimes migrate upward in soil from shallow groundwater. When precipitation is insufficient to leach ions from the soil profile, salts accumulate in soil resulting in a high salinity (Blaylock 1994).

Soil salinity in agriculture refers to high concentration of soluble salts in soil moisture of the root zone (Diby and Harshad 2014). According to Ghassemi et al. (1995), soil salinization may be either a natural process, also known as primary salinization, or a human-induced (anthropogenic) process, known as secondary salinization.

The major causes of naturally induced salinity are intrusion of highly salinized water in coastal (oceans and sea) or continental (fossil salt aquifers) regions, aeolian by wind-borne salt from salt surface waters (oceans, lakes) deposited inland, and dissolution of parent rock material. In contrast, secondary salinization is mainly caused by some agricultural practices (irrigation with brackish water, fertigation, application of inorganic/organic soil amendments, and improper cropping patterns and rotations), deforestation, construction of reservoirs, salt farming, etc. (Mitsuchi et al. 1986; Rengasamy 2006; Biggs et al. 2010).

In 2006, Rengasamy, based on soil and groundwater problem all over the world, classified the salinity into three major types, differently to the classification of "primary" or "secondary" salinity as defined by Ghassemi et al. (1995). According to this classification, salinity may be associated to three sources:

(i) Groundwater-associated salinity (GAS), when water exits from ground to the soil surface with upward movement, it brings dissolved salts in it, and at soil surface, water will evaporate, and salts will be accumulated. It was reported by Talsma (1963) that salt accumulation is high when the water table is less than 1.5 m below the soil surface.
(ii) Non-groundwater-associated salinity (NAS), in this case, salts are introduced by rain, weathering, and aeolian deposits and stored within the soil solum.
(iii) Irrigation-associated salinity (IAS), because of insufficient leaching, salts introduced by irrigation water are stored within root zone.

Additionally, about one-third of the world's irrigated lands contributing with more than 30% of global agricultural production are salt affected (Munns and Tester 2008), because irrigation modifies soil hydration balance by generating an extra supply of water; this supply is always associated with added salts and consequently results in soil becoming salty and unproductive (Wu et al. 2008).

Soil salinity is measured by passing an electrical current through a soil solution extracted from soil sample, and the ability of this "saturated solution" to carry current is called electrical conductivity (EC) (Shin et al. 2016).

In general, soils are considered to be saline if they have concentration sufficient to interfere with the growth of most of crop species (Bui 2013); however, US Department of Agriculture (USDA) (1954) defines a saline soil as one in which EC measured in units of decisiemens/meter or millisiemens/cm of the saturation extract (ECe) in the root zone exceeds 4 dS m^{-1} (approximately 40 mM NaCl) at 25° and an exchangeable sodium percentage (ESP) of 15 or more.

2.3 Effects of Salinity on Soil, Plants, and Microorganisms

2.3.1 Effect on Soil Quality

Soil salinity is not a recent issue; it has been a major concern to global agriculture throughout human history. It is considered as a major factor reducing crop productivity and a main cause to abandon lands designed for agricultural purposes (Dodd and Perez-Alfocea 2012).

In addition, salinity not only decreases agricultural production of most crops but also affects soil physicochemical as well as biological properties (soil quality) and ecological balance of area (Benlloch-Gonzalez et al. 2005; Shrivastava and Kumar 2015; Arshadullah et al. 2017); it is also responsible for low economic returns (Hu and Schmidhalter 2002).

It is well documented that excess accumulation of salt leads to negative effects on soil structure, increase in osmotic pressure, and decrease in water potential and declines turgor potential of cells (plants and microbes). The most soil processes that have immediate consequences for soil biological activity or soil conservation occur within the soil pores (either in the pore space or on the surface of particles forming the pores); for this reason, pores are considered as central physical features in soil structure (Kay 1990; Rengasamy and Olsson 1991). However, the elevation of sodium (Na^+) concentration causes soil dispersion, swelling of the clay platelets, and disruption of the forces involved in binding of clay particles. These changes cause clay particles to plug soil pores (especially in fine-textured soil horizons) (Burrow et al. 2002) leading to reduction in soil aeration (prevalence of anaerobic conditions), water infiltration (deteriorating soil hydraulic properties), and also soil compaction (Abu-Sharar et al. 1987). Thus, excessive sodium concentration raises soil pH, soil erosion, and water runoff (Ondrasek et al. 2010; Diby and Harshad 2014). Thereby, the presence of multivalent cations can increase the sorption of organic matter to soil particles (Mikutta et al. 2007; Mavi et al. 2012) and thus reduces organic matter amount available for decomposition (Oades 1988; Six et al. 2000). Because of these modifications, soil becomes unsuitable for proper root growth and plant development.

2.3.2 Effect on Plant Growth Activities

Salinity constitutes a most critical constraint, which hinders and causes a significant decrease in agricultural production in many areas worldwide (Tank and Saraf 2010). It is a limiting growth factor for many types of plants, especially non-halophytic ones (Saghafi et al. 2013). Most of the widely used crops in human or animal nutrition such as cereals, forages, or horticultural crops are susceptible to excessive salt concentration either dissolved in irrigation water or already present naturally in rhizosphere soil (Ondrasek et al. 2010). It was reported that vegetables are more sensitive to salt than forages and grains (Waller and Yitayew 2016).

All soils contain some water-soluble salts; many of them constitute plant nutrients, but an excessive accumulation of them can cause many deleterious effects for plant and also suppress its growth (Paul 2012, Shrivastava and Kumar 2015). It is not easy to quantify plants' salt tolerance because it varies appreciably with many environmental factors (soil fertility, soil physical conditions, distribution of salt in soil profile, irrigation regime, and climate) and plant factors (growth stage, root stock, and variety of "genera, species, and cultivars") (Juan et al. 2005; Niu et al. 2012).

Salinity affects almost all plant development aspects including germination, vegetative growth, and reproductive development (Bartels and Sunkar 2005). Rahman et al. (2000) and Jamil et al. (2006) described that seed germination and early seedling growth are the most salt-sensitive plant growth stages, because seedling root is in direct contact with salt-affected soil. Some plants like barley, rice, wheat, and corn showed a resistance in their first growth stages (Baniaghil et al. 2013).

High concentration of salts in root rhizosphere badly affects plant growth through complex interactions (Arbona et al. 2005). Salt stress affects plants in two ways: causing water deficit referred also as "drought stress" and triggering ion toxicity resulting from the excessive uptake of less demanded elements, mainly Cl^- and Na^+ (Setia et al. 2013; Bharti et al. 2016; Hingole and Pathak 2016). Excessive accumulation of sodium (Na^+) in cell walls can rapidly lead to osmotic stress and cell death (Munns 2002). Uptake and accumulation of Cl^- may disrupt photosynthetic function through the inhibition of nitrate reductase activity (Xu et al. 2000). In rhizosphere, under excessive sodium and chloride concentrations, there are competitive interactions with other nutrient ions (K^+, NO_3^-, and $H_2PO_4^-$) for binding sites and transport proteins in root cells and partitioning within plant cells (Grattan and Grieve 1999; Tester and Davenport 2003). Once the capacity of cells to store salts is exhausted, salts build up in the intercellular space leading to cell dehydration and death (White and Broadley 2001; Sheldon et al. 2004). Increasing salt concentrations in the environment elevate the osmolarity outside the root zone; as the external salt concentrations rise, cells loose water, and the water absorption becomes very difficult (Baniaghil et al. 2013). As mentioned above, many salts are also plant nutrients, but high salt levels in the soil can obstruct nutrient balance in plant (reducing intake N, Ca, K, P, Fe, Zn) or interfere with some nutrients' uptake and cause metabolic disturbances in processes where low Na^+ and high K^+ or Ca^{2+} are required for optimum function (Blaylock 1994; Diby and Harshad 2014). It has been stated by Shaheen et al. (2013) that salinity stress causes perturbation in gaseous exchange plant processes like photosynthesis, sub-stomatal CO_2 concentration, and net CO_2 assimilation and transpiration rate. Salinity reduces photosynthesis capacity mainly through complete or partial stomata closure and osmotic stress (Meloni et al. 2003). Further, plant responses to osmotic stress are reduction in leaf area and chlorophyll content, defoliation, and carbon assimilation reduction (Shannon and Grieve 1999). Soil salinity significantly reduces plant phosphorus (P) uptake because phosphate ions precipitate with Ca ions (Bano and Fatima 2009). Like other abiotic stresses, salinity leads to oxidative stress due to increased production and accumulation of reactive oxygen species (ROS) such as singlet oxygen, superoxide ion, hydrogen peroxide, and hydroxyl radical on the cell, which are detrimental to plant survival and can damage the biomembranes, proteins, nucleic acids, and enzymes (Azevedo Neto et al. 2008; Mishra et al. 2009; Shahbazi et al. 2011). Salinity triggers alteration of metabolic processes, membrane disorganization, reduction of cell division and expansion, and genotoxicity, resulting in decreased growth and premature activation of programmed cell death (PCD) (Flowers 2004; Carillo et al. 2011). Salinity adversely affects reproductive development by inhabiting microsporogenesis and stamen filament elongation, enhancing programmed cell death in some tissue types, ovule abortion, and senescence of fertilized embryos (Shrivastava and Kumar 2015). It has an adverse effect on cell cycle and differentiation; it arrests the cell cycle by reducing the expression and activity of cyclins and cyclin-dependent kinases that result in fewer cells in the meristem, thus limiting growth (Javid et al. 2011). Salinity can increase the rate of ethylene biosynthesis via elevated levels of 1-aminocyclopropane-1-carboxylic acid (ACC), which may lead to physiological changes in plant tissues (Zapata et al. 2004; Tank and Saraf 2010). Ethylene is an important gaseous phytohormone, which is required for many physiological processes like seed germination, root elongation, ripening of fruit, and organ senescence (Bleecker and Kende 2000) but up to certain concentrations (more than required level); it acts as a negative plant growth regulator (Holguin and Glick 2001; Huang et al. 2003); it exacerbates leaf and petal abscission plus organ senescence causing early death (Mayak et al. 2004; Cheng et al. 2007); then it acts as a stress signaling molecule (Hahn and March 2009). It has also been reported that in various plant species ethylene acts as a negative regulator of nodulation (Schaller 2012). Salinity is also known to affect phytohormone synthesis, other plant growth-stimulating factors (Xiong and Zhu 2002), protein synthesis, lipid metabolism (Parida and Das 2005), and energy production by plants (Larcher 1980). Plants with perturbed nutrient relations caused by salinity are more susceptible to invasion of different pathogenic microorganisms (Romic et al. 2008).

Many studies have demonstrated that salinity inhibits seed germination of various crops such as rice (Xu et al. 2011), wheat (Egamberdieva 2009), maize (Khodarahmpour et al. 2012), faba bean (Rabie and Almadini 2005), soybean (Essa 2002), cabbage, and tomato (Bojovic et al. 2010). It has been reported that increased salinity leads to decreased root length of groundnut (Mensah et al. 2006) and chickpea (Al-Mutawa 2003); these reports are in line with results observed by Egamberdieva (2011), where increasing salt content reduced shoot length (50%) and root length (7%) of bean seedling grown in a gnotobiotic sand system. Inhibitory effect of salinity on photosynthesis has been confirmed in cowpea (Taffouo et al. 2010) and sugar beet (Dadkhah 2011). A similar study reported an inverse relationship between salt concentration and chlorophylls "a" and "b" and total chlorophyll content for bean plant (Qados 2011). In basil, Golpayegani and Tilebeni (2011) confirmed that salinity decreases photosynthesis, stomatal conductance, chlorophyll content, and mineral uptake. In eggplant, accumulation of Cl^- in leaves was accompanied with decreased concentration of $NO3^-$ (Savvas and Lanz 2000). Kapoor and Srivastava (2010) confirmed decrease in protein content in black gram plants treated with different salt concentrations. In tomato, Takagi et al. (2009) reported decreased whole plant biomass with reduced leaf photosynthesis and transport of carbon assimilates as an effect of salinity.

2.3.3 Effect on Rhizospheric Bacteria

Even though soil microorganisms constitute less than 0.5% (w/w) of soil mass, they play a key role in maintaining soil fertility (Tate 2000). Microbial biomass is an important labile fraction of soil organic matter, functioning both as an agent of transformation and recycling of organic matter and soil nutrients, by participating in many processes such as nitrification, ammonification, nitrogen fixation, and other ones leading to decomposition of soil organic matter and hence to the transformation of nutrients. These microbes influence plant growth and development directly or indirectly (Diby and Harshad 2014). Because of that, any factor affecting the rhizospheric microbes and its functions influences the nutrient availability and plant growth (Diby and Harshad 2014).

Salinity not only inhibits plant growth and development but also negatively affects microbial functions and microbial biomass and changes microbial community structure which interferes with soil fertility and reduces crops productivity (Tripathi et al. 2007; Andronov et al. 2012), especially rhizospheric bacteria (Ofek et al. 2006). These microbial modifications are considered as an indirect effect of salinity on plant growth (Chowdhury et al. 2011).

Soluble salts in soil increase the osmotic potential, drawing water out of microbial cells. As for plant roots, low osmotic potential makes it difficult for microbes to remove water from the soil (Oren 1999). Microorganisms in hypertonic environments (low water activity) either die or remain dormant, lowering the activity of various degrading enzymes released by microbes and eventually decreasing fertility (Tripathi et al. 2007; Egamberdieva et al. 2010). Cytoplasmic membranes of organisms are permeable to water but not to other metabolites. Therefore, hyper- or hypo-osmotic shock exerted on cells causes a concomitant decrease or increase in the cytoplasmic volume leading to plasmolysis and cell death (Nabti et al. 2010). However, soil microbes can adapt to low osmotic potential. This process is highly bioenergetically taxing, because they must maintain an osmotic equilibrium between the cytoplasm and the surrounding medium while excluding sodium ions from cell interior, leading to reduction in growth and activity of the surviving microbes (Jiang et al. 2007; Ibekwe et al. 2010). In addition, the toxicities of Na^+ and other accompanying ions (Cl^- and CO_3^{2-}), along with very high pH, also inhibit microbial growth and activity (Zahran 1997).

Nelson and Mele (2007) showed that sodium chloride affects the rhizosphere microbial community structure through its influence on the quantity and/or quality of root exudates. Nabti et al. (2010) reported that salinity inhibits bacterial nodulation and nitrogen fixation activity by causing alteration of proteins involved in the initial attachment steps (adsorption and anchoring) of bacteria to plant roots in symbiotic interaction, in addition to alteration of exopolysaccharide (EPS) and lipopolysaccharide (LPS) composition of the bacterial cell surface, impairment of molecular signal exchange between bacteria and their plant host due to the alteration of membrane glucan contents, and inhibition of bacterial mobility and chemotaxis toward plant roots. Singleton and Bohlool (1984) have also demonstrated the disturbance effect of symbiotic interaction between legumes and rhizobia by soil salinity. Rabie et al. (2005) reported decrease in nodulation and nitrogen fixation with reduced nitrogenase activity in legumes such as soybean, common bean, and faba bean. High salinity inhibited nitrogen fixation at the level of *nifH* gene expression and nitrogenase activity in *Azospirillum* sp. (Tripathi et al. 2002). Some studies showed that soil respiration decreased with increasing soil salinity (Yuan et al. 2007; Wong et al. 2009; Setia et al. 2010). Many enzyme activities such as urease, amylase, invertase, catalase, alkaline phosphatase, and β-glucosidase were strongly inhibited by salinity (Frankenberger and Bingham 1982; Ghollaratta and Raiesi 2007; Pan et al. 2013). With an increase in the salinity level above 5%, the total count of bacteria and *Actinobacteria* were drastically reduced in studies conducted by Omar et al. (1994) and Wichern et al. (2006).

2.4 Halotolerant PGPRs as a New Strategy to Alleviate Salt Stress in Plant

It must be noticed that salinity is one of the major abiotic stresses affecting plant growth and productivity in arid and semiarid regions (Yadav et al. 2011); it is an important growth-limiting factor for most non-halophytic plants. According to FAO, more than 831 million hectares of land are affected by salinity throughout the world (Martinez-Beltran and Manzur 2005).

Moreover, a number of authors observed that salt effect on growth of different crops such as pepper, tomato, soybean, wheat, canola, and lettuce was alleviated by using halotolerant PGPRs. Similarly, it was reported that PGPRs' increased shoot and

Fig. 2.1 Effect of PGPRs inoculation on wheat plant growth under nonsaline and saline conditions. Plants were grown in soil under glasshouse conditions and supplemented with 150 mM NaCl through irrigation and harvested 60 days after germination. (Bharti et al. 2016)

root length and dry weight of soybean (Naz et al. 2009) and wheat (Fig. 2.1) (Bharti et al. 2016) under salt stress. *P. extremorientalis* TSAU20 and *P. chlororaphis* TSAU13 were shown to improve growth of various crops under saline conditions (Egamberdieva and Kucharova 2009). Jha and Subramanian (2013) cited the role of PGPRs in the regulation of ion concentration, nutrient uptake, and antioxidant enzymes in plant cultivated under salt stress. Moreover, the results obtained by Sapsirisopa et al. (2009) revealed that treatments of rice with *Bacillus megaterium* A12ag significantly increased N and P content in shoots. PGPR strains can also enhance chlorophyll, carotenoid, nitrogen, and phosphorus concentrations in plant (Jha and Subramanian 2013; Saghafi et al. 2013). In addition, co-inoculation with *Azospirillum lipoferum* and *Pseudomonas fluorescens* increases meaningfully chlorophyll a and b and carotenoid contents in wheat (Saghafi et al. 2013); similar results were also reported by Nadeem et al. (2006) who showed that inoculation of maize seeds with rhizobacterial strains augmented chlorophyll a and b and carotenoid contents of fresh leaves at 10 dS m^{-1}. On the other hand, PGPRs can induce antioxidant enzyme production in salt-stressed plant, which is an essential mechanism to detoxify reactive oxygen species (ROS). Kumar et al. (2003) and Jha and Subramanian (2013) reported that the antioxidant activity is more significant when PGPRs are present. PGPR application could also increase seed germination; it is the case of inoculation with *P. pseudoalcaligenes* and *Bacillus pumilus* showing an increase of paddy seed germination under saline conditions (Jha and Subramanian 2013). Likewise, they showed that *B. megaterium* A12ag improved significantly the rice seed germination at EC 10 dS/m. The same strain had similar effect on tomato seeds at NaCl concentration between 30 and 90 mM. It also augmented seedling dry weight at NaCl up to 120 mM (Chookietwattana and Maneewan 2012).

Due to their promising effect on plant growth under saline stress, the use of PGPRs is becoming prevalent worldwide. In addition to alleviating salinity stress in

Table 2.1 Salinity stress alleviation by PGPR species in various crops

PGPRs	Crops	Effect	References
Brevibacterium epidermidis RS15, *Micrococcus yunnanensis* RS222, and *Bacillus aryabhattai* RS341	Canola	Enhanced plant growth and reduced level of ethylene	Siddikee et al. (2010)
Rhizobium and *Pseudomonas*	Mung bean	Improved plant growth and increased salt tolerance index of plant	Ahmad et al. (2013)
Pseudomonas putida R4 and *P. chlororaphis* R5	Cotton	Improved seed germination and seedlings growth and alleviated salt stress in cotton seedlings	Egamberdieva et al. (2015)
Dietzia natronolimnaea STR1	Wheat	Promoted salinity tolerance of plant and enhanced gene expression of antioxidant enzymes	Bharti et al. (2016)
Bacillus megaterium and *Enterobacter* sp.	Okra	Improved plant growth, increased antioxidant enzyme activities, and upregulated ROS pathway genes	Habib et al. (2016)
Bacillus sp.	Wheat	Improved shoot and root length	Orhan (2016)
Bacillus mojavensis K78	Wheat	Improved plant growth, increased chlorophyll content and nutrient uptake, and reduced levels of ethylene	Pourbabaee et al. (2016)
Agrobacterium tumefaciens and *Pseudomonas* sp.	Peanut	Enhanced plant growth, maintained ion homeostasis, and reduced reactive oxygen species (ROS) level	Sharma et al. (2016)
Bacillus cereus Y5, *Bacillus* sp. Y14, and *B. subtilis* Y16	Wheat	Improved plant growth	Khan et al. (2017)

plants (Table 2.1), several reports suggest that PGPRs can also enhance plant tolerance toward other abiotic stresses such as drought (Sandhya et al. 2009; Timmusk and Wagner 1999), chilling injury (Barka et al. 2006), metal toxicity (Dell' Amico et al. 2008; Bensidhoum et al. 2016), and elevated temperature stress (Ali et al. 2009).

2.4.1 Mechanisms of Plant Promotion by Halotolerant PGPRs in Saline Soil

Potential of halotolerant PGPRs to increase crop productivity under stressful environment involves various mechanisms. These are similar to those implicated in halosensitive bacteria for plant growth promotion under elevated salinity. These mechanisms are summarized in Fig. 2.2.

2.4.1.1 1-Aminocyclopropane-1-carboxylic Acid (ACC) Deaminase

Ethylene is a gaseous plant hormone involved in plant growth and development as well as in plant response to a wide range of abiotic and biotic stresses (Abeles et al. 1992). The presence of ethylene may be stimulatory or inhibitory depending on its concentration, the nature of the physiological process, and the growth phase of plant (Shaharoona et al. 2006). Under stress conditions such as salinity, plant hormone ethylene endogenously regulates plant homoeostasis resulting in reduced root and shoot growth (Shrivastava and Kumar 2015; Vurukonda et al. 2016). A number of PGPRs promote plant growth by lowering stress ethylene concentrations through their 1-aminocyclopropane-1-carboxylic acid (ACC) deaminase enzymes (Glick 1995; Penrose et al. 2001; Glick 2014). This enzyme can hydrolyze the endogenous levels of ACC, which is an immediate precursor of ethylene biosynthesis in plants, into ammonia and α-ketobutyrate resulting in reduced level of ethylene (Hontzeas et al. 2004; Glick et al. 2007; Nascimento et al. 2014; Santoyo et al. 2016; Khan et al. 2017). PGPRs producing this enzyme protect plant from deleterious effect of salinity and drought (Glick et al. 1998; Saleem et al. 2007; Zahir et al. 2009). By studying the role of ACC deaminase enzyme-producing bacteria, numerous authors reported the implication of ACC deaminase in promoting growth and health status of plant growing in salt-affected soils. El-Tarabily et al. (2008) showed that ACC deaminase producing *Streptomyces filipinensis* 15 and *S. atrovirens* 26 can positively influence plant growth by reducing tomato endogenous ACC levels in both roots and shoots. Similarly, ethylene production by tomato seedlings was reduced by the bacterium *Achromobacter piechaudii* as reported by Mayak et al. (2004). They revealed the significant increase of fresh and dry weights of tomato grown in NaCl (up to 172 mM). Nadeem et al. (2006) have also mentioned that inoculation of canola by ACC deaminase producing *P. syringae* and *Pseudomonas* sp. improves yield and growth under salinity (12 dS m^{-1}). ACC deaminase-containing PGPRs enhance chlorophyll contents of canola, lettuce, as well as maize (Glick et al. 1997; Han and Lee 2005; Nadeem et al. 2006). Saravanakumar and Samiyappan (2007) exhibited that *P. fluorescens* containing ACC deaminase activity enhanced saline resistance in groundnut plants and increased yield as compared to *Pseudomonas* strains lacking ACC deaminase activity. The role of PGPRs containing ACC deaminase in improving nodulation in legumes has been demonstrated by Shaharoona et al. (2006) showing that inoculation with PGPR containing ACC deaminase promoted nodule formation by indigenous competitive *Bradyrhizobium*. Thereby, these bacteria could be a promising approach for successful nodulation in legumes (Zafar-ul-Hye et al. 2013).

2.4.1.2 Phytohormone Biosynthesis

Plant hormones play a crucial role in controlling the way in which plants grow and develop (Davies 1995). The commonly recognized classes of phytohormones, regarded as the "classical five," are the auxins, gibberellins, cytokinins, abscisic acid, and ethylene (Baca and Elmerich 2007). It is well documented that the most important mechanisms adopted by PGPRs to improve plant growth involve phytohormone production, which can control several plant growth parameters and development both under normal and stressed environment. Numerous authors showed that salt stress affects negatively endogenous levels of phytohormones. Indeed, exogenous application of plant growth regulators alleviated salt effect in several plants, e.g., wheat (Nabti 2007; Egamberdieva and Kucharova 2009), maize (Khalid et al. 2013; Kaya et al. 2013), cotton (Egamberdieva et al. 2015), and mung bean (Ahmad et al. 2013). Indole-3-acetic acid (IAA) is one of the most important phytohormones that regulate many aspects of plant growth and development, e.g., cell division, seed germination, differentiation of tissues, leaf expansion, root elongation, and resistance to stressful conditions (Egamberdieva 2008; Maleki et al. 2010; Martínez-Viveros et al. 2010). Egamberdieva et al. (2015) reported that IAA-producing bacterial strains *P. putida* R4 and *P. chlororaphis* R5 improved seed germination up to 64 and 73% in response of NaCl stress. These two strains have significantly increased root and shoot length of cotton seedling at 100 mM NaCl compared to the uninoculated seedlings. Khalid et al. (2013) reported also the role of exogenous application of IAA in alleviating the deleterious effect of salt stress in maize plant. It was found that IAA-overproducing PGPR strains are able to alleviate different stress conditions by increasing proline amount in *Medicago* plants (Saghafi et al. 2013). The ability of halotolerant bacteria to modulate plant stress level by producing IAA was previously reported for bacteria isolated from halophyte root in Argentina (Sgroy et al. 2009), coastal soil (Siddikee et al. 2010), rhizosphere of *C. annuum* growing in desert areas (Marasco et al. 2012), and *Salicornia* plant rhizosphere (Mapelli et al. 2013). Besides its role in stimulating plant growth, IAA produced by PGPRs can be absorbed by plants, and in combination with endogenous plant, IAA can stimulate ACC synthase (Penrose and Glick 2001; Gerhardt et al. 2006). This enzyme converts S-adenosylmethionine (SAM) to ACC which is the immediate precursor to ethylene (Kende 1993). Much of this ACC is exuded from seeds and roots then taken up by PGPRs and subsequently converted by ACC deaminase to ammonia and α-ketobutyrate (Glick et al. 1998; Penrose 2000; Holguin and Glick 2001; Gerhardt et al. 2006). Therefore, ACC reduction leads to lowering ethylene content in plant, thus alleviating the inhibitory effect of high ethylene concentration on plant development (Fig. 2.2). High concentration of IAA, which reflects a higher synthesis rate of ACC, might result in a negative impact on root growth and seed germination (Siddikee et al. 2011; Nakbanpote et al. 2014). Siddikee et al. (2010) revealed that halotolerant bacteria producing both ACC deaminase and IAA enhance root elongation and dry weight of canola to a greater extent than the strains producing solely ACC deaminase. Glick et al. (2007) suggested that IAA and ACC deaminase stimulate root growth in a coordinated fashion. Since IAA stimulates ACC production which is converted into ethylene and since ACC deaminase hydrolyzes ACC and decreases ethylene level, the final ethylene effect or root growth depends on the balance of IAA and ACC deaminase (Arora et al. 2012). Certainly as cited by Glick (2014), synergy between ACC deaminase-producing bacteria and IAA is indispensable to enhance plant growth and development under salt stress.

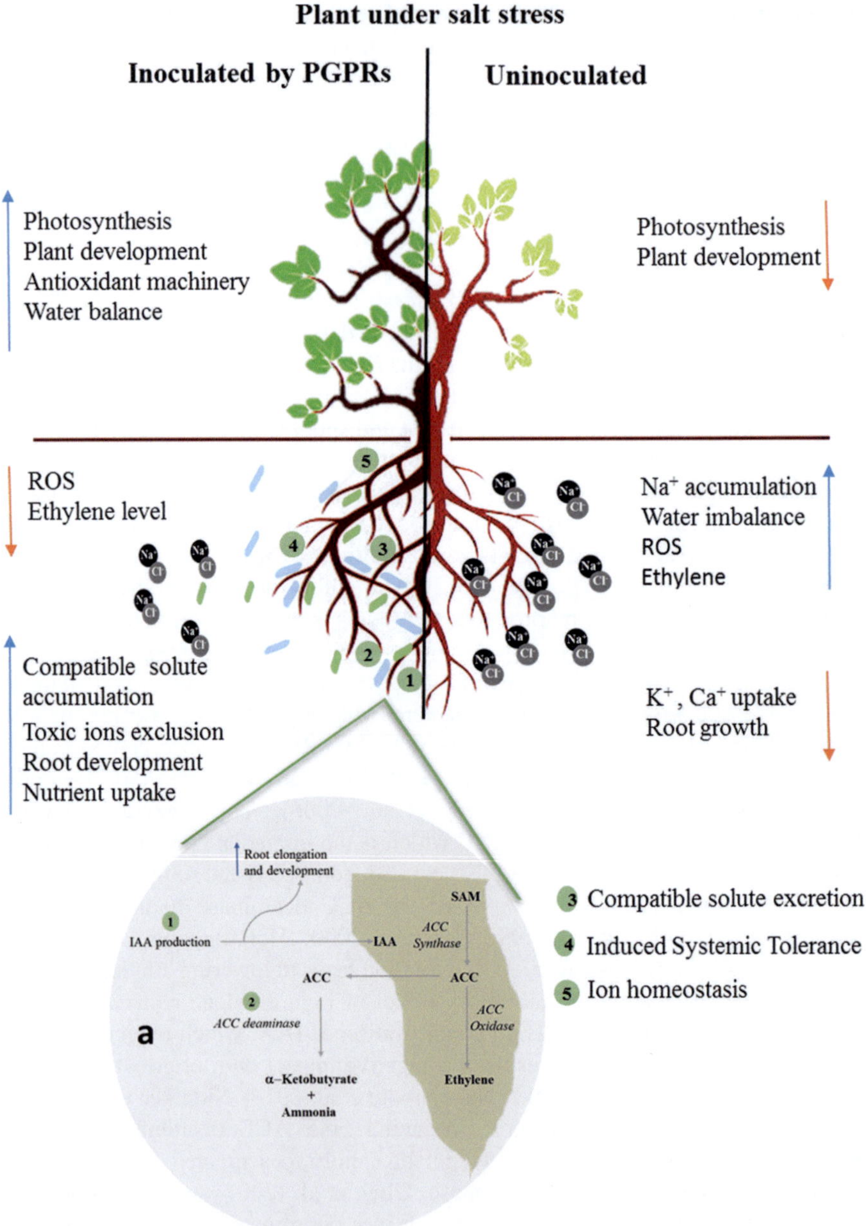

Fig. 2.2 Mechanisms involved by PGPRs to alleviate salt stress in plant. (**a**) ACC deaminase enzyme produced by PGPR reduces the ethylene level in plant. *ACC* 1-aminocyclopropane-1-carboxylic acid, *IAA* indole-3-acetic acid, *SAM* S-adenosyl methionine, *ROS* reactive oxygen species

2.4.1.3 Induced Systemic Tolerance (IST): Antioxidant Enzymes

As sessile organisms, plants are frequently exposed to several biotic and abiotic stresses. The application of PGPRs may help plant to cope with these two types of stress. Unlike direct mechanisms involved by PGPRs to improve plant growth (e.g., phytohormones synthesis, phosphate solubilization, and nitrogen fixation), indirect mechanisms lead to protect plant from both biotic and abiotic stress. Jha and Saraf (2012) suggested that PGPRs can induce defense programs such as systemic acquired resistance (SAR) and induced systemic resistance (ISR), thus reducing phytotoxic microbial communities. Recently, many studies about the role of PGPRs in eliciting plants to tolerate abiotic stresses, e.g., drought, salt, cold, and high temperature, were published. The term "induced systemic tolerance" (IST) was proposed for PGPRs that induced physical and chemical changes in plants that result in enhanced tolerance to abiotic stresses (Mantelin and Touraine 2004). Among the mechanisms involved by PGPR plant tolerance induction under salt stress, antioxidant enzymes can be mentioned.

Under abiotic stress such as salinity, plant can suffer from membrane damage due to the generation of reactive oxygen species (ROS) including superoxide radical (O_2^-), hydroxyl radical (OH^-), and hydrogen peroxide (H_2O_2) which causes oxidative damage to biomolecules such as lipids and proteins and eventually leads to plant death (Del Rio et al. 2003). Perturbation of the cellular redox state is adjusted by the activation of antioxidant system in plant (Gill and Tuteja 2010; Jha and Subramanian 2013), by production of antioxidant enzymes, e.g., superoxide dismutase (SOD), peroxidase (POX), and catalase (CAT), and nonenzymatic antioxidants such as ascorbate, glutathione, and a-tocopherol (Del Rio et al. 2003). Numerous reports were edited recently on the ability of PGPRs to activate plant antioxidant machinery and to reduce salt stress. Han and Lee (2005) mentioned the ability of PGPRs to induce antioxidant enzymes in lettuce; they said that the inoculation with two PGPR strains, *Serratia* sp. and *Rhizobium* sp., into saline soils alleviated salinity effects on the antioxidant enzymes ascorbate, peroxidase (APX), and glutathione reductase (GR). Kohler et al. (2009) showed that the induction of antioxidant enzymes by PGPR strains increased the tolerance of lettuce grown under sever salt stress. PGPR strain *Bacillus cereus* Pb25 increased significantly antioxidant enzyme (POD, SOD, and CAT) activities in *Vigna radiata* (mung bean) (Islam et al. 2015). To monitor the putative changes in the expression of genes encoding the ROS-scavenging enzymes, Gururani et al. (2013) subjected the PGPR-inoculated and non-inoculated potato plants to different stresses. They showed that the mRNA expression of SOD and APX, in bacteria-inoculated plants growing under stress conditions, increased considerably in comparison to non-inoculated stressed plants. Similarly, mRNA expression levels of genes encoding for other antioxidative enzymes such as CAT and GR also increased in bacteria-treated plants. Therefore, inoculation with selected PGPRs could serve as a useful tool for alleviating oxidative stress in plants. According to Bharti et al. (2016), salt-stressed plants applied with halotolerant PGPR showed an improvement in gene expression levels as compared to uninoculated salt-stressed plants. They noticed that shoot of APX, CAT, and POD of PGPR-inoculated salt-stressed plants recorded a threefold increase in gene expression as compared to uninoculated non-stressed plants. Plants' salt-stressed roots inoculated with PGPRs recorded a two- to threefold increase in gene expression of antioxidants in comparison with the non-stressed control plants.

2.4.1.4 Ion Homeostasis and Nutrient Uptake

Salt stress affects plant development either through perturbation of water balance or imbalance in ratio of ion homeostasis. Numerous reports showed that salinity increased Na^+ accumulation and consequently decreased Ca^+ and K^+ uptake (Yildirim et al. 2006; Zhang et al. 2008; Tank and Saraf 2010; Shukla et al. 2011; Rojas-Tapias et al. 2012; Aamir et al. 2013; Jha and Subramanian 2013; Younesi et al. 2013; Sharma et al. 2016). Accumulation of toxic ions (Na^+ and Cl^-) perturbs water balance and ion homeostasis, which eventually disturbs metabolic processes and hormonal status and affects enzymes, photosynthetic function, transpiration, and translocation of nutrients (Munns 2002). Maintenance of low Na^+/K^+ is one of the predominant mechanisms involved by PGPRs to improve plant growth under high salinity (Munns and Tester 2008). They have the ability to increase mineral nutrient exchange and to alleviate nutrient imbalance caused by the high influx of Na^+ and Cl^- ions. According to Hamdia et al. (2004) and Ilangumaran and Smith (2017), PGPRs promote growth of salt-affected plant by changing the selectivity of cation (Na^+, K^+, and Ca^+), maintaining a higher K^+/Na^+ ratio, reducing Na^+ and Cl^- accumulation in leaves, increasing Na^+ exclusion via roots, and boosting the activity of high-affinity K^+ transporters. Several authors highlighted the PGPRs' role in reducing Na^+/K^+ ratio in plant under salt stress. Peanut seedlings treated with three isolates (MBE01, MBE02, and MBE05) showed lower shoot Na^+/K^+ ratio than non-inoculated seedlings under salinity stress (Sharma et al. 2016). Han and Lee (2005) showed that PGPR strains increased the tolerance of maize plant by lowering Na^+/K^+ ratio. Plants inoculated by PGPR strains *Brachybacterium saurashtrense* and *Brevibacterium casei* achieved the maximum-reduced Na^+ concentration in addition to the Na^+/K^+ ratio lowered at both 0 and 100 mM NaCl (Shukla et al. 2011). Salt stress affects also gene expression involved in ion homeostasis. Pinedo et al. (2015) clarified the deleterious effect of salt stress on the expression of ion homeostasis-related gene (*Arabidopsis* K^+ transporter 1, high-affinity K^+ transporter 1, sodium hydrogen exchanger 2, and *Arabidopsis* salt overly sensitive 1). They displayed that PGPR strain *Burkholderia phytofirmans* PsJN regulates the expression of important ion homeostasis-related genes after short- and long-term exposure to salt stress. A tissue-specific regulation of HKT1 by the soil bacterium *B. subtilis* GB03 confers salt tolerance in *Arabidopsis thaliana*. Under salt stress, this strain concurrently downregulates HKT1 expression in roots and upregulates its expression in shoots, resulting in lower Na^+ accumulation throughout the plant compared to the control (Zhang et al. 2008). Recently, Niu et al. (2015) reported that *B. subtilis* GB03 strain upregulated *PtSOS1* gene expression under salt stress (200 mM NaCl), which is consistent with improved salt tolerance and reduced Na^+ accumulation in shoots, roots, and whole plants of *P. tenuiflora*.

Salinity reduces plant nutrient uptake, which leads to reduction of plant growth and development. Salt stress causes competitive interactions between toxic ions and nutrient ions for binding sites and transport proteins in root cells and thereafter for translocation, deposition, and partitioning within the plant (Tester and Davenport 2003). Consequently, to this interaction, the concentration of nutrient elements will be reduced in whole plant. Numerous researches exhibited the PGPR ability to provide plants with nutrient elements under salt stress (Yildirim et al. 2006; Zhang et al. 2008; Shukla et al. 2011; Aamir et al. 2013; Jha and Subramanian 2013; Niu et al. 2015; Sharma et al. 2016). Jha and Subramanian (2013) found that plant inoculated by PGPR strains increased N, P, and K concentration and reduced that of Na and Ca. Similarly, Aamir et al. (2013) observed that N, P, and K concentration and protein content in grain increased by inoculation and co-inoculation with *Rhizobium* and plant growth-promoting rhizobacteria.

2.4.1.5 Osmoprotective Compounds: Compatible Solutes

Under salt stress and to be adapted to the fluctuation in osmolarity, bacteria apply two main strategies: "salt in" mechanism and production of organic osmolytes (Korcan et al. 2015; Mohammadipanah et al. 2015), which is the major mechanism involved by both prokaryote and eukaryote organisms to alleviate salt stress. The osmoadaptation strategy in halophilic and halotolerant bacteria consists on the accumulation of K^+ ions and glutamate (first response) (Sleator et al. Hill 2001; Le Rudulier et al. 2002); if the osmotic shock persists or increases, bacteria accumulate compatible solutes (second response) (Lucht et al. Bremer 1994; Bartlett et al. Roberts 2004). Compatible solutes are so named because, even at high concentration in the cytoplasm, they do not interfere with the vital cellular processes (Kempf and Bremer 1998) and they can be synthesized by halophiles organisms or accumulated by uptake from surrounding medium (Mohammadipanah et al. 2015). These organic osmolytes harbor different classes of molecules including amino acids and their derivatives (proline, glutamate, glutamine, ectoine, and taurine), sugars (sucrose, trehalose), methylamines (glycine betaine), and polyols (mannitol, erythritol, and glycerol) (Oren 2002; Yancey 2005). The presence of halotolerant bacteria in saline soils may play an important role in interaction with plants (Nabti et al. 2015). They can improve plant growth by realizing these molecules in medium. Such organic compounds, taken up and accumulated by plants unable to synthesize them de novo and able to improve growth under inhibitory osmolarities, are called osmoprotectants. Thus, the concept of osmoprotectant supposes an ecological cycle in which compatible solutes are transported from producers to consumers injured by osmotic stress (Arora et al. 2012). A number of studies on halophilic and halotolerant PGPRs demonstrated their ability to increase compatible solutes in plants under stress conditions. Sandhya et al. (2010) observed that *Pseudomonas* strain GAP 45 increased compatible solutes and antioxidant status of maize plants under water-stressed conditions. Proline accumulation is a very common response in plants exposed to abiotic stress. Saghafi et al. (2013) showed that PGPR strains increased proline contents in wheat cultivars. Similarly, Naz et al. (2009) reported that PGPR isolates increase significantly proline contents of shoots and roots of soybean plant under salt stress. It was noted that *Medicago* plant inoculated by IAA-overproducing PGPR strain is able to cope with different stress conditions and accumulates high level of proline (Saghafi et al. 2013). Increased proline and total soluble sugar contents in the PGPR-treated wheat plants contributed significantly to their osmotolerance (Upadhyay et al. 2012). Similarly, trehalose metabolism in rhizobia also seems important for improving plant growth, yield, and adaptation to abiotic stress of leguminous plants (Suarez et al. 2008). In addition to their ability to restore plant growth by providing compatible solutes, halotolerant PGPRs can improve plant growth under salt stress by the same mechanisms used by PGPRs in normal conditions.

2.5 Halotolerant PGPR Algae: A Promising Consortium to Alleviate Salt Stress in Plant

The mechanisms used by PGPRs to improve plant growth and health under salt stress can be inhibited or weakened if these bacteria are unable to overcome this stress. As reported previously, the accumulation of exogenous compatible solutes protects and stimulates bacterial growth in hyperosmotic environment. Compatible solute accumulation by uptake from the medium is preferred over biosynthesis, since uptake is energetically less expensive than biosynthesis (Imhoff 1986). Nabti et al. (2010) highlighted that bacteria belonging to the genus *Azospirillum* accumulate compatible solutes such as glutamate, proline, glycine betaine, and trehalose in order to adapt to salt stress.

By their composition rich in phytohormones, polysaccharides, nutrient elements, and vitamins, algae can protect and improve plant growth in normal and stressed environment. Besides, they are the major source for compatible solute for plant and bacteria in a stressed environment. Ghoul et al. (1995) showed that algal extracts contain high amount of compatible solutes as betaines, amino acids, and DMSP. They noticed that algal extracts convey significant osmoprotection to *E. coli* under salt stress. Tolerance to NaCl of *A. brasilense* NH was increased up to 600 mmol/l in the presence of glycine betaine and *Ulva lactuca* extracts (Nabti et al. 2007). Pichereau et al. (1998) demonstrated also the role of algae to stimulate bacterial growth in a stressed environment, by their organic osmolytes GB (N, N, N-trimethylglycine) and 3-dimethylsulfoniopropionate (DMSP) which acts as osmoprotectants. Few studies have been performed on the role of the consortium algae/PGPRs in plant growth stimulation. These studies consist of the simultaneous application of PGPRs and algae extracts under stress conditions, where the algae extracts were used as source of osmoprotectants. The main role of this combination is to restore bacterial growth and improve their interaction with plants and thus ameliorates plant productivity in stressed environment. As reported by Nabti et al. (2007), algal extracts enhance bacteria performance and thus support their plant growth-promoting effect. Arif (2016) showed that *Ulva lactuca* extracts restored germination, biomass, chlorophyll contents, and protein of wheat inoculated by *P. fluorescens* under salt stress. Similarly, Nabti et al. (2007) exhibited that the PGPR strain *A. brasilense* restored wheat growth under salt stress even better in the presence of *Ulva lactuca* extracts.

2.6 Conclusion

Improvement of plant growth in saline soils by inoculation with halotolerant PGPRs could be a good solution as a suitable alternative to improve crop growth and yield in this hostile environment. Furthermore, these bacteria showed a high capacity to enhance plant growth under saline conditions by using multiple and variable mechanisms, such as the strategy of osmoregulation. In addition, stimulation of plant growth in saline soils seems to involve the same mechanisms like other halosensitive bacteria.

References

Abeles FB, Morgan PW, Saltveit ME Jr (1992) Ethylene in plant biology, 2nd edn. Academic, New York, p 414

Abu-Sharar TM, Bingham FT, Rhoades JD (1987) Stability of soil aggregates as affected by electrolyte concentration and composition. Soil Sci Soc Am J 51:309–314

Ahmad M, Zahir AZ, Nazli F, Akram F, Arshad M, Khalid M (2013) Effectiveness of halo- tolerant, auxin producing *Pseudomonas* and *Rhizobium* strains to improve osmotic stress tolerance in mung bean (*Vigna radiata* L). Braz J Microbiol 44:1341–1348

Ali SZ, Sandhya V, Grover M, Rao LV, Kishore VN, Venkateswarlu B (2009) Pseudomonas sp. strain AKM-P6 enhances tolerance of sorghum seedlings to elevated temperatures. Biol Fertil Soils 46:45–55

Al-Mutawa MM (2003) Effect of salinity on germination and seedling growth of chick pea (Cicer arietinum L.) genotypes. Int J Agron Biol 5:227–229

Amir M, Aslem A, Khan MY, Jamshaid MU, Ahmed M, Asghar HN, Zahir ZA (2013) Co-inoculation with *Rhizobium* and plant growth promoting Rhizobacteria (PGPR) for inducing salinity tolerance in mung bean under field condition of semi-arid climate. Asian J Biol 1:17–22

Andronov EE, Petrova SN, Pinaev AG, Pershina EV, Rakhimgalieva SZ, Akhmedenov KM, Gorobets AV, Sergaliev NK (2012) Analysis of the structure of microbial community in soils with different degrees of salinization using T-RFLP and real time PCR techniques. Eurasian Soil Sci 45:147–156

Anonymous (2015) Soils are endangered, but the degradation can be rolled back. Food and Agriculture Organization of United Nations, Rome

Arbona V, Macro AJ, Iglessias DJ, Lopez-Climent MF, Talon M, Gomez-Cadnas A (2005) Carbohydrate depletion in roots and leaves of salt- affected potted *Citrus clementina* L. *Plant Growth Regul* 46:153–160

Arif F (2016) Effet du stress salin et d´osmoprotecteurs naturels sur la germination de blé dur (*Triticumdurum*) inoculé par *Pseudomonas fluorescens*. Thèse de Doctorat, Université Ferhat Abbas, Sétif 1. Faculté des Sciences de la Nature et de la Vie, Sétif, 156p

AroraNK TS, Singh S, Lal N, Maheshwari DK (2012) PGPR for protection of plant health under saline conditions. In: Maheshwari DK (ed) Bacteria in agrobiology-stress management. Springer, Berlin/Heidelberg, pp 239–258

Arshadullah M, Hyder SI, Mahmood IA, Sultan T, Naveed S (2017) Mitigation of salt stress in wheat plant (Triticum aestivum) by plant growth promoting rhizobacteria for ACC deaminase. Int J Adv Res Biol Sci 4:41–46. ISSN: 2348–8069

Ashraf M, Ahmad MSA, Öztürk M, Aksoy A (2012) Crop improvement through different means: challenges and prospects. In: Ashraf M et al (eds) Crop production for agricultural improvement. Springer, Dordrecht, pp 1–15

Azevedo Neto AD, Gomes-Filho E, Prisco JT (2008) Salinity and oxidative stress. In: Khan NA, Sarvajeet S (eds) Abiotic stress and plant responses. IK International, New Delhi, pp 58–82

Baca BE, Elmerich C (2007) Microbial production of plant hormones. In: Elmerich C, Newton WE (eds) Associative and endophytic nitrogen-fixing bacteria and cyanobacterial associations. Springer, Dordrecht, pp 113–143

Baniaghil N, Arzanesh MH, Ghorbanli M, Shahbazi M (2013) The effect of plant growth promoting rhizobacteria on growth parameters, antioxidant enzymes and microelements of canola under salt stress. J Appl Environ Biol Sci 3:17–27

Bano A, Fatima M (2009) Salt tolerance in *Zea mays* (L) following inoculation with *Rhizobium* and *Pseudomonas*. Biol Fertil Soils 45:405–413

Barka EA, Nowak J, Clement C (2006) Enhancement of chilling resistance of inoculated grapevine plantlets with a plant growth-promoting rhizobacterium *Burkholderia phytofirmans* strain PsJN. Appl Environ Microbiol 72:7246–7252

Bartels D, Sunkar R (2005) Drought and salt tolerance in plants. Crit Rev Plant Sci 24:23–58

Bartlett DH, Roberts M (2004) Osmotic stress. In: Schaechter M (ed) The desk encyclopedia of microbiology. Elsevier/Academic Press, Amsterdam/Boston, pp 754–766

Benlloch-Gonzalez M, Fournier JM, Ramos J, Benlloch M (2005) Strategies underlying salt tolerance in halophytes are present in *Cynara cardunculus*. Plant Sci 168:653–659

Bensidhoum L, Nabti E, Tabli N, Kupferschmied P, Weiss A, Rothballer M, Schmid M, Keel C, Hartmann A (2016) Heavy metal tolerant *Pseudomonas protegens* isolates from agricultural well water in northeastern Algeria with plant growth promoting, insecticidal and antifungal activities. Eur J Soil Biol 75:38–46

Bharti N, Pandey SS, Barnawal D, Patel VK, Kalra A (2016) Plant growth promoting rhizobacteria *Dietzia natronolimnaea* modulates the expression of stress responsive genes providing protection of wheat from salinity stress. Sci Rep 6:1–16. https://doi.org/10.1038/srep34768

Biggs AJW, Watling KM, Cupples N, Minehan K (2010) Salinity risk assessment for the Queensland Murray-Darling Region. Queensland Department of Environment and Resource Management, Toowoomba

Blaylock AD (1994) Soil salinity, salt tolerance and growth potential of horticultural and landscape plants. Co-operative Extension Service, University of Wyoming, Department of Plant, Soil and Insect Science, College of Agriculture, Laramie

Bleecker AB, Kende H (2000) Ethylene: a gaseous signal molecule in plants. Annu Rev Cell Dev Biol 16:1–18

Bojovic B, Delic G, Topuzovic M, Stankovic M (2010) Effects of NaCl on seed germination in some species from families *Brassicaceae* and *Solanaceae*. Kragujevac J Sci 32:83–87

Bui EN (2013) Soil salinity: a neglected factor in plant ecology and biogeography. J Arid Environ 92:14–25

Buringh P (1978) Food production potential of the world. In: Sinha R (ed) The world food problem: consensus and conflict. Pergamon Press, Oxford, p 477e485

Burrow DP, Surapaneni A, Rogers ME, Olsson KA (2002) Groundwater use in forage production: the effect of saline-sodic irrigation and subsequent leaching on soil sodicity. Aust J Exp Agric 42:237–247

Carillo P, Annunziata MG, Pontecorvo G, Fuggi A, Woodrow P (2011) Salinity stress and salt tolerance. In: Arun S (ed) Abiotic stress in plants—mechanisms and adaptations. InTech, Croatia, pp 22–38

Chedlly A, Ozturk M, Ashraf M, Grignon C (2008) Biosaline agriculture and high salinity tolerance. Birkhauser Verlag-AG (Springer Science), Basel, p 367

Cheng Z, Park E, Glick BR (2007) 1-Aminocyclopropane-1-carboxylic acid deaminase from Pseudomonas putida UW4 facilitates the growth of canola in the presence of salt. Can J Microbiol 53:912–918

Chookietwattana K, Maneewan K (2012) Screening of efficient halotolerant solubilizing bacterium and its effect on promoting plant growth under saline conditions. World Appl Sci J 16:1110–1117

Chowdhury N, Marschner P, Burns RG (2011) Soil microbial activity and community composition: impact of changes in matric and osmotic potential. Soil Biol Biochem 43:1229–1236

CISEAU (2006) Extent of salinization and strategies for salt-affected land prevention and rehabilitation. Background Paper, In Electronic Conference on Salinization Organized and Coordinated by IPTRID 2006

Dadkhah A (2011) Effect of salinity on growth and leaf photosynthesis of two sugar beet (*Beta vulgaris* L.) cultivars. J Agric Sci Technol 13:1001–1012

Davies PJ (1995) Plant hormones physiology, biochemistry and molecular biol, 2nd edn. Kluwer Academic Publishers, Dordrecht, p 833

Del Rio LA, Corpas FJ, Sandalio LM, Palma JM, Barroso JB (2003) Plant peroxisomes, reactive oxygen metabolism and nitric oxide. IUBMB Life 55:71–81

Dell'Amico E, Cavalca L, Andreoni V (2008) Improvement of *Brassica napus* growth under cadmium stress by cadmium resistant rhizobacteria. Soil Biol Biochem 40:74–84

Diby P, Harshad L (2014) Plant-growth-promoting rhizobacteria to improve crop growth in saline soils: a review. Agron Sustain Dev 34:737–752

Dikilitas M, Karakas S (2012) Behavior of plant pathogens for crops under stress during the determination of physiological, biochemical, and molecular approaches for salt stress tolerance. In: Ashraf M et al (eds) Crop production for agricultural improvement. Springer, Dordrecht, pp 417–441

Dodd IC, Perez-Alfocea F (2012) Microbial alleviation of crop salinity. J Exp Bot 63:3415–3428

Egamberdieva D (2008) Plant growth promoting properties of rhizobacteria isolated from wheat and pea grown in loamy sand soil. Turk J Biol 32:9–15

Egamberdieva D (2009) Alleviation of salt stress by plant growth regulators and IAA producing bacteria in wheat. Acta Phys Plant 31:861–864

Egamberdieva D (2011) Survival of *Pseudomonas extremorientalis* TSAU20 and *P. chlororaphis* TSAU13 in the rhizosphere of common bean (*Phaseolus vulgaris*) under saline conditions. Plant Soil Environ 57:122–127

Egamberdieva D, Kucharova Z (2009) Selection for root colonizing bacteria stimulating wheat growth in saline soils. BiolFertil Soils 45:563–571

Egamberdieva D, Renella G, Wirth S, Islam R (2010) Secondary salinity effects on soil microbial biomass. Biol Fertil Soils 46:445–449

Egamberdieva D, Jabborova D, Hashem A (2015) *Pseudomonas* induces salinity tolerance in cotton (*Gossypium hirsutum*) and resistance to *Fusarium* root rot through the modulation of indole-3-acetic acid. Saudi J Biol Sci 22:773–779

El-Tarabily KA, Nassar AH, Sivasithamparam K (2008) Promotion of growth of bean (*Phaseolus vulgaris* L.) in a calcareous soil by a phosphate-solubilizing, rhizosphere-competent isolate of *Micromonospora endolithica*. Appl Soil Ecol 39:161–171

Essa TA (2002) Effect of salinity stress on growth and nutrient composition of three soybean (Glycine max (L.) Merrill) cultivars. J Agron Crop Sci 188:86–93

FAO (2005) Global network on integrated soil management for sustainable use of salt-affected soils. FAO, Rome. ftp://ftp.fao.org/agl/agll/docs/misc23.pdf

FAO: The Food and Agriculture Organization of the United Nations (2002) Crops and drops: making the best use of water for agriculture. FAO, Rome, http://www.fao.org/docrep/w5146e/w5146e0a.htm

Flowers TJ (2004) Improving crop salt tolerance. J Exp Bot 55:307–319

Frankenberger WT, Bingham FT (1982) Influence of salinity on soilless enzyme activities. Soil Sci Soc Am J 46:1173–1177

Gerhardt KE, Greenberg BM, Glick BR (2006) The role of ACC deaminase in facilitating the phytoremediation of organics, metals and salt. Curr Trends Microbiol 2:61–73

Ghassemi F, Jakeman AJ, Nix HA (1995) Salinization of land and water resources: human causes, extent, management and case studies. Center for resource and environmental studies. The Australian National University, Canberra

Ghollaratta M, Raiesi F (2007) The adverse effects of soil salinization on the growth of *Trifolium alexandrinum* L. and associated microbial and biochemical properties in a soil from Iran. Soil Biol Biochem 39:1699–1702

Ghoul M, Minet J, Bernard T, Dupray E, Cornier M (1995) Marine macroalgae as a source for osmoprotection for *Escherichia coli*. Microb Ecol 30:171–181

Gill SS, Tuteja N (2010) Reactive oxygen species and antioxidant machinery in abiotic stress. Plant Physiol Biochem 48:909–930

Glick BR (1995) The enhancement of plant growth by free-living bacteria. Can J Microbiol 41:109–117

Glick BR (2014) Bacteria with ACC deaminase can promote plant growth and help to feed the world. Microbiol Res 169:30–39

Glick BR, Liu C, Ghosh S, Dumbrof EB (1997) Early development of canola seedlings in the presence of the plant growth-promoting rhizobacterium Pseudomonas putida GR12-2. Soil Biol Biochem 29:1233–1239

Glick BR, Penrose DM, Li J (1998) A model for the lowering of plant ethylene concentrations by plant growth promoting bacteria. J Theor Biol 190:63–68

Glick BR, Chenz Z, Czarny J, Duan J (2007) Promotion of plant growth by ACC-deaminase producing soil bacteria. Eur J Plant Pathol 119:329–339

Golpayegani A, Tilebeni HG (2011) Effect of biological fertilizers on biochemical and physiological parameters of Basil (Ociumum basilicm L.). Med Plant Am–Eur J Agric Environ Sci 11:411–416

Grattan SR, Grieve CM (1999) Salinity-mineral nutrient relations in horticultural crops. Sci Hortic 78:127–157

Gray EJ, Smith DL (2005) Intracellular and extracellular PGPR: commonalities and distinctions in the plant-bacterium signaling processes. Soil Biol Biochem 37:395–412

Griggs D, Stafford-Smith M, Gaffney O, Rockström J, Ohman MC, Shyamsundar P, Steffen W, Glaser G, Kanie N, Noble I (2013) Policy: sustainable development goals for people and planet. Nature 495:305–307

Gururani MA, Upadhyaya CP, Baskar V, Venkatesh J, Nookaraju A (2013) Plant growth-promoting rhizobacteria enhance abiotic stress tolerance in *Solanum tuberosum* through inducing changes in the expression of ROS-scavenging enzymes and improved photosynthetic performance. J Plant Growth Regul 32:245–258

Habib SH, Kausar H, Saud HM (2016) Plant growth-promoting rhizobacteria enhance salinity stress tolerance in Okra through ROS-scavenging enzymes. Biomed Res. https://doi.org/10.1155/2016/6284547

Hahn A, March HK (2009) Mitogen activated protein kinase cascades and ethylene: signaling, biosynthesis, or both? Plant Physiol 149:1207–1210

Hamdia MA, Shaddad MAK, Doaa MM (2004) Mechanisms of salt tolerance and interactive effects of *Azospirillum brasilense* inoculation on maize cultivars grown under salt stress conditions. Plant Growth Regul 44:165–174

Han HS, Lee KD (2005) Plant growth promoting rhizobacteria effect on antioxidant status, photosynthesis, mineral uptake and growth of lettuce under soil salinity. Res J Agric Biol Sci 1:210–215

Hingole HS, Pathak AP (2016) Saline soil microbiome: a rich source of halotolerant PGPR. J Crop Sci Biotechnol 19:231–239

Holguin G, Glick BR (2001) Expression of ACC deaminase gene from *Enterobacter cloacae* UW4 in *Azospirillum brasilense*. Microb Ecol 41:81–288

Hontzeas N, Saleh SS, Glick BR (2004) Changes in gene expression in canola roots by ACC deaminase-containing plant growth-promoting bacteria. Mol Plant-Microbe Interact 17:865–871

Hu Y, Schmidhalter U (2002) Limitation of salt stress to plant growth. In: Hock B, Elstner CF (eds) Plant toxicology. Marcel Dekker, New York, pp 91–224

Huang Y, Hutchison LH, Laskey CE, Kieber JJ (2003) Biochemical and functional analysis of CTR1, a protein kinase that negatively regulates ethylene signaling in *Arabidopsis*. Plant J 33:221–233

Ibekwe AM, Poss JA, Grattan SA, Grieve CM, Suarez D (2010) Bacterial diversity in cucumber (*Cucumis sativus*) rhizosphere in response to salinity, soil pH, and boron. Soil Biol Biochem 42:567–575

Ilangumaran G, Smith DL (2017) Plant growth promoting rhizobacteria in amelioration of salinity stress: a systems biology perspective. Front Plant Sci 8:1–14. https://doi.org/10.3389/fpls.2017.01768

Imhoff JF (1986) Osmoregulation and compatible solutes in eubacteria. FEMS Microbiol Rev 39:57–66

Islam F, Yasmeen T, Arif MS, Ali S, Ali B, Hameed S, Zhou W (2015) Plant growth promoting bacteria confer salt tolerance in Vigna radiata by up-regulating antioxidant defense and biological soil fertility. Plant Growth Regul. https://doi.org/10.1007/s10725-015-0142-y

Jamil M, Lee DB, Jung KY, Ashraf M, Lee SC, Rhal ES (2006) Effect of salt (NaCl) stress on germination and early seedling growth of four vegetables species. J Cent Eur Agric 7(273):282

Jamil A, Riaz S, Ashraf M, Foolad MR (2011) Gene expression profiling of plants under salt stress. Crit Rev Plant Sci 30:435–458

Javid MG, Sorooshzadeh A, Moradi F, Sanavy Seyed AMM, Allahdadi I (2011) The role of phytohormones in alleviating salt stress in crop plants. AJCS 5:726–734

Jha CK, Saraf M (2012) Hormonal signaling by PGPR improves plant health under stress conditions. In: Maheshwar DK (ed) Bacteria in agrobiology: stress management. Springer, Berlin/Heidelberg, pp 119–140

Jha Y, Subramanian RB (2013) Paddy plants inoculated with PGPR show better growth physiology and nutrient content under saline conditions. Chil J Agric Res 73:213–219

Jiang HC, Dong HL, Yu BS, Liu XQ, Li YL, Ji SS, Zhang CL (2007) Microbial response to salinity change in Lake Chaka, a hypersaline lake on Tibetan plateau. Environ Microbiol 9:2603–2621. https://doi.org/10.1111/j.1462-2920.2007.01377.x

Juan M, Rivero RM, Romero L, Ruíz JM (2005) Evaluation of some nutritional and biochemical indicators in selecting salt-resistant tomato cultivars. Environ Exp Bot 54:193–201

Kapoor K, Srivastava A (2010) Assessment of salinity tolerance of *Vigna mungo* var. Pu-19 using ex vitro and in vitro methods. Asian J Biotechnol 2:73–85

Kasim WA, Gaafar RM, Abou-Ali RM, Omar MN, Hewait HM (2016) Effect of biofilm forming plant growth promoting rhizobacteria on salinity tolerance in barley. AOAS 61:217–227

Kay BD (1990) Rates of change of soil structure under different cropping systems. Adv Soil Sci 12:1–52

Kaya C, Ashraf M, Dikilitas M, Tuna AL (2013) Alleviation of salt stress induced adverse effects on maize plants by exogenous application of indole acetic acid (IAA) and inorganic nutrients –a field trial. Aust J Crop Sci 7:249–254

Kempf B, Bremer E (1998) Uptake and synthesis of compatible solutes as microbial stress responses to high-osmolality environments. Arch Microbiol 170:319–330

Kende H (1993) Ethylene biosynthesis. Annu Rev Plant Physiol Plant Mol Biol 44:283–307

Khalid S, Parvaiz M, Nawaz K, Hussain H, Arshad A, Shawaka S, Sarfaraz ZN, Waheed T (2013) Effect of indole acetic acid (IAA) on morphological, biochemical and chemical attributes of two varieties of maize (Zea mays L.) under salt stress. World Appl Sci J 26:1150–1159

Khan MY, Zahir ZA, Asghar HN, Waraich EA (2017) Preliminary investigations on selection of synergistic halotolerant plant growth promoting rhizobacteria for inducing salinity tolerance in wheat. Pak J Bot 49:1541–1551

Khodarahmpour Z, Ifar M, Motamedi M (2012) Effects of NaCl salinity on maize (Zea mays L.) at germination and early seedling stage. Afr J Biotechnol 11:298–304

Kohler J, Hernandezb A, Caravacaa F, Roldan A (2009) Induction of antioxidant enzymes is involved in the greater effectiveness of a PGPR versus AM fungi with respect to increasing the tolerance of lettuce to severe salt stress. Environ Exp Bot 65:245–252

Korcan SE, Konuk M, Erdoğmuş SF (2015) Beneficial usages of halophilic microorganisms. In: Maheshwari DK, Saraf M (eds) Halophile: biodiversity and sustainable exploitation. Springer, Cham, pp 261–276

Kumar SG, Reddy AM, Sudhakar C (2003) NaCl effects on proline metabolism in two high yielding genotypes of mulberry (*Morus alba* L.) with contrasting salt tolerance. Plant Sci 165:1245–1251

Lafe O (2013) Agriculture. In: Lafe O (ed) Abulecentrism, rapid development of society catalyzed at the local community level. Springer, Cham, pp 127–132

Larcher W (1980) Physiological plant ecology: ecophysiology and stress physiology of functional groups, 2nd edn. Springer, Berlin

Le Rudulier D, Mandon K, Dupont L, Trinchant JC (2002) Salinity effects on physiology of soil microorganisms. In: Bitton G (ed) Encyclopedia of environmental microbiology. Wiley-Interscience, Canada, pp 2774–2789

Lucht JM, Bremer E (1994) Adaptation of Escherichia coli to high osmolarity environments: osmoregulation of the high-affinity glycine betaine transport system ProU. FEMS Microbiol Rev 14:3–20

Maleki M, Mostafaee S, Mokhtarnejad L, Farzaneh M (2010) Characterization of *Pseudomonas fluorescens* strain CV6 isolated from cucumber rhizosphere in Varamin as a potential biocontrol agent. AJCS 4:676–683

Mantelin S, Touraine B (2004) Plant growth-promoting bacteria and nitrate availability: impacts on root development and nitrate uptake. J Exp Bot 55:27–34

Mapelli F, Marasco R, Rolli E, Barbato M, Cherif H, Guesmi A, Ouzari I, Daffonchio D, Borin S (2013) Potential for plant growth promotion of rhizobacteria associated with Salicornia growing in Tunisian hypersaline soils. Biomed Res Int. https://doi.org/10.1155/2013/248078

Marasco R, Rolli E, Ettoumi B, Vigani G, Mapelli F, Borin S, Abou-Hadid AF, El-Behairy UA, Sorlini C, Cherif A, Zocchi G, Daffonchio D (2012) A drought resistance-promoting microbiome is selected by root system under desert farming. PLoS ONE 7:e48479. https://doi.org/10.1371/journal.pone.0048479

Martinez-Beltran J, Manzur CL (2005) Overview of salinity problems in the world and FAO strategies to address the problem In: Proceedings of the international salinity forum, Riverside, California. USDA-ARS Salinity Lab Riverside, p 311–313

Martinez-Viveros O, Jorquera MA, Crowley DE, Gajardo G, Mora ML (2010) Mechanisms and practical considerations involved in plant growth promotion by rhizobacteria. J Soil Sci Plant Nutr 10:293–319

Mavi MS, Sanderman J, Chittleborough DJ, Cox JW, Marschner P (2012) Sorption of dissolved organic matter in salt-affected soils: effect of salinity, sodicity and texture. Sci Total Environ 435:337–344

Mayak S, Tirosh T, Glick BR (2004) Plant growth-promoting bacteria that confer resistance in tomato and pepper to salt stress. Plant Physiol Biochem 42:565–572

Meloni DA, Oliva MA, Martinez CA, Cambraia J (2003) Photosynthesis and activity of superoxide dismutase, peroxidase and glutathione reductase in cotton under salt stress. Environ Exp Bot 49:69–76. https://doi.org/10.1016/S0098-8472(02)00058-8

Mensah JK, Akomeah PA, Ikhajiagbe B, Ekpekurede EO (2006) Effects of salinity on germination, growth and yield of five groundnut genotypes. Afr J Biotechnol 5:1973–1979

Mikutta R, Mikutta C, Kalbitz K, Scheel T, Kaiser K, Jahn R (2007) Biodegradation of forest floor organic matter bound to minerals via different binding mechanisms. Geochim Cosmochim Acta 71:2569–2590

Mishra M, Mishra PK, Kumar U, Prakash V (2009) NaCl phytotoxicity induces oxidative stress and response of antioxidant systems in *Cicer arietinum* L. CV. Abrodhi. Bot Res Int 2:74–82

Mitsuchi M, Wichaidit P, Jeungnijnirund S (1986) Outline of soils of the North-East plateau, Thailand: their characteristics and constraints. Technical p. 131, paper No. 1. Khon-Kaen: Agricultural Development Research Center in the North-East.

Mohammadipanah F, Hamedi J, Dehhaghi M (2015) Halophilic bacteria: potentials and applications in biotechnology. In: Maheshwari DK, Saraf M (eds) Halophile: biodiversity and sustainable exploitation. Springer, Cham, pp 277–321

Mona BH, Soha SM, Eman AES, Mostafa AA, Fayez M (2016) Cereal-PGPR interweave in salt-affected environments: towards plant persistence and growth promotion. IJSER 7:1774–1814
Munns R (2002) Comparative physiology of salt and water stress. Plant Cell Environ 25:239–250
Munns R, Tester M (2008) Mechanisms of salinity tolerance. Annu Rev Plant Biol 59:651–681
Nabti E (2007) Restauration de la croissance de *Azospirillum brasilense* et de Blé dur et leur osmoprotection par *Ulva lactuca* en Milieux Salés. Thèse de Doctorat en Science Biologique. Université Abderrahmane Mira, Faculté des sciences de la nature et de la vie, Bejaia, algérie 147p
Nabti E, Sahnoune M, Adjrad S, Van Dommelen A, Ghoul M, Schmid M, Hartmann A (2007) A halophilic and osmotolerant Azospirillum brasilense strain from Algerian soil restores wheat growth under saline conditions. Eng Life Sci 7:354–360
Nabti E, Sahnoune M, Ghoul M, Fischer D, Hofmann A, Rothballer M, Schmid M, Hartmann A (2010) Restoration of growth of durum wheat (Triticum durum var. Waha) under saline conditions due to inoculation with the rhizosphere bacterium Azospirillum brasilense NH and extracts of the marine alga Ulva lactuca. J Plant Growth Regul 29:6–22
Nabti E, Schmid M, Hartmann A (2015) Application of halotolerant bacteria to restore plant growth under salt stress. In: Maheshwari DK, Saraf M (eds) Halophile: biodiversity and sustainable exploitation. Springer, Cham, pp 235–359
Nadeem SM, Hussain I, Naveed M, Asghar HN, Zahir ZA, Arshad M (2006) Performance of plant growth promoting rhizobacteria containing ACC-deaminase activity for improving growth of maize under salt-stressed conditions. Pak J Sci 43:114–120
Nakbanpote W, Panitlurtumpai N, Sangdee A, Sakulpone N, Sirisom P, Pimthong A (2014) Salt-tolerant and plant growth-promoting bacteria isolated from Zn/Cd contaminated soil: identification and effect on rice under saline conditions. J Plant Interact 9:379–387
Nascimento FX, Rossi MJ, Soares CRFS, McConkey BJ, Glick BR (2014) New insights into 1-aminocyclopropane-1-carboxylic acid (ACC) deaminase phylogeny, evolution and ecological significance. PLoS ONE 9:e99168. https://doi.org/10.1371/journal.pone.0099168
Naz I, Bano A, Ul-Hassan T (2009) Isolation of phytohormones producing plant growth promoting rhizobacteria from weeds growing in Khewra salt range, Pakistan, and their implication in providing salt tolerance to Glycine max (L.). Afr J Biotechnol 8:5762–5766
Nelson DR, Mele PM (2007) Subtle changes in the rhizosphere microbial community structure in response to increased boron and sodium chloride concentrations. Soil Biol Biochem 39:340–351
Niu G, Xu W, Rodriguez D, Sun Y (2012) Growth and physiological responses of maize and sorghum genotypes to salt stress. ISRN Agron 2012:145072., 12 pages. https://doi.org/10.5402/2012/145072
Niu SQ, Li HR, Paré PW, Aziz M, Wang SM, Shi H, Li J, Han QQ, Guo SQ, Li J, Guo Q, Ma Q, Zhang JL (2015) Induced growth promotion and higher salt tolerance in the halophyte grass *Puccinellia tenuiflora* by beneficial rhizobacteria. Plant Soil. https://doi.org/10.1007/s11104-015-2767-z
Oades JM (1988) The retention of organic-matter in soils. Biogeochemistry 5:35–70
Ofek M, Ruppel S, Waisel Y (2006) Effects of salinity on rhizosphere bacterial communities associated with different root types of Vicia faba L. In: Ozturk M, Waisel Y, Khan A, Gork G (eds) Biosaline agriculture and salinity tolerance in plants. Birkhauser, Basel, pp 1–21
Omar SA, Abdel-Sater MA, Khallil AM, Abdalla MH (1994) Growth and enzyme activities of fungi and bacteria in soil salinized with sodium chloride. Folia Microbiol 39:23–28. https://doi.org/10.1007/BF02814524
Ondrasek G, Rengel Z, Romic D, Savic R (2010) Environmental salinisation processes in agro-ecosystem of Neretva river estuary. Novenytermeles 59:223–226
Oren A (1999) Bioenergetic aspects of halophilism. Microbiol Mol Biol Rev 63:334–340
Oren A (2002) Diversity of halophilic microorganisms: environments, phylogeny, physiology, and applications. J Ind Microbiol Biotechnol 28:56–63

Orhan F (2016) Alleviation of salt stress by halotolerant and halophilic plant growth-promoting bacteria in wheat (*Triticum aestivum*). BJM 4(7):621–627

Pan CC, Liu CA, Zhao HL, Wang Y (2013) Changes of soil physicochemical properties and enzyme activities in relation to grassland salinization. Eur J Soil Biol 55:13–19

Parida AK, Das AB (2005) Salt tolerance and salinity effects on plants: a review. Ecotoxicol Environ Safe 60:324–349. https://doi.org/10.1016/j.ecoenv.2004.06.010

Patel BB, Patel Bharat B, Dave RS (2011) Studies on infiltration of saline–alkali soils of several parts of Mehsana and Patan districts of north Gujarat. J Appl Technol Environ Sanitation 1:87–92

Paul D (2012) Osmotic stress adaptations in rhizobacteria. J Basic Microbial 52:1–10

Penrose DM (2000) The role of ACC deaminase in plant growth promotion. PhD Thesis, University of Water Loo, Canada.

Penrose DM, Glick BR (2001) Levels of 1-aminocyclopropane-1-carboxylic acid (ACC) in exudates and extracts of canola seeds treated with plant growth promoting bacteria. Can J Microbiol 41:368–372

Penrose DM, Moffatt BA, Glick BR (2001) Determination of 1-aminocyclopropane-l-carboxylic acid (ACC) to assess the effects of ACC deaminase-containing bacteria on roots of canola seedlings. Can J Microbiol 47:77–80

Pessarakli M (1991) Dry matter yield nitrogen-15 absorption and water uptake by green bean under sodium chloride stress. Crop Sci 31:1633–1640

Pichereau V, Pocard JA, Hamelin J, Blanco C, Bernard T (1998) Differential effects of dimethylsulfoniopropionate, dimethylsulfonioacetate, and other S-methylated compounds on the growth of *Sinorhizobium meliloti* at low and high osmolarities. Appl Environ Microbiol 64:1420–1429

Pinedo I, Ledger T, Greve M, Poupin MJ (2015) *Burkholderia phytofirmans* PsJN induces long-term metabolic and transcriptional changes involved in *Arabidopsis thaliana* salt tolerance. Front Plant Sci. https://doi.org/10.3389/fpls.2015.00466

Pourbabaee AA, Bahmani E, Alikhani HA, Emami S (2016) Promotion of Wheat Growth under Salt Stress by Halotolerant Bacteria Containing ACC deaminase. J Agric Sci Technol 18:855–864

Qados AMSA (2011) Effect of salt stress on plant growth and metabolism of bean plant *Vicia faba* (L.). J Saudi Soc Agric Sci 10:7–15

Rabie GH, Almadini AM (2005) Role of bioinoculants in development of salt-tolerance of Vicia faba plants under salinity stress. Afr J Biotechnol 4:210–222

Rabie GH, Aboul-Nasr MB, Al-Humiany A (2005) Increase salinity tolerance of cowpea plants by dual inoculation of Am fungus *Glomus clarum* and nitrogen-fixer *Azospirillum brasilense*. Mycobiology 33:51–61

Rahman MS, Matsumuro T, Miyake H, Takeoka Y (2000) Salinity-induced ultrastructural alternations in leaf cells of rice (Oryza sativa L.). Plant Prod Sci 3:422–429

Rengasamy P (2006) World salinization with emphasis on Australia. J Exp Bot 57:1017–1023

Rengasamy P, Olsson KA (1991) Sodicity and soil structure. Aust J Soil Res 29:935–952

Rojas-Tapias D, Moreno-Galvan A, Pardo-Diaz S, Obando M, Rivera D, Bonilla R (2012) Effect of inoculation with plant growth promoting bacteria (PGPB) on amelioration of saline stress in maize (Zea mays). Appl Soil Ecol 61:264–272

Romic D, Ondrasek G, Romic M, Josip B, Vranjes M, Petosic D (2008) Salinity and irrigation method affect crop yield and soil quality in watermelon (Citrullus Lanatus L.) growing. Irrig Drain 57:463–469. https://doi.org/10.1002/ird.358

Saghafi K, Ahmadi J, Asgharzadeh A, Bakhtiari S (2013) The effect of microbial inoculants on physiological responses of two wheat cultivars under salt stress. Int J Adv Biol Biomed Res 4:421–431

Saleem M, Arshad M, Hussain S, Bhatti AS (2007) Perspective of plant growth promoting rhizobacteria (PGPR) containing ACC deaminase in stress agriculture. J Ind Microbiol Biotechnol 34:635–648

Sandhya V, Ali S, Grover M, Reddy G, Venkateswarlu B (2009) Alleviation of drought stress effects in sunflower seedlings by the exopolysaccharides producing Pseudomonas putida strain GAP-P45. Biol Fertil Soils 46:17–26
Sandhya V, Ali SZ, Grover M, Reddy G, Venkateswaralu B (2010) Effect of plant growth promoting Pseudomonas spp. on compatible solutes antioxidant status and plant growth of maize under drought stress. Plant Growth Regul. https://doi.org/10.1007/s10725-010-9479-4
Santoyo G, Moreno-Hagelsieb G, Orozco-Mosqueda MC, Glick BR (2016) Plant growth-promoting bacterial endophytes. Microbiol Res 183:92–99
Sapsirisopa S, Chookietwattana K, Maneewan K, Khaengkhan P (2009) Effect of salt-tolerant *Bacillus* inoculum on rice KDML 105 cultivated in saline soil. Asian J Food Ag-Ind Special Issue S69–S74
Saranga Y, Menz M, Jiang CX, Robert JW, Yakir D, Andrew HP (2001) Genomic dissection of genotype X environment interactions conferring adaptation of cotton to arid conditions. Genome Res 11:1988–1995
Saravanakumar D, Samiyappan R (2007) ACC deaminase from Pseudomonas fluorescens mediated saline resistance in groundnut (Arachis hypogea) plants. J Appl Microbiol 102:1283–1292
Savvas D, Lanz F (2000) Effects of NaCl or nutrient-induced salinity on growth, yield, and composition of eggplant grown in rockwool. Sci Hortic 84:37–47
Schaller GE (2012) Ethylene and the regulation of plant development. BioMed Cent Biol 10:9
Selvakumar G, Kim K, Hu S, Sa T (2014) Effect of salinity on plants and the role of arbuscular mycorrhizal fungi and plant growth-promoting rhizobacteria in alleviation of salt stress. In: Ahmad P, Wani MR (eds) Physiological mechanisms and adaptation strategies in plants under changing environment. Springer, New York, pp 115–144
Setia R, Marschner P, Baldock J, Chittleborough D (2010) IsCO2 evolution in saline soils affected by an osmotic effect and calcium carbonate? Biol Fertil Soils 46:781–792
Setia R, Gottschalk P, Smith P, Marschner P, Baldock J, Setia D, Smith J (2013) Soil salinity decreases global soil organic carbon stocks. Sci Total Environ 465:267–272
Sgroy V, Cassán F, Masciarelli O, Papa MFD, Lagares A, Luna V (2009) Isolation and characterization of endophytic plant growth-promoting (PGPB) or stress homeostasis-regulating (PSHB) bacteria associated to the halophyte Prosopis strombulifera. Appl Microbiol Biotechnol 85:371–381
Shaharoona B, Arshad M, Zahir ZA, Khalid A (2006) Performance of *Pseudomonas* spp. containing ACC-deaminase for improving growth and yield of maize (*Zea mays* L.) in the presence of nitrogenous fertilizer. Soil Biol Biochem 38:2971–2975
Shahbazi M, Arzani A, Saeidi G (2011) Effect of NaCl treatments of seed germination and antioxidant activity of Canola (*Brassica Napus*) Cultivars. Bangladesh J Bot 41:67–73
Shaheen S, Naseer S, Ashraf M, Akram NA (2013) Salt stress affects water relations, photosynthesis, and oxidative defense mechanisms in *Solanum melongena* L. J. Plant Interact 8:85–96
Shannon MC, Grieve CM (1999) Tolerance of vegetable crops to salinity. Sci Hortic 78(5):38
Sharma S, Kulkarni J, Jha B (2016) Halotolerant rhizobacteria promote growth and enhance salinity tolerance in peanut. Front Microbiol 7:1600. https://doi.org/10.3389/fmicb.2016.01600
Sheldon A, Menzies NW, Bing SH, Dalal RC (2004) The effect of salinity on plant available water. Supersoil 2004: 3rd Australian New Zealand soil conference
Shilpi M, Girdhar K, Tuteja PN (2008) Calcium and salt-stress signaling in plants: shedding light on SOS pathway. Arch Biochem Biophys 471:146–158
Shin W, Siddikee MA, Joe MM, Benson A, Kim K, Selvakumar G, Kang Y, Jeon S, Samaddar S, Chatterjee P, Walitang D, Chanratana M, Sa T (2016) Halotolerant plant growth promoting bacteria mediated salinity stress amelioration in plants. Korean J Soil Sci Fertil 49:355–367
Shrivastava P, Kumar R (2015) Soil salinity: a serious environmental issue and plant growth promoting bacteria as one of the tools for its alleviation. Saudi J Biol Sci 22:123–131
Shukla PS, Agarwal PK, Jha B (2011) Improved salinity tolerance of *Arachis hypogaea* (L.) by the interaction of halotolerant plant-growth-promoting rhizobacteria. J Plant Growth Regul. https://doi.org/10.1007/s00344-011-9231-y

Siddikee MA, Chauhan PS, Anandham R, Gwang-Hyun H, Tongmin S (2010) Isolation, characterization, and use for plant growth promotion under salt stress of ACC deaminase-producing halotolerant bacteria derived from coastal soil. J Microbiol Biotechnol 20:1577–1584

Siddikee MA, Glick BR, Chauhan PS, Sa T (2011) Enhancement of growth and salt tolerance of red pepper seedlings (*Capsicum annuum* L.) by regulating stress ethylene synthesis with halotolerant bacteria containing 1-aminocyclopropane-1-carboxylic acid deaminase activity. Plant Physiol Biochem 49:427–434

Singh PK, Kumar V, Maurya N, Choudhary H, Kumar V, Gupta AK (2014) Grassroots solutions to overcome abiotic and biotic environmental stress in agriculture. In: Singh SB et al (eds) Translational research in environmental and occupational stress. Springer, New Delhi, pp 11–16

Singleton PW, Bohlool BB (1984) Effect of salinity on nodule formation by soybean. Plant Physiol 74:72–76

Six J, Paustian K, Elliott ET, Combrink C (2000) Soil structure and organic matter: I. Distribution of aggregate-size classes and aggregate-associated carbon. Soil Sci Soc Am J 64:681–689

Sleator RD, Hill C (2001) Bacterial osmoadaptation: the role of osmolytes in bacterial stress and virulence. FEMS Microbiol Rev 26:49–71

Stanković MS, Petrović M, Godjevac D, Stevanović ZD (2015) Screening inland halophytes from the central Balkan for their antioxidant activity in relation to total phenolic compounds and flavonoids: are there any prospective medicinal plants? J Arid Environ 120:26–32

Suarez R, Wong A, Ramirez M, Barraza A, Orozco MD, Cevallos MA, Lara M, Hernandez G, Iturriaga G (2008) Improvement of drought tolerance and grain yield in common bean by overexpressing trehalose-6-phosphate synthase in rhizobia. Mol Plant-Microbe Interact 21:958–966

Taffouo VD, Wamba OF, Yombi E, Nono GV, Akoe A (2010) Growth, yield, water status and ionic distribution response of three bambara groundnut (*Vigna subterranean*) (L.) verdc. landraces grown under saline conditions. Int J Bot 6:53–58

Takagi M, El-Shemy HA, Sasaki S, Toyama S, Kanai S, Saneoka H, Fujita K (2009) Elevated CO2 concentration alleviates salinity stress in tomato plant. Acta Agric Scand B—S P 59:87–96

Talsma T (1963) The control of saline groundwater. Thesis for the degree of doctor in land technology, University of Wageningen. Repr Bull Univ Wagening 63:1–68

Tank N, Saraf M (2010) Salinity-resistant plant growth promoting rhizobacteria ameliorates sodium chloride stress on tomato plants. J Plant Interact 5:51–58

Tate RL III (2000) Soil microbiology. Wiley, New York

Tester M, Davenport R (2003) Na^+ tolerance and Na^+ transport in higher plants. Ann Bot 91:503–527

Timmusk S, Wagner GH (1999) The plant-growth-promoting rhizobacterium *Paenibacillus polymyxa* induces changes in *Arabidopsis thaliana* gene expression: a possible connection between biotic and abiotic stress responses. Mol Plant-Microbe Interact 12:951–959

Tokala RK, Strap JL, Jung CM, Crawford DL, Salove H, Deobald LA, Bailey FJ, Morra MJ (2002) Novel plant microbe rhizosphere interaction involving *S. lydicus* WYEC108 and the pea plant (*Pisum sativum*). Appl Environ Microbiol 68:2161–2171

Tripathi AK, Verma SC, Ron EZ (2002) Molecular characterization of a salt-tolerant bacterial community in the rice rhizosphere. Res Microbiol 153:579–584

Tripathi S, Chakraborty A, Chakrabartia K, Bandyopadhyayc BK (2007) Enzyme activities and microbial biomass in coastal soils of India. Soil Biol Biochem 39:2840–2848

United Nations, Department of Economic and Social Affairs, Population Division (2004) World population prospects: world population to 2300. Working Paper No. ST/ESA/SER.A/236. New York

United States Department of Agriculture (1954) Diagnosis and improvement of saline and alkali soils. In: Richards LA (ed) Agriculture hand book no. 60. Oxford & IBH, New Delhi, pp 1–160

Upadhyay SK, Maurya SK, Singh DP (2012) Salinity tolerance in free-living plant growth promoting rhizobacteria. Indian J Res 3:73–78

Vacheron J, Desbrosses G, Bouffaud ML, Touraine B, Moënne-Loccoz Y, Muller D, Legendre L, Wisniewski-Dyé F, Prigent-Combaret C (2013) Plant growth promoting rhizobacteria and root system functioning. Front Plant Sci 4:1–19

Vurukonda SSKP, Vardharajula S, Shrivastava M, SkZ A (2016) Enhancement of drought stress tolerance in crops by plant growth promoting rhizobacteria. Microbiol Res 184:13–24

Waller P, Yitayew M (2016) Water and salinity stress. In: Waller P, Yitayew M (eds) Irrigation and drainage engineering. Springer, Cham, pp 51–65

White PJ, Broadley MR (2001) Chloride in soils and its uptake and movement within the plant: a review. Ann Bot 88:967–988. https://doi.org/10.1006/anbo.2001.1540

Wichern J, Wichern F, Joergensen RG (2006) Impact of salinity on soil microbial communities and the decomposition of maize in acidic soils. Geoderma 137:100–108

Wild A (2003) Soils, land and food: managing the land during the twenty-first century. Cambridge University Press, Cambridge

Wong VNL, Dalal RC, Greene RSB (2009) Carbon dynamics of sodic and saline soils following gypsum and organic material additions: a laboratory incubation. Appl Soil Ecol 41:29–40

Wu J, Vincent B, Yang J, Bouarfa S, Vidal A (2008) Remote sensing monitoring of changes in soil salinity: a case study in Inner Mongolia, China. Sensors 8:7035–7049

Xiong L, Zhu JK (2002) Salt-stress signal transduction. In: Scheel D, Wasternack C (eds) Plant signal transduction. Frontiers in molecular biology series. Oxford University Press, Oxford, pp 165–197

Xu ZH, Saffigna PG, Farquhar GD, Simpson JA, Haines RJ, Walker S, Osborne DO, Guinto D (2000) Carbon isotope discrimination and oxygen isotope composition in clones of the F (1) hybrid between slash pine and Caribbean pine in relation to tree growth, water-use efficiency and foliar nutrient concentration. Tree Physiol 20:1209–1217

Xu GY, Rocha PS, Wang ML, Xu ML, Cui YC, Li LY, Zhu YX, Xia X (2011) A novel rice calmodulin-like gene, OsMSR2, enhances drought and salt tolerance and increases ABA sensitivity in Arabidopsis. Planta 234:47–59

Yadav S, Irfan M, Ahmad A, Hayat S (2011) Causes of salinity and plant manifestations to salt stress: a review. J Environ Biol 32:667–685

Yancey PH (2005) Organic osmolytes as compatible, metabolic and counteracting cytoprotectants in high osmolarity and other stresses. J Exp Biol 208:2819–2830

Yao LX, Wu ZS, Zheng YY, Kaleem I, Li C (2010) Growth promotion and protection against salt stress by Pseudomonas putida Rs-198 on cotton. Eur J Soil Biol 46:49–54. https://doi.org/10.1016/j.ejsobi.2009.11.002

Yildirim E, Taylor AG, Spittler TD (2006) Ameliorative effects of biological treatments on growth of squash plants under salt stress. Sci Hortic 111:1–6

Younesi O, Chaichi MR, Postini K (2013) Salt tolerance in alfalfa following inoculation with Pseudomonas. Middle-East J Sci Res 16:101–107

Yuan BC, Xu XG, Li ZZ, Gao TP, Gao M, Fan XW, Deng HM (2007) Microbial biomass and activity in alkalized magnesic soils under arid conditions. Soil Biol Biochem 39:3004–3013

Zafar-ul-Hye M, Ahmed M, Shahzad SM (2013) Synergistic effect of Rhizobia and plant growth promoting rhizobacteria on the growth and nodulation of lentil seedlings under axenic conditions. Soil Environ 32:79–86

Zahir ZA, Ghani U, Naveed M, Nadeem SM, Asghar HN (2009) Comparative effectiveness of *Pseudomonas* and *Serratia* sp. containing ACC-deaminase for improving growth and yield of wheat (*Triticum aestivum* L.) under salt-stressed conditions. Arch Microbiol 191:415–424

Zahran HH (1997) Diversity, adaptation and activity of the bacterial flora in saline environments. Biol Fertil Soils 25:211–223

Zapata PJ, Serrano M, Pretel MT, Amoros A, Botella MA (2004) Polyamines and ethylene changes during germination of different plant species under salinity. Plant Sci 167:781–788

Zhang H, Kim M-S, Dowd SE, Shi H, Pare PW (2008) Soil bacterial confer plant tolerance by tissue-specific regulation of the sodium transporter HKT1. Mol Plant-Microbe Interact 21:737–744

Halotolerant Rhizobacteria: A Promising Probiotic for Saline Soil-Based Agriculture

3

Ankita Alexander, Avinash Mishra, and Bhavanath Jha

Abstract

Soil salinity is a serious threat to sustainable agriculture, and a number of research are going on to improve saline-resistant crops by using various breeding methods and genetic engineering tools. These methods are time-consuming, often face yield penalties, and many other ethical issues. There is a need to explore other more stable, environmentally friendly methods for the sustainable agriculture. Exploration of plant growth-promoting rhizobacteria (PGPR) associated with salt-tolerant plants (halophytes) and their use as probiotics for saline soil agriculture are a promising substitute for classical approaches. Salinity is one of the major abiotic stress reported from arid and semiarid regions which causes a major loss in the agriculture productivity. Halophytes are adapted to the saline environment because of their genetic makeup and associated microbiome. These microbiomes have potential to survive in the saline condition, but they are not thoroughly explored. Several studies showed that bacteria associated with halophytes, directly and indirectly, support the plant growth and yield in saline conditions; thus, these bacteria can be used as probiotics for salt-sensitive plants (glycophytes) grown in the salt-affected area to enhance the productivity. PGPR induce many morphological, physiological, and genetic changes in a plant which compensate the pressure of salt stress. The genetic level changes in plants due to application or the presence of PGPR are known as induced systemic resistance (ISR). PGPR secrete some beneficial elements like organic solutes, siderophores, etc. to survive in harsh conditions. PGPR also help plants to maintain their

A. Alexander · A. Mishra (✉) · B. Jha (✉)
CSIR-Central Salt and Marine Chemicals Research Institute, Bhavnagar, Gujarat, India

Academy of Scientific and Innovative Research, Council of Scientific and Industrial Research, Ghaziabad, India
e-mail: avinash@csmcri.res.in; jha.bhavanath@gmail.com

M. Kumar et al. (eds.), *Saline Soil-based Agriculture by Halotolerant Microorganisms*, https://doi.org/10.1007/978-981-13-8335-9_3

osmotic pressure and nutrient balance. The presence of PGPR also affects the level of various phytohormones in plants which play a major role in growth, development, and stress response of the plant.

Keywords
Halophytes · Salt-sensitive plants · Halotolerant PGPR · Salinity stress · Crop plants

3.1 Introduction

As human probiotics, plant-probiotic microorganisms (PPM) are beneficial microorganisms which can act as bio-protectants (reduce pathogen infection), bio-fertilizers (increase yield), and biostimulants, additionally also reduce various biotic and abiotic plant stresses (de Souza Vandenberghe et al. 2017). According to FAO/WHO expert consultation report, probiotics are "live microorganisms which when administered in adequate amounts confer a health benefit on the host" (Hill et al. 2014).

Plant growth is affected by various abiotic factors like abundance of heavy metals, soil toxicity, salinity, cold, drought, temperature and flood. Among these, soil salinity is a major cause which adversely affects the plant growth and productivity (Príncipe et al. 2007). The soil is considered saline when electrical conductivity (EC) of the saturation extract of the rhizosphere soil (root zone) exceeds 4 dS m^{-1} which is equivalent to approximately 40 mM NaCl at 25 °C; commonly, a higher EC limits the crop productivity (Munns and Tester 2008). Primary salinization is caused by the salts that are already present in the soil, and secondary salinization is a result of excessive use of chemical fertilizers, soil amendments, and inadequate irrigation management (Carillo et al. 2011). According to reports of FAO (2016), 20% of total agriculture land has been severely affected by excessive soil salinity.

Soil salinity adversely affects the plant health at a physiological, biochemical, and molecular level which decreases the growth and yield of crop plants (Roy et al. 2014). The development of salt stress-resistant transgenic plants is considered as an option for saline soil agriculture; however, there is a need for environmentally friendly, ethical, and natural procedure for sustainable agriculture (Ashraf 2009). Halophytic plant species are adapted to saline soils, naturally. These plants can survive in this harsh environment by their specific genetic makeup and their root-associated microbiome having very diverse rhizobacteria (Mishra and Tanna 2017; Mukhtar et al. 2017). Rhizobacteria isolated from microbiome of halophytes may help the glycophytic plant to grow in saline soil and enhance the productivity of glycophytic crop plants. These rhizobacteria can be used for the improvement of the soil quality.

3.2 Effect of Salt on Plant Growth

Plant growth is affected by salt at multiple levels. The primary effect of soil salinity on plant growth and productivity is briefly discussed as follows:

3.2.1 Osmotic Stress

During stress, water intake by root is very difficult which results in reduced transpiration rate and leaf expansion, stomatal closure, water retention, water use efficiency, and increased rate of senescence of older leaves (Munns and Tester 2008). Osmotic stress also leads to the production of reactive oxygen species (ROS) which results in autophagy (Han et al. 2011).

3.2.2 Nutrient Imbalance

Soil salinity profoundly influences the nutrient imbalance and causes nutrients deficiency in plants (Blaylock 1994). In saline soil, it is difficult to take phosphorus (P) because phosphate ions precipitate with Ca ions and become unavailable for plant utilization (Bano and Fatima 2009).

3.2.3 Decrease in Photosynthesis

Salt stress damages the photosynthetic capability of the plant. During salt stress, growth of the plant is affected, and it also causes a reduction in leaf area, degradation of chlorophyll and carotene content, low level of stomatal conductance, and harmful effects on PSII and PSI which altogether result in the decrease of photosynthesis ability of plant (Netondo et al. 2004).

3.2.4 Effect on Reproductivity

Salinity badly disturbs reproductive machinery of the affected plant. Salt stress inhibits microsporogenesis (formation of microspores), floral development, ovule abortion, embryo arrest, and senescence of fertilized embryos, but induces apoptosis (programmed cell death) in some reproductive tissues (Shrivastava and Kumar 2015).

3.2.5 Ion Toxicity

Ion toxicity is the major problem of soil salinity. During salinity, excess sodium ion causes leaf burn and defoliation (Podmore 2009). Chloride toxicity causes leaf bronzing and necrotic spots in plant species (Rahnama et al. 2010). A high level of Cl^- disturbs NO_3^- uptake in leaves which results in the decrease of nitrate reductase (NR) activity, affecting the nitrogen assimilation in plants (Baki et al. 2000). Excessive salt causes replacement of K^+ by Na^+ which promotes conformational changes in many proteins and enzymes, and these affect many metabolic and molecular alterations in plants (Zhu 2002; Chinnusamy et al. 2006).

3.2.6 Oxidative Stress

Salinity induces the generation of reactive oxygen species (ROS) and free radicals which causes oxidative stress and reduces the level of antioxidant enzymes such as catalase (CAT), superoxide dismutase (SOD), peroxidase (POX), glutathione (GSH), ascorbate peroxidase (APX), glutathione reductase (GR), glutathione-S-transferase (GST), and monodehydroascorbate reductase in plants (AbdElgawad et al. 2016). ROS are harmful to cell components and damage lipids, protein, and nucleic acids; furthermore, long-term exposure causes the death of host species (del Rio et al. 2003).

3.2.7 Production of Excess Ethylene

Plants, when subjected to salinity, produce excessive ethylene, which severely retards root development (Mahajan and Tuteja 2005). Under biotic and abiotic stress conditions, ethylene causes a decrease in root growth and senescence in crop plants (Ma et al. 2003). In general, salt stress adversely affects the cell cycle and differentiation by reducing the expression and activity of cyclins and cyclin-dependent kinases. Some reports also showed that salinity also reduces seed germination, seedling growth, and enzyme activity (Seckin et al. 2009), which ultimately affects plant yield.

3.3 Need for Plant Probiotic

For the sustainable agriculture toward biotic and abiotic stress, transgenic approaches are commonly used, but there are many shortcomings and limitation of transgenic plants. These approaches need a long time and need much manpower from lab to field and after that have less success rate (Coleman- Derr and Tringe 2014). Transgenic approaches caused unstable progeny due to the involvement of multiple genes in abiotic stress responses and also the uncertainty of gene flow in the next generation (Jewell et al. 2010). Transgenic crops are facing the difficulty of public acceptance as well as many ethical issues (Fedoroff 2010). Molecular techniques which are involved in the development of transgenic crops are most of the time not applicable or responsive in case of tetraploid and hexaploid species. Saline and alkaline stresses are typically linked in nature (Saslis-Lagoudakis et al. 2014), and the tolerance of transgenic toward high soil alkalinity (pH) and/or salt alkaline-mixed conditions could not fully obtain (Yamaguchi and Blumwald 2005). Sometimes transgenic salt-tolerant crops, especially in which transcription factors have been genetically modified, resulting in the low yield of the plant (Roy et al. 2014). When a plant is facing salt stress, many genes, proteins, and metabolites are activated at the same time, but transgenic plants are concentrated on either only one gene or one promoter only, so protection against salt stress is not obtained completely or up to desired level (Bhatnagar-Mathur et al. 2008).

To overcome these problems and for the fulfillment of increasing food demand in an environment-friendly manner, there is a need of exploring more useful alternative strategies for sustainable agriculture. The health of the plant and its response toward any stress are affected by not only its genome but also microbiome of its surroundings (Munns and Gilliham 2015; Vannier et al. 2015). Plant root system releases a large number of nutrients, known as rhizodeposits (e.g., exudates, border cells, and mucilage). Rhizodeposits influence the rhizosphere by regulating the microbial diversity and activity on plant roots (Cook et al. 1995). The rhizosphere is a narrow zone of soil including surrounding plant roots and it can contain up to 10^{11} microbial cells per gram of the root (Egamberdieva 2008).

The microbes have coevolved with their corresponding hosts and adapted their community structure for specific environmental stresses (Lau and Lennon 2012). Thus, interacting network among root, soil, and microorganisms plays an imperative role in supporting healthy growth and defense mechanism under unfavorable conditions (biotic and abiotic stress) for both the host and its associated organisms. Some of these microbes support their host plant regarding growth, yield, in biotic and abiotic stress (Fig. 3.1). Due to plant growth-promoting traits of these bacteria, they are called plant growth-promoting bacteria (PGPB), and PGPB which are found in the rhizosphere and stimulate plant growth are called plant growth-promoting rhizobacteria (PGPR) (Kloepper and Schroth 1981). PGPR can be classified as extracellular and intracellular. Extracellular PGPR (ePGPR) exist in the rhizosphere, on the rhizoplane, in the spaces between root cortex cells, while intracellular PGPR (iPGPR) exist inside root cells, generally in specialized nodules (Gray and Smith 2005).

3.4 Rhizobacteria as Probiotic

The term induced systemic tolerance (IST) has been proposed for PGPR-induced physical and chemical changes that result in enhanced abiotic stress tolerance (Shrivastava and Kumar 2015). Halotolerant rhizobacteria have many characteristics which improve the plant health (Fig. 3.2), including nitrogen fixation, phosphate solubilization, production of phytohormones and exopolysaccharide, aminocyclopropane-1-carboxylate deaminase (ACCD) production, and biolytic enzyme secretion (de Souza et al. 2015).

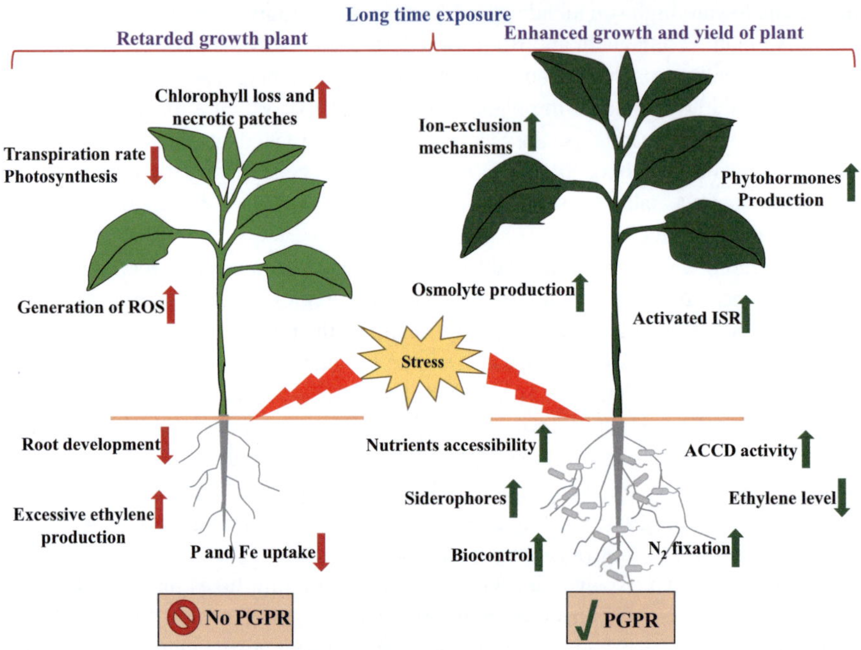

Fig. 3.1 Role of PGPR in the plant growth promotion

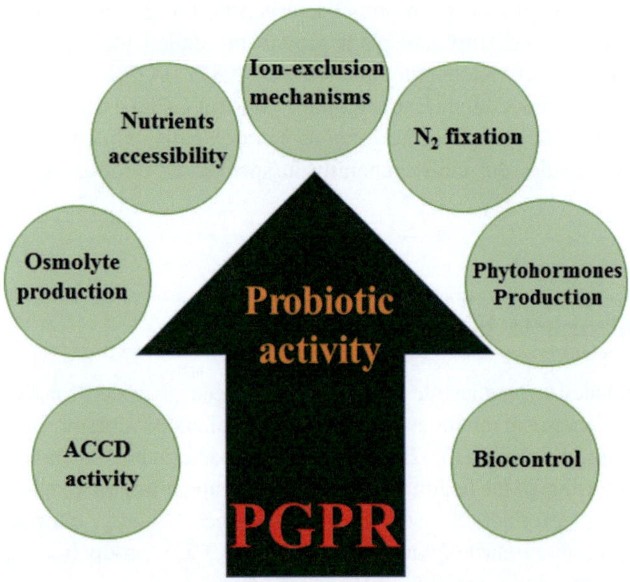

Fig. 3.2 Probiotic activity of PGPR for plant growth promotion

3.4.1 ACCD Activity

Many halotolerant bacterial isolates showed the ACCD activity which required for plant health (Santoyo et al. 2016). Plants naturally select ACCD-producing bacteria under stress conditions. PGPR produce the enzyme ACC deaminase which cleaves the plant ethylene precursor ACC into ammonia and α-ketobutyrate and decrease the level of ethylene in seedlings and plants. Additionally, ACCD stimulates IAA production and thus enhances the growth and survival of plant under abiotic stress (Glick 2014; Mayak et al. 2004a, b).

3.4.2 Osmolyte Production

Osmolytes are highly soluble substances with no net charge at physiological pH, so they do not interact with proteins and compatible with many cellular functions such as DNA–protein interactions, protein–protein interactions, cellular metabolism, and osmotic balance (Lippert and Galinski 1992; Welsh 2000). The most studied osmolytes, produced and secreted by halotolerant PGPR, are trehalose, glucosylglycerol (GG), proline, glycine betaine, and choline (Qurashi and Sabri 2013). Trehalose is nonreducing disaccharide which is not common in vascular plants but acts as an osmoprotectant. It stabilizes dehydrated enzymes and membranes, thus protects them from desiccation. Trehalose metabolism is important for plant growth, and yield of leguminous plants under abiotic stress conditions (Suarez et al. 2008). *Stenotrophomonas rhizophila* is halotolerant rhizobacteria which can grow at 4% NaCl and produce the trehalose in response to salt stress (Roder et al. 2005).

3.4.3 Ion-Exclusion Mechanisms

Rhizobacteria can alter root uptake of toxic ions by slight changes in the physiology of host plant (by regulating ion transporters) and by modification in physical barriers around the roots (Siddikee et al. 2011) or by directly reducing accumulation of toxic ions (Na^+ and Cl^-). Rhizobacteria also improve the nutritional status of the plant by maintaining the essential level of both macro (N, P, and K) and micronutrients (Zn, Fe, Cu, and Mn). Zahir et al. (2009) showed higher K^+/Na^+ ratios in salinized crop plants like wheat after inoculation with *Pseudomonas* and *Serratia* sp.

3.4.4 Increment in Nutrients Accessibility

Phosphorus (P) is abundantly available in soils in both organic and inorganic form, but it is not easily available for plants because the majority of soil P is found in insoluble forms. Microbial-induced changes in rhizosphere pH (organic acid excretion) convert insoluble phosphorus in soil into the soluble state (El-Tarabily and

Youssef 2010). In soil, iron occurs as Fe^{3+} and forms insoluble hydroxides which becomes inaccessible to plants. PGPR also secret some organic molecules called as siderophores. These siderophores help plants to take iron because they chelate Fe^{3+} (Rajkumar et al. 2010). It is shown that siderophores produced by *Enterobacter* sp. 638 and *Pseudomonas* sp. enhance the growth of poplar and tomato, respectively (Taghavi et al. 2009; Nishma et al. 2014).

3.4.5 N_2 Fixation

Nitrogen fixation is the process of conversion of atmospheric N_2 to NH_3. Microorganisms involved in N_2 fixation are rhizosphere associated either symbiotic or endophytic. Rhizosphere-associated bacteria use carbon compounds from root exudates for N_2 fixation. All these microorganisms contain specific genes for N_2 fixation, for example, a *nifH* gene that encodes for dinitrogenase reductase enzyme which converts N_2 to NH_3 and makes it available for plants (Raymond et al. 2004).

3.4.6 Production of Phytohormones

Rhizobacteria also produce a variety of phytohormones like cytokinin, indole-3-acetic acid (IAA), abscisic acid (ABA), and gibberellic acid (GA) (Glick 2012). Phytohormones produced by PGPR play an important role in increasing root surface area, and numbers of root tip in many plants, thus increasing the accessibility of plants to soil nutrients (Narula et al. 2006). Similarly, GA enhances the shoot growth and total biomass of the plant (Boiero et al. 2007). ABA regulates the stomatal activity of plants hence indirectly affects the photosynthesis response to the saline soil (Dodd 2003). Many PGPR, isolated from saline soil, produced ABA which can enhance the growth of plants under saline condition (Naz et al. 2009).

3.4.7 Biocontrol Agent

PGPR also secret some antibiotic compounds which play a major role in the reduction of phytopathogens and deleterious rhizobacteria (Stutz et al. 1986). By acting as a biocontrol agent, rhizobacteria stimulate plant growth indirectly. They synthesize extracellular enzymes to hydrolyze the cell wall of pathogens and compete with harmful bacteria for niches within the rhizosphere (Zahir et al. 2004; van Loon 2007). PGPR also activate induced systemic resistance (ISR) in host plant to fight with pathogens at the molecular level (Chandler et al. 2008).

3.5 Halotolerant Rhizobacteria

Bacteria which are associated with halophytic plants can tolerate salt level up to 4–30% (Yuan et al. 2016) and play an essential role in the survival of halophytes in the harsh environment and are called as halophyte rhizosphere-associated microbes (HRAMs). These bacteria are endowed with specific metabolic signals which are essential for plant's survival under salt stress (Szymanska et al. 2013). The plant–microbes interaction effectively improves the soil quality and fertility (Yuan et al. 2016). HRAMs have all PGPR activity with high salt tolerance, so they are the most suitable candidate for growth and yield enhancement in glycophytic plants in salt-affected areas in most environmentally friendly manner.

3.6 Halotolerant Bacteria as Plant Probiotics for Saline Soil-Based Agriculture

Proper osmotic regulation is a crucial point for plants growing in the saline area; the absence of this results in loss of turgidity, cell dehydration, failure of photosynthetic machinery, nutritional disorders, toxicities, less crop productivity, and ultimately death of cells (Ashraf 2004; Shrivastava and Kumar 2015).

Halotolerant bacteria survive under salt stress conditions because they modify the osmolality of the rhizosphere (Egamberdieva 2011). These modified osmolytes are made available for utilization of plant and act as free radical scavengers, a regulator of the photosynthetic apparatus, and stabilizer of subcellular structures (Yang et al. 2009), hence affecting the plant growth (Miller and Wood 1996), so, they are considered as an effective candidate for plant growth promotion.

Previously, many studies showed that halotolerant PGPR are capable to ameliorate the detrimental effect of salinity on the growth of basil, canola, maize, and tomato plants (Kang et al. 2014; Rojas-Tapias et al. 2012). Karuppasamy et al. (2011) showed that the growth of tree legumes *Samanea saman* could be improved by the application of stress-tolerant rhizobia. Halotolerant bacterial consortia isolated from avocado trees could ameliorate salinity stress in other plant species such as wheat (Barra et al. 2016). Upadhyay et al. (2009) isolated 130 rhizobacterial strains from the rhizosphere of wheat plants grown in a saline zone and showed that out of 130, 24 were tolerant to relatively high levels (up to 8%) of NaCl. All of the 24 salt-tolerant isolates were able to produce IAA, 10 isolates solubilized phosphorus, eight isolates produced siderophores, 6 were involved in gibberellin production, and 2 isolates contained the *nifH* gene, which indicates that the whole consortia have PGPR ability.

In another study, Nadeem et al. (2010) evaluate a number of axenic cultures of rhizobacterial strains for their plant–growth potential under varying salinity stress conditions. At a high salinity level (15 dS m^{-1}), the PGPR–inoculants improved overall plant growth including plant height, root length, plant biomass, and also grain yield by 37%, 70%, 116%, and 111%, respectively. It was also observed that treated plants also had higher K^+/Na^+ ratios, relative water, and chlorophyll contents along with lower proline contents. Potential halotolerant rhizobacteria reported for sustainable saline soil agriculture are listed in Table 3.1.

Table 3.1 Potential halotolerant rhizobacteria with different plant growth-promoting traits

Name of bacteria	Isolated from	Plant growth-promoting trait observed	References
Achromobacter piechaudii ARV8	Rhizosphere of arid and saline environment	Increase in the biomass of tomato seedlings under salt stress	Mayak et al. (2004a, b)
Bacillus pumilus, *B. aquimaris*, *B. arsenicus*, *B. sporothermodurances*, *Arthrobacter* sp., *B. cereus*, *Pseudomonas medicona*, and *B. subtilis*	Rhizosphere of 30 days old wheat plants grown in a saline zone	Production of indole-3-acetic acid (IAA), phosphorus solubilization, siderophore production, gibberellin production, and the presence of *nifH* gene	Upadhyay et al. (2009)
Acinetobacter, *Alcaligenes faecalis*, *Bacillus cereus*, *Enterobacter hormaechei*, *Pantoae*, and *Pseudomonas aeruginosa*	Rhizobacteria isolated from wheat and pea grown in loamy sand soil	Improved plant growth and nutrition under salt stress	Egamberdieva (2008)
Stenotrophomonas rhizophila strain DSM14405T	Highly salinated soils	Increase of germination rate, shoot, and root growth of wheat, tomato, lettuce, sweet pepper, melon, celery, and carrot	Egamberdieva (2011)
Bacillus megaterium	Degraded soil	Increase in root hydraulic conductance in *Zea mays*	Marulanda et al. (2010)
Pseudomonas mendocina	Rhizosphere	Enhanced growth and nutrient uptake of *Lactuca sativa*	Kohler et al. (2009)
Azotobacter chroococcum	Rhizosphere	Enhanced oxidative stress tolerance in sugar beet	Stajner et al. (1997)
Pseudomonas putida Rs-198	Alkaline soil	Increased the K^+/Na^+ ratios	Yao et al. (2010)
Consortia of plant growth-promoting rhizobacteria	Rhizosphere of weeds (*Chrysopogon aucheri*, *Lactuca dissecta*, *Solanum surattense*, and *Sonchus arvensis*) grown in saline soils	Increase in salt tolerance in *Glycine max* L.	Naz et al. (2009)
Bacillus pumilus with *P. pseudoalcaligenes*	Root tissue of rice as well as the	Induction of osmoprotectant and	Jha et al. (2011)

(continued)

Table 3.1 (continued)

Name of bacteria	Isolated from	Plant growth-promoting trait observed	References
	rhizospheric soil of rice	antioxidant proteins in paddy	
Azospirillum strains	Saline or non-saline soil	Increased shoot dry weight and grain yield in wheat	Nia et al. (2012)
Halobacillus sp. and *Bacillus halodenitrificans*	Rhizosphere of saline habitats	Ameliorated salt stress in wheat seedlings and increased root length	Ramadoss et al. (2013)
Virgibacillus marismortui and *Terribacillus halophilus*	Dessert soil	Improved the stem growth in tomato	Essghaier et al. (2014)
Promicromonospora sp. SE188	Field soil sample	Improved growth of *Solanum lycopersicum*	Kang et al. (2012)
Group of rhizobacteria	Roots of *Salicornia* grown in hypersaline soils	Increased plant growth	Mapelli et al. (2013)
Bacillus subtilis, B. atrophaeus, B. spharicus, Staphylococcus kloosii and *Kocuria Erythromyxa*	Rhizosphere of plants naturally grown high salty soils	Promoted yield and growth of strawberry and lettuce under saline conditions	Karlidag et al. (2010) and Yildirim et al. (2011)
Zhihengliuella halotolerans, Staphylococcus succinus, Bacillus gibsonii, Oceanobacillus oncorhynchi, Halomonas sp., and *Thalassobacillus* sp.	Salt-affected soils	Improved growth in *Triticum aestivum* L. under salt stress	Orhan (2016)
Azospirillum spp.	Mangrove rhizosphere	Growth promotion in *Salicornia bigelovii*	Bashan et al. (2000)
Endophytic bacteria *Pseudomonas* sp. ISE-12 and *Xanthomonadales* sp. CSE-34	Roots of halophyte *Salicornia europaea*	Growth promotion in *Beta vulgaris* under saline conditions	Piernik et al. (2017)
Consortia od rhizobacteria mainly *Pseudomonas* sp.	Roots of coastal sand dune plants	Increase in plant growth	Shin et al. (2007)
Klebsiella oxytoca Rs-5	Salinized cotton fields	Salt stress relief and growth promotion in cotton	Yue et al. (2007)
Pseudomonas aeruginosa PF23	Rhizosphere of different plants	Plant growth promotion, biocontrol, and stress	Tewari and Arora (2014)

(continued)

Table 3.1 (continued)

Name of bacteria	Isolated from	Plant growth-promoting trait observed	References
		amelioration in sunflower under saline conditions	
Bacillus, *Pseudomonas*, *Klebsiella*, *Serratia*, *Arthrobacter*, *Streptomyces*, *Isoptericola*, and *Microbacterium*	Roots of halophyte *Limonium sinense*	Stimulated the growth of the *Limonium sinense* seed germination and seedling growth under salinity stress	Qin et al. (2014)
Novosphingobium pokkalii sp. nov	Rhizosphere of saline-tolerant pokkali rice	Increase in growth of pokkali rice plants (*VTL-2* variety)	Krishnan et al. (2017)
Bacillus sp. and *Pseudomonas lini*	Rhizospheric soil and roots of *Distichlis spicata*	Growth of *Arabidopsis thaliana*, *Cucumis sativus*, and *Citrullus lanatus*	Palacio-Rodríguez et al. (2017)
Enterobacter asburiae, *Bacillus thuringiensis*, *Moraxella pluranimalium*, *Bacillus thuringiensis*, and *Pseudomonas stutzeri*	*Acacia arabia* grown in a saline soil	Alleviation of salt stress of *Triticum aestivum*	Raheem and Ali (2015)
Flavobacterium johnsoniae, *Pseudomonas putida*, *Achromobacter xylosoxidans*, and *Azotobacter chroococcum*	Agricultural sites used for wheat cultivation	Remarkable enhancement in germination of wheat seeds	Abdelwahab et al. (2017)
Azospirillum brasilense	Rhizosphere of many kinds of grasses	Alleviation of saline stress in *Trifolium repens*	Khalid et al. (2017)
Montagnulaceae sp. Jp-root-44	Rhizosphere of halotolerant seep weed *Suaeda salsa*	Plant growth promotion of *Cucumis sativus* and *Oryza sativa* in normal and in salt stress	Yuan et al. (2016)
Pseudomonas syringae and *Pseudomonas fluorescens* in different combination of *Rhizobium* strain	Rhizosphere and nodules of mung bean grown in salt-affected fields	Improved total dry weight and salt tolerance of mung bean	Ahmad et al. (2013)
Enterobacter sp. EJ01	Rhizosphere soil of *Dianthus japonicus*, a halophyte, in reclaimed land	Promoted growth and increased the salt resistance of tomato seedlings	Kim et al. (2014)

(continued)

Table 3.1 (continued)

Name of bacteria	Isolated from	Plant growth-promoting trait observed	References
Brevibacterium epidermidis RS15, *Micrococcus yunnanensis RS222*, and *Bacillus aryabhattai RS341*,	Rhizosphere of six different naturally grown halophytic plants of the saline coastal region	Enhanced growth of canola seedlings under saline stress	Siddikee et al. (2010)
Pseudomonas fluorescens, *P. aeruginosa*, and *P. stutzeri*	Rhizosphere of tomato plants grown in agricultural land	Enhanced fresh weight as well as dry weight of tomato when grown in the presence of salt	Tank and Saraf (2010)
Cellulosimicrobium sp. *S16*	Salt-affected agricultural rhizospheric soil	Increase in growth of barley seedlings as well as increase in biological control abilities	Nabti et al. (2014)
Bacillus pumilus, *Pseudomonas mendocina*, *Arthrobacter* sp., *Halomonas* sp., and *Nitrinicola lacisaponensis*	High saline habitats	Induced salt tolerance in *Triticum aestivum* L.	Tiwari et al. (2011)
Hartmannibacter diazotrophicus	Rhizosphere of *Plantago winteri* from a natural salt meadow	Increase in growth of *Hordeum vulgare* under salt stress	Suarez et al. (2015)
Azospirillum brasilense	Rhizosphere of salt-affected wheat field soil	Enhanced growth of durum wheat (*Triticum durum* var. waha) under saline soil conditions	Nabti et al. (2010)
Streptomyces isolate	Rhizosphere of wheat	Increase in growth and development of a wheat plant in normal and saline conditions	Sadeghi et al. (2012)
Alcaligenes faecalis, *Bacillus pumilus*, and *Ochrobactrum* sp.	Soils from root region of rice grown in coastal soils	Positive impacts on different growth parameters of rice under salinity stress	Bal et al. (2013)
Bacillus subtilis and *Arthrobacter* sp.	Rhizosphere of 30 days old wheat plants grown in a saline zone	Alleviation of the adverse effects of soil salinity on wheat growth	Upadhyay et al. (2012)
Azospirillum sp.	Nonsaline and saline soil	Positive impact on wheat seedlings under salinity stress	Zarea et al. (2012)
Pseudomonas pseudoalcaligenes and *P. putida*	Saline areas	Growth promotion in chickpea under saline stress	Patel et al. (2012)

(continued)

Table 3.1 (continued)

Name of bacteria	Isolated from	Plant growth-promoting trait observed	References
Pseudomonas sp.	Rhizosphere soil samples of eggplants (*Solanum melongena* L.) grown in the coastal saline soil	Facilitated growth of eggplants seedlings under salt stress	Fu et al. (2010)
Bacillus sp. and *Ochrobactrum* sp.	Saline soils	Growth promotion of *Zea mays* and *Agropyron elongatum* under salinity conditions	Príncipe et al. (2007)
Pseudomonas fluorescens	Rhizosphere of coastal and forestry ecosystems	Enhanced the saline resistance in groundnut plants	Saravanakumar and Samiyappan (2007)
Azospirillum	Maize rhizosphere	Enhanced growth of two cultivars of maize in saline stress	Hamdia et al. (2004)
Micrococcus yunnanensis, *Planococcus rifietoensis*, and *Variovorax paradoxus*	Saline habitats at the edge desert	Enhanced salt stress tolerance remarkably in *Beta vulgaris* L.	Zhou et al. (2017)
Consortia of *Enterobacter*, *Serratia*, *Serratia*, *Pseudomonas*, *Microbacterium* and *Achromobacter*,	Rhizospheric soils of avocado trees located near the sea	Ameliorated the effect of salt (NaCl) stress on the emergence, growth, and biomass of wheat seedlings	Barra et al. (2016)
Pseudomonas alcaligenes, *P. denitrificans*, *Bacillus polymyxa*, and *Mycobacterium phlei*	Soil of the root zone of melon	Significantly enhanced plant growth and nutrient uptake in cotton and pea in the pot experiments	Egamberdiyeva and Hoflich (2004)
Bacillus megaterium and *Enterobacter* sp.	Crop fields	Increase in salinity stress tolerance in okra	Habib et al. (2016)
Achromobacter piechaudii ARV8	Arid and saline environment	Increased in the biomass of tomato in saline stress	Mayak et al. (2004b)
Pseudomonas putida, *Enterobacter cloacae*, *Serratia ficaria*, and *Pseudomonas fluorescens*	Saline soil	Enhanced the germination percentage, germination rate in wheat at high salinity	Nadeem et al. (2013)

3.7 Conclusion

Halotolerant rhizobacteria-based agriculture is more efficient and environment-friendly than plant breeding and modification at the genetic level. The plant microbiome is considered as a pan–genome (second plant genome) having the capability to enhance host stress tolerance. It is well established that multiple plant growth-promoting traits are present in the bacterial communities of a halophytic rhizosphere. A number of halotolerant bacteria were isolated from the rhizosphere of halophytes, and their potential was studied to enhance plant growth under saline environments. Although single PGPR strain can enhance plant growth and tolerance to abiotic stress, several reports suggest that plants often supported by more than a single type of symbiotic microbe to combat salinity. A co-culture system containing two or more microbes is proven to be more efficient for the plant health. However, the use of PGPR in agriculture needs more extensive research and awareness. In addition to the study on plant physiology or microbiome of the host plant, it is also needed to concentrate on plant-microbe interaction to understand the whole mechanism during biotic and abiotic stress. Identification of the plant–microbes exudates, plant–microbes signals, and key players in the rhizosphere microbiome will provide unique metabolic and physiological markers which can elucidate the mechanism of plant–growth promotion by PGPR, additionally can also reveal whether and how plants recruit and stimulate beneficial (micro)organisms or its result of any stress.

Acknowledgment CSIR-CSMCRI Communication No.: PRIS-42/2018.

References

AbdElgawad H, Zinta G, Hegab MM, Pandey R, Asard H, Abuelsoud W (2016) High salinity induces different oxidative stress and antioxidant responses in maize seedlings organs. Front Plant Sci 7:276

Abdelwahab RAI, Cherif A, Cristina C, Nabti E (2017) Extracts from seaweeds and Opuntia ficus-indica cladodes enhance diazotrophic-PGPR halotolerance, their enzymatic potential, and their impact on wheat germination under salt stress. Pedosphere

Ahmad M, Zahir ZA, Nazli F, Akram F, Arshad M, Khalid M (2013) Effectiveness of halo-tolerant, auxin producing Pseudomonas and Rhizobium strains to improve osmotic stress tolerance in mung bean (Vigna radiata L.). Braz J Microbiol 44(4):1341–1348

Ashraf M (2004) Some important physiological selection criteria for salt tolerance in plants. Flora-Morphology, Distrib, Funct Ecol Plants 199(5):361–376

Ashraf M (2009) Biotechnological approach of improving plant salt tolerance using antioxidants as markers. Biotechnol Adv 27(1):84–93

Baki GK, Siefritz F, Man HM, Weiner H, Kaldenhoff R, Kaiser WM (2000) Nitrate reductase in Zea mays L. under salinity. Plant Cell Environ 23(5):515–521

Bal HB, Nayak L, Das S, Adhya TK (2013) Isolation of ACC deaminase producing PGPR from rice rhizosphere and evaluating their plant growth promoting activity under salt stress. Plant and Soil:1–13

Bano A, Fatima M (2009) Salt tolerance in Zea mays (L). following inoculation with Rhizobium and Pseudomonas. Biol Fertil Soils 45(4):405–413

Barra PJ, Inostroza NG, Acuña JJ, Mora ML, Crowley DE, Jorquera MA (2016) Formulation of bacterial consortia from avocado (Persea americana Mill.) and their effect on growth, biomass and superoxide dismutase activity of wheat seedlings under salt stress. Appl Soil Ecol 102:80–91

Bashan Y, Moreno M, Troyo E (2000) Growth promotion of the seawater-irrigated oilseed halophyte Salicornia bigelovii inoculated with mangrove rhizosphere bacteria and halotolerant Azospirillum spp. Biol Fertil Soils 32(4):265–272

Bhatnagar-Mathur P, Vadez V, Sharma KK (2008) Transgenic approaches for abiotic stress tolerance in plants: retrospect and prospects. Plant Cell Rep 27(3):411–424

Blaylock AD (1994) Soil salinity, salt tolerance, and growth potential of horticultural and landscape plants. University of Wyoming, Cooperative Extension Service, Department of Plant, Soil, and Insect Sciences, College of Agriculture

Boiero L, Perrig D, Masciarelli O, Penna C, Cassán F, Luna V (2007) Phytohormone production by three strains of Bradyrhizobium japonicum and possible physiological and technological implications. Appl Microbiol Biotechnol 74(4):874–880

Carillo P, Annunziata MG, Pontecorvo G, Fuggi A, Woodrow P (2011) Salinity stress and salt tolerance. In: Abiotic stress in plants-mechanisms and adaptations. InTech

Chandler D, Davidson G, Grant W, Greaves J, Tatchell G (2008) Microbial biopesticides for integrated crop management: an assessment of environmental and regulatory sustainability. Trends Food Sci Technol 19(5):275–283

Chinnusamy V, Zhu J, Zhu JK (2006) Gene regulation during cold acclimation in plants. Physiol Plant 126(1):52–61

Coleman-Derr D, Tringe SG (2014) Building the crops of tomorrow: advantages of symbiont-based approaches to improving abiotic stress tolerance. Front Microbiol 5:283

Cook RJ, Thomashow LS, Weller DM, Fujimoto D, Mazzola M, Bangera G, Kim D-S (1995) Molecular mechanisms of defense by rhizobacteria against root disease. Proc Natl Acad Sci 92 (10):4197–4201

de Souza R, Ambrosini A, Passaglia LM (2015) Plant growth-promoting bacteria as inoculants in agricultural soils. Genet Mol Biol 38(4):401–419

de Souza Vandenberghe LP, Garcia LMB, Rodrigues C, Camara MC, de Melo Pereira GV, de Oliveira J, Soccol CR (2017) Potential applications of plant probiotic microorganisms in agriculture and forestry. AIMS Microbiol 3(3):629–648

Del Río LA, Corpas FJ, Sandalio LM, Palma JM, Barroso JB (2003) Plant peroxisomes, reactive oxygen metabolism and nitric oxide. IUBMB Life 55(2):71–81

Dodd IC (2003) Hormonal interactions and stomatal responses. J Plant Growth Regul 22(1):32–46

Egamberdieva D (2008) Plant growth promoting properties of rhizobacteria isolated from wheat and pea grown in loamy sand soil. Turk J Biol 32(1):9–15

Egamberdieva D (2011) Survival of Pseudomonas extremorientalis TSAU20 and P. chlororaphis TSAU13 in the rhizosphere of common bean (Phaseolus vulgaris) under saline conditions. Plant Soil Environ 57(3):122–127

Egamberdiyeva D, Höflich G (2004) Effect of plant growth-promoting bacteria on growth and nutrient uptake of cotton and pea in a semi-arid region of Uzbekistan. J Arid Environ 56 (2):293–301

El-Tarabily KA, Youssef T (2010) Enhancement of morphological, anatomical and physiological characteristics of seedlings of the mangrove Avicennia marina inoculated with a native phosphate-solubilizing isolate of Oceanobacillus picturae under greenhouse conditions. Plant Soil 332(1–2):147–162

Essghaier B, Dhieb C, Rebib H, Ayari S, Boudabous ARA, Sadfi-Zouaoui N (2014) Antimicrobial behavior of intracellular proteins from two moderately halophilic bacteria: strain J31 of Terribacillus halophilus and strain M3-23 of Virgibacillus marismortui. J Plant Pathol Microbiol 5(1):1

FAO (2016) FAO Soil Portal. http://www.fao.org/soils-portal/en/

Fedoroff NV (2010) The past, present and future of crop genetic modification. New Biotechnol 27 (5):461–465

Fu Q, Liu C, Ding N, Lin Y, Guo B (2010) Ameliorative effects of inoculation with the plant growth-promoting rhizobacterium Pseudomonas sp. DW1 on growth of eggplant (Solanum melongena L.) seedlings under salt stress. Agric Water Manag 97(12):1994–2000

Glick BR (2012) Plant growth-promoting bacteria: mechanisms and applications. Scientifica 2012:963401. https://doi.org/10.6064/2012/963401

Glick BR (2014) Bacteria with ACC deaminase can promote plant growth and help to feed the world. Microbiol Res 169(1):30–39

Gray EJ, Smith DL (2005) Intracellular and extracellular PGPR: commonalities and distinctions in the plant–bacterium signaling processes. Soil Biol Biochem 37(3):395–412

Habib SH, Kausar H, Saud HM (2016) Plant growth-promoting rhizobacteria enhance salinity stress tolerance in okra through ROS-scavenging enzymes. BioMed Res Int 2016:6284547

Hamdia MAE-S, Shaddad M, Doaa MM (2004) Mechanisms of salt tolerance and interactive effects of Azospirillum brasilense inoculation on maize cultivars grown under salt stress conditions. Plant Growth Regul 44(2):165–174

Han S, Yu B, Wang Y, Liu Y (2011) Role of plant autophagy in stress response. Protein Cell 2 (10):784–791

Hill C, Guarner F, Reid G, Gibson GR, Merenstein DJ, Pot B, Morelli L, Canani RB, Flint HJ, Salminen S (2014) Expert consensus document: The International Scientific Association for Probiotics and Prebiotics consensus statement on the scope and appropriate use of the term probiotic. Nat Rev Gastroenterol Hepatol 11(8):506

Jewell MC, Campbell BC, Godwin ID (2010) Transgenic plants for abiotic stress resistance. In: Transgenic crop plants. Springer, pp 67–132

Jha Y, Subramanian R, Patel S (2011) Combination of endophytic and rhizospheric plant growth promoting rhizobacteria in Oryza sativa shows higher accumulation of osmoprotectant against saline stress. Acta Physiol Plant 33(3):797–802

Kang S-M, Khan AL, Hamayun M, Hussain J, Joo G-J, You Y-H, Kim J-G, Lee I-J (2012) Gibberellin-producing Promicromonospora sp. SE188 improves Solanum lycopersicum plant growth and influences endogenous plant hormones. J Microbiol 50(6):902

Kang S-M, Khan AL, Waqas M, You Y-H, Kim J-H, Kim J-G, Hamayun M, Lee I-J (2014) Plant growth-promoting rhizobacteria reduce adverse effects of salinity and osmotic stress by regulating phytohormones and antioxidants in Cucumis sativus. J Plant Interact 9(1):673–682

Karlidag H, Esitken A, Yildirim E, Donmez MF, Turan M (2010) Effects of plant growth promoting bacteria on yield, growth, leaf water content, membrane permeability, and ionic composition of strawberry under saline conditions. J Plant Nutr 34(1):34–45

Karuppasamy K, Nagaraj S, Kathiresan K (2011) Stress tolerant rhizobium enhances the growth of Samanea saman (JECQ) Merr. Afr J Basic Appl Sci 3(6):278–284

Khalid M, Bilal M, Hassani D, Iqbal HM, Wang H, Huang D (2017) Mitigation of salt stress in white clover (Trifolium repens) by Azospirillum brasilense and its inoculation effect. Bot Stud 58(1):5

Kim K, Jang Y-J, Lee S-M, Oh B-T, Chae J-C, Lee K-J (2014) Alleviation of salt stress by Enterobacter sp. EJ01 in tomato and Arabidopsis is accompanied by up-regulation of conserved salinity responsive factors in plants. Mol Cells 37(2):109

Kloepper J, Schroth M (1981) Relationship of in vitro antibiosis of plant growth-promoting rhizobacteria to plant growth and the displacement of root microflora. Phytopathology 71 (10):1020–1024

Kohler J, Hernández JA, Caravaca F, Roldán A (2009) Induction of antioxidant enzymes is involved in the greater effectiveness of a PGPR versus AM fungi with respect to increasing the tolerance of lettuce to severe salt stress. Environ Exp Bot 65(2):245–252

Krishnan R, Menon RR, Busse H-J, Tanaka N, Krishnamurthi S, Rameshkumar N (2017) Novosphingobium pokkalii sp nov, a novel rhizosphere-associated bacterium with plant beneficial properties isolated from saline-tolerant pokkali rice. Res Microbiol 168(2):113–121

Lau JA, Lennon JT (2012) Rapid responses of soil microorganisms improve plant fitness in novel environments. Proc Natl Acad Sci 109(35):14058–14062

Lippert K, Galinski EA (1992) Enzyme stabilization be ectoine-type compatible solutes: protection against heating, freezing and drying. Appl Microbiol Biotechnol 37(1):61–65

Ma W, Guinel FC, Glick BR (2003) Rhizobium leguminosarum biovar viciae 1-aminocyclopropane-1-carboxylate deaminase promotes nodulation of pea plants. Appl Environ Microbiol 69(8):4396–4402

Mahajan S, Tuteja N (2005) Cold, salinity and drought stresses: an overview. Arch Biochem Biophys 444(2):139–158

Mapelli F, Marasco R, Rolli E, Barbato M, Cherif H, Guesmi A, Ouzari I, Daffonchio D, Borin S (2013) Potential for plant growth promotion of rhizobacteria associated with Salicornia growing in Tunisian hypersaline soils. BioMed Res Int 2013:248078

Marulanda A, Azcón R, Chaumont F, Ruiz-Lozano JM, Aroca R (2010) Regulation of plasma membrane aquaporins by inoculation with a Bacillus megaterium strain in maize (Zea mays L.) plants under unstressed and salt-stressed conditions. Planta 232(2):533–543

Mayak S, Tirosh T, Glick BR (2004a) Plant growth-promoting bacteria confer resistance in tomato plants to salt stress. Plant Physiol Biochem 42(6):565–572

Mayak S, Tirosh T, Glick BR (2004b) Plant growth-promoting bacteria that confer resistance to water stress in tomatoes and peppers. Plant Sci 166(2):525–530

Miller KJ, Wood JM (1996) Osmoadaptation by rhizosphere bacteria. Annu Rev Microbiol 50 (1):101–136

Mishra A, Tanna B (2017) Halophytes: potential resources for salt stress tolerance genes and promoters. Front Plant Sci 8:829

Mukhtar S, Ishaq A, Hassan S, Mehnaz S, Mirza MS, Malik KA (2017) Comparison of microbial communities associated with halophyte (Salsola stocksii) and non-halophyte (Triticum aestivum) using culture-independent approaches. Pol J Microbiol 66(3):353–364

Munns R, Gilliham M (2015) Salinity tolerance of crops–what is the cost? New Phytol 208 (3):668–673

Munns R, Tester M (2008) Mechanisms of salinity tolerance. Annu Rev Plant Biol 59:651–681

Nabti E, Sahnoune M, Ghoul M, Fischer D, Hofmann A, Rothballer M, Schmid M, Hartmann A (2010) Restoration of growth of durum wheat (Triticum durum var. waha) under saline conditions due to inoculation with the rhizosphere bacterium Azospirillum brasilense NH and extracts of the marine alga Ulva lactuca. J Plant Growth Regul 29(1):6–22

Nabti E, Bensidhoum L, Tabli N, Dahel D, Weiss A, Rothballer M, Schmid M, Hartmann A (2014) Growth stimulation of barley and biocontrol effect on plant pathogenic fungi by a Cellulosimicrobium sp. strain isolated from salt-affected rhizosphere soil in northwestern Algeria. Eur J Soil Biol 61:20–26

Nadeem SM, Zahir ZA, Naveed M, Asghar HN, Arshad M (2010) Rhizobacteria capable of producing ACC-deaminase may mitigate salt stress in wheat. Soil Sci Soc Am J 74(2):533–542

Nadeem SM, Zahir ZA, Naveed M, Nawaz S (2013) Mitigation of salinity-induced negative impact on the growth and yield of wheat by plant growth-promoting rhizobacteria in naturally saline conditions. Ann Microbiol 63(1):225–232

Narula N, Deubel A, Gans W, Behl R, Merbach W (2006) Paranodules and colonization of wheat roots by phytohormone producing bacteria in soil. Plant Soil Environ 52(3):119

Naz I, Bano A, Ul-Hassan T (2009) Isolation of phytohormones producing plant growth promoting rhizobacteria from weeds growing in Khewra salt range, Pakistan and their implication in providing salt tolerance to Glycine max L. Afr J Biotechnol 8(21):5762–5768

Netondo GW, Onyango JC, Beck E (2004) Sorghum and salinity. Crop Sci 44(3):797–805

Nia SH, Zarea MJ, Rejali F, Varma A (2012) Yield and yield components of wheat as affected by salinity and inoculation with Azospirillum strains from saline or non-saline soil. J Saudi Soc Agric Sci 11(2):113–121

Nishma K, Adrisyanti B, Anusha S, Rupali P, Sneha K, Jayamohan N, Kumudini B (2014) Induced growth promotion under in vitro salt stress tolerance on solanum lycopersicum by fluorescent pseudomonads associated with rhizosphere. IJASER 3:422–430

Orhan F (2016) Alleviation of salt stress by halotolerant and halophilic plant growth-promoting bacteria in wheat (Triticum aestivum). Braz J Microbiol 47(3):621–627

Palacio-Rodríguez R, Coria-Arellano JL, López-Bucio J, Sánchez-Salas J, Muro-Pérez G, Castañeda-Gaytán G, Sáenz-Mata J (2017) Halophilic rhizobacteria from Distichlis spicata promote growth and improve salt tolerance in heterologous plant hosts. Symbiosis 73 (3):179–189

Patel D, Jha CK, Tank N, Saraf M (2012) Growth enhancement of chickpea in saline soils using plant growth-promoting rhizobacteria. J Plant Growth Regul 31(1):53–62

Piernik A, Hrynkiewicz K, Wojciechowska A, Szymańska S, Lis MI, Muscolo A (2017) Effect of halotolerant endophytic bacteria isolated from Salicornia europaea L. on the growth of fodder beet (Beta vulgaris L.) under salt stress. Arch Agron Soil Sci 63(10):1404–1418

Podmore C (2009) Irrigation salinity–causes and impacts. Primefact 937(1):1–4

Príncipe A, Alvarez F, Castro MG, Zachi L, Fischer SE, Mori GB, Jofré E (2007) Biocontrol and PGPR features in native strains isolated from saline soils of Argentina. Curr Microbiol 55 (4):314–322

Qin S, Zhang Y-J, Yuan B, Xu P-Y, Xing K, Wang J, Jiang J-H (2014) Isolation of ACC deaminase-producing habitat-adapted symbiotic bacteria associated with halophyte Limonium sinense (Girard) Kuntze and evaluating their plant growth-promoting activity under salt stress. Plant Soil 374(1–2):753–766

Qurashi AW, Sabri AN (2013) Osmolyte accumulation in moderately halophilic bacteria improves salt tolerance of chickpea. Pak J Bot 45:1011–1016

Raheem A, Ali B (2015) Halotolerant rhizobacteria: beneficial plant metabolites and growth enhancement of Triticum aestivum L. in salt-amended soils. Arch Agron Soil Sci 61 (12):1691–1705

Rahnama A, James RA, Poustini K, Munns R (2010) Stomatal conductance as a screen for osmotic stress tolerance in durum wheat growing in saline soil. Funct Plant Biol 37(3):255–263

Rajkumar M, Ae N, Prasad MNV, Freitas H (2010) Potential of siderophore-producing bacteria for improving heavy metal phytoextraction. Trends Biotechnol 28(3):142–149

Ramadoss D, Lakkineni VK, Bose P, Ali S, Annapurna K (2013) Mitigation of salt stress in wheat seedlings by halotolerant bacteria isolated from saline habitats. SpringerPlus 2(1):6

Raymond J, Siefert JL, Staples CR, Blankenship RE (2004) The natural history of nitrogen fixation. Mol Biol Evol 21(3):541–554

Roder A, Hoffmann E, Hagemann M, Berg G (2005) Synthesis of the compatible solutes glucosylglycerol and trehalose by salt-stressed cells of Stenotrophomonas strains. FEMS Microbiol Lett 243(1):219–226

Rojas-Tapias D, Moreno-Galván A, Pardo-Díaz S, Obando M, Rivera D, Bonilla R (2012) Effect of inoculation with plant growth-promoting bacteria (PGPB) on amelioration of saline stress in maize (Zea mays). Appl Soil Ecol 61:264–272

Roy SJ, Negrão S, Tester M (2014) Salt resistant crop plants. Curr Opin Biotechnol 26:115–124

Sadeghi A, Karimi E, Dahaji PA, Javid MG, Dalvand Y, Askari H (2012) Plant growth promoting activity of an auxin and siderophore producing isolate of Streptomyces under saline soil conditions. World J Microbiol Biotechnol 28(4):1503–1509

Santoyo G, Moreno-Hagelsieb G, del Carmen Orozco-Mosqueda M, Glick BR (2016) Plant growth-promoting bacterial endophytes. Microbiol Res 183:92–99

Saravanakumar D, Samiyappan R (2007) ACC deaminase from Pseudomonas fluorescens mediated saline resistance in groundnut (Arachis hypogea) plants. J Appl Microbiol 102(5):1283–1292

Saslis-Lagoudakis CH, Hua X, Bui E, Moray C, Bromham L (2014) Predicting species' tolerance to salinity and alkalinity using distribution data and geochemical modelling: a case study using Australian grasses. Ann Bot 115(3):343–351

Seckin B, Sekmen AH, Türkan I (2009) An enhancing effect of exogenous mannitol on the antioxidant enzyme activities in roots of wheat under salt stress. J Plant Growth Regul 28(1):12
Shin D-S, Park M-S, Jung S, Lee M-S, Lee K-H, Bae K-S, Kim S-B (2007) Plant growth-promoting potential of endophytic bacteria isolated from roots of coastal sand dune plants. J Microbiol Biotechnol 17(8):1361–1368
Shrivastava P, Kumar R (2015) Soil salinity: a serious environmental issue and plant growth promoting bacteria as one of the tools for its alleviation. Saudi J Biol Sci 22(2):123–131
Siddikee MA, Chauhan PS, Anandham R, Han G-H, Sa T (2010) Isolation, characterization, and use for plant growth promotion under salt stress, of ACC deaminase-producing halotolerant bacteria derived from coastal soil. J Microbiol Biotechnol 20(11):1577–1584
Siddikee MA, Glick BR, Chauhan PS, Jong Yim W, Sa T (2011) Enhancement of growth and salt tolerance of red pepper seedlings (Capsicum annuum L.) by regulating stress ethylene synthesis with halotolerant bacteria containing 1-aminocyclopropane-1-carboxylic acid deaminase activity. Plant Physiol Biochem 49(4):427–434
Štajner D, Kevrešan S, Gašić O, Mimica-Dukić N, Zongli H (1997) Nitrogen and Azotobacter chroococcum enhance oxidative stress tolerance in sugar beet. Biol Plant 39(3):441
Stutz EW, Défago G, Kern H (1986) Naturally occurring fluorescent pseudomonads involved in suppression of black root rot of tobacco. Phytopathology 76(2):181–185
Suárez R, Wong A, Ramírez M, Barraza A, Orozco MC, Cevallos MA, Lara M, Hernández G, Iturriaga G (2008) Improvement of drought tolerance and grain yield in common bean by overexpressing trehalose-6-phosphate synthase in rhizobia. Mol Plant-Microbe Interact 21 (7):958–966
Suarez C, Cardinale M, Ratering S, Steffens D, Jung S, Montoya AMZ, Geissler-Plaum R, Schnell S (2015) Plant growth-promoting effects of Hartmannibacter diazotrophicus on summer barley (Hordeum vulgare L.) under salt stress. Appl Soil Ecol 95:23–30
Szymańska S, Piernik A, Hrynkiewicz K (2013) Metabolic potential of microorganisms associated with the halophyte Aster tripolium L. in saline soils. Ecol Quest 18(1):9–19
Taghavi S, Garafola C, Monchy S, Newman L, Hoffman A, Weyens N, Barac T, Vangronsveld J, van der Lelie D (2009) Genome survey and characterization of endophytic bacteria exhibiting a beneficial effect on growth and development of poplar trees. Appl Environ Microbiol 75 (3):748–757
Tank N, Saraf M (2010) Salinity-resistant plant growth promoting rhizobacteria ameliorates sodium chloride stress on tomato plants. J Plant Interact 5(1):51–58
Tewari S, Arora NK (2014) Multifunctional exopolysaccharides from Pseudomonas aeruginosa PF23 involved in plant growth stimulation, biocontrol and stress amelioration in sunflower under saline conditions. Curr Microbiol 69(4):484–494
Tiwari S, Singh P, Tiwari R, Meena KK, Yandigeri M, Singh DP, Arora DK (2011) Salt-tolerant rhizobacteria-mediated induced tolerance in wheat (Triticum aestivum) and chemical diversity in rhizosphere enhance plant growth. Biol Fertil Soils 47(8):907
Upadhyay SK, Singh DP, Saikia R (2009) Genetic diversity of plant growth promoting rhizobacteria isolated from rhizospheric soil of wheat under saline condition. Curr Microbiol 59(5):489–496
Upadhyay SK, Singh JS, Saxena AK, Singh DP (2012) Impact of PGPR inoculation on growth and antioxidant status of wheat under saline conditions. Plant Biol 14(4):605–611
Van Loon L (2007) Plant responses to plant growth-promoting rhizobacteria. Eur J Plant Pathol 119 (3):243–254
Vannier N, Mony C, Bittebière A-K, Vandenkoornhuyse P (2015) Epigenetic mechanisms and microbiota as a toolbox for plant phenotypic adjustment to environment. Front Plant Sci 6:1159
Welsh DT (2000) Ecological significance of compatible solute accumulation by micro-organisms: from single cells to global climate. FEMS Microbiol Rev 24(3):263–290
Yamaguchi T, Blumwald E (2005) Developing salt-tolerant crop plants: challenges and opportunities. Trends Plant Sci 10(12):615–620

Yang J, Kloepper JW, Ryu C-M (2009) Rhizosphere bacteria help plants tolerate abiotic stress. Trends Plant Sci 14(1):1–4

Yao L, Wu Z, Zheng Y, Kaleem I, Li C (2010) Growth promotion and protection against salt stress by Pseudomonas putida Rs-198 on cotton. Eur J Soil Biol 46(1):49–54

Yildirim E, Turan M, Ekinci M, Dursun A, Cakmakci R (2011) Plant growth promoting rhizobacteria ameliorate deleterious effect of salt stress on lettuce. Sci Res Essays 6 (20):4389–4396

Yuan Z, Druzhinina IS, Labbé J, Redman R, Qin Y, Rodriguez R, Zhang C, Tuskan GA, Lin F (2016) Specialized microbiome of a halophyte and its role in helping non-host plants to withstand salinity. Sci Rep 6:32467

Yue H, Mo W, Li C, Zheng Y, Li H (2007) The salt stress relief and growth promotion effect of Rs-5 on cotton. Plant Soil 297(1–2):139–145

Zahir ZA, Arshad M, Frankenberger WT (2004) Plant growth promoting rhizobacteria: applications and perspectives in agriculture. Adv Agron 81:98–169

Zahir ZA, Ghani U, Naveed M, Nadeem SM, Asghar HN (2009) Comparative effectiveness of Pseudomonas and Serratia sp. containing ACC-deaminase for improving growth and yield of wheat (Triticum aestivum L.) under salt-stressed conditions. Arch Microbiol 191(5):415–424

Zarea M, Hajinia S, Karimi N, Goltapeh EM, Rejali F, Varma A (2012) Effect of Piriformospora indica and Azospirillum strains from saline or non-saline soil on mitigation of the effects of NaCl. Soil Biol Biochem 45:139–146

Zhou N, Zhao S, Tian CY (2017) Effect of halotolerant rhizobacteria isolated from halophytes on the growth of sugar beet (Beta vulgaris L.) under salt stress. FEMS Microbiol Lett 364(11). https://doi.org/10.1093/femsle/fnx091

Zhu J-K (2002) Salt and drought stress signal transduction in plants. Annu Rev Plant Biol 53 (1):247–273

Development of Salt Tolerance in Crops Employing Halotolerant Plant Growth–Promoting Rhizobacteria Associated with Halophytic Rhizosphere Soils

4

Ashok Panda and Asish Kumar Parida

Abstract

Increasing levels of salinity in agricultural lands is one of the most serious environmental concerns that pose a risk to the food security of the growing human population of the world. According to the United Nations Environment Program, the total areas of salt-stressed agricultural lands and croplands have increased by approximately 20% and 50%, respectively, worldwide. The total land area that cannot be used as agricultural land is increasing by 1–2% every year as a result of soil salinization, mostly in dry areas. Increasing soil salinity is becoming the prime reason for substantial decreases in agricultural yield due to inhibitory effects of salinity on growth, photosynthesis, protein synthesis, lipid metabolism, and many other metabolic processes of plants. Production of salt-tolerant crop varieties is a prerequisite for meeting increasing food demands and creating sustainable agriculture practices. The halophytic rhizosphere is a reservoir of plant growth–promoting rhizobacteria (PGPRs), which can enhance plant adaptation and growth under high salinity. Among free-living soil bacteria, PGPRs play an essential role in promoting plant growth even in stress conditions. PGPRs have both direct and indirect effects on plant growth. The direct mechanisms involve biosynthesis of phytohormones, enhanced nitrogen fixation, and higher levels of phosphate solubilization. The indirect mechanisms involve inhibition of phytopathogens that reduce plant growth. Various studies have illustrated that salinity-tolerant PGPRs obtained from rhizosphere soils of various halophytic species have potential for use in development of glycophytic salt-tolerant crops in salt-dominated agricultural lands through their use as

A. Panda · A. K. Parida (✉)
Division of Biotechnology and Phycology, Council of Scientific and Industrial Research–Central Salt and Marine Chemicals Research Institute (CSIR-CSMCRI), Bhavnagar, Gujarat, India

Academy of Scientific and Innovative Research, Council of Scientific and Industrial Research–Central Salt and Marine Chemicals Research Institute (CSIR-CSMCRI), Bhavnagar, Gujarat, India
e-mail: asishparida@csmcri.org; asishparida@csmcri.res.in

M. Kumar et al. (eds.), *Saline Soil-based Agriculture by Halotolerant Microorganisms*, https://doi.org/10.1007/978-981-13-8335-9_4

bioinoculants. To accomplish this goal, PGPRs adapt various mechanisms such as modulation of phytohormones, gene expression, protein function, and metabolite synthesis. PGPRs modulate synthesis of 1-aminocyclopropane-1-carboxylate (ACC) deaminase along with indoleacetic acid (IAA), which function in stress signaling and induce various stress-responsive pathways. Implementation of PGPR inoculation in the advancement of agriculture to increase global food security is desirable. This chapter focuses on the salinity tolerance mechanisms of PGPRs and the roles of PGPRs in developing salt tolerance in various glycophytic crop species.

Keywords
ACC deaminase · Halophyte · IAA · PGPR · Phytohormones · Salinity · Salt-tolerant crop · Siderophore

4.1 Introduction

Salinization now affects more than one billion hectares of land worldwide and is considered a serious environmental threat to many coastal countries (Egamberdieva and Lugtenberg 2014). The major factors responsible for the increase in salinity are poor agricultural practices, irrigation with saline water, and low precipitation. Gradual decreases in the availability of fresh water result in use of salt water for irrigation (Egamberdieva and Lugtenberg 2014).

Natural salinity is defined as addition of salt to the soil or surface water bodies by natural means. Intrusion of salt water into coastal lands and deposition of salt by winds are the major natural sources of salt accumulation (Manchanda and Garg 2008). Moreover, addition of salts to agricultural land from the ocean occurs through transfer by winds or by rainfall precipitation of ocean water. In addition, salt can be added to soil by weathering of mineral rocks (Paul and Lade 2014).

Anthropogenic causes of increasing soil salinity include excessive irrigation with saline water, which makes most agricultural land unfit for agriculture by increasing the level of dissolved salts (Egamberdieva and Lugtenberg 2014). Other causative factors include excessive supplementation with inorganic fertilizers and various changes in the soil structure. All of these factors continuously contribute to land degradation, representing a large burden on natural ecosystems (Paul and Lade 2014). The main impacts of high soil salinity are degradation of the soil structure through excessive accumulation of Na^+ along with deterioration of the hydraulic properties of the soil. Apart from this, high levels of soil Na^+ tend to increase the soil pH and increase soil erosion. Most importantly, Na^+ plays a decisive role in destruction of other clay minerals by dispersion. The process of dispersion occurs through replacement of Ca^{2+} and Mg^{2+} in the inner layer of soil coagulates by intruding Na^+ (Paul and Lade 2014). Agricultural lands with less rainfall and higher temperatures due to extreme climate change convert into saline land more rapidly.

This results in occurrence of more arid regions, favoring even more salt accumulation (Shanker and Venkateswarlu 2011). As mentioned earlier, salinization is defined as accumulation of different kinds of salt ions, including magnesium (Mg^{2+}), potassium (K^+), chloride (Cl^-), calcium (Ca^{2+}), carbonate (CO_3^{2-}), sulfate (SO_4^{2-}), sodium (Na^+), and bicarbonate (HCO_3^-). Consequently, the total area of land that is affected by salinity and the total area of abandoned agricultural land are increasing day by day (Paul and Lade 2014). Salinity stress generates reactive oxygen species (ROS), which damage DNA, RNA, and proteins (Habib et al. 2016). ROS also cause decreases in chlorophyll levels and the activity of root meristems, hindering root growth. On the other hand, various antioxidant enzymes such as superoxide dismutase (SOD), catalase (CAT), and ascorbate peroxidase (APX) have the ability to scavenge ROS and maintain the level of ROS within the tolerance limit (Jaleel et al. 2009). SOD is a metalloenzyme that protects cells and various cellular organs from damage caused by oxidative stress, by neutralizing superoxide radicals. APX plays a significant role in breakdown of H_2O_2 produced by SOD. Furthermore, H_2O_2 produced by the action of SOD is neutralized by catalase into H_2O and O_2 (Habib et al. 2016).

Development of salinity-tolerant crops is becoming a more desirable scientific achievement in view of the increase in soil salinity in agricultural lands and the decrease in the rate of agricultural productivity due to salinization. Well-organized soil management practices, which involve efficient cost management and formulation of methods that are easily adaptable, are a challenge that must be met for successful reclamation of saline soils (Munns and Tester 2008). In addition, in contrast to conventional agriculture, it is essential to institute sustainable management practices that provide more support to meet the future needs of agriculture. Various microorganisms that play helpful roles in agriculture in an eco-friendly manner have been shown to be of great benefit (Rodriguez and Redman 2008). Moreover, various researchers have demonstrated that the associations between rhizospheric microbes and stressed plants help the plants to adapt to their microenvironment (Paul and Lade 2014). A variety of agricultural crops grown at a wide range of salinity levels have been shown to respond to such rhizospheric microorganisms, which enhance the growth of these crops. This approach make it possible to develop salinity-tolerant varieties of crops, although this was previously believed to be impossible. Stress alleviation in crops that are habituated to saline soils occurs by implementation of beneficial rhizobacteria that colonize the rhizosphere soil of the plants. Successful implementation of this approach is possible because of the wide range of tolerance of these microorganisms, their productive interactions with plants, and use of prospective implementation methods. The percentage of rhizobacteria in inoculations that generally play beneficial roles in plant growth under adverse environmental conditions is about 2–5%. Plant growth–promoting rhizobacteria (PGPRs) make soil ecosystems nutrient rich and make crop yields sustainable (Hayat et al. 2010). The various ways in which microbes stimulate plant growth and alleviate stress responses include (1) efficient nutrient mobilization in the soil, (2) protection from phytopathogens, and (3) improvement of soil structure and quality by sequestration of toxic heavy metals. Microbes can also aid degradation of various xenobiotic compounds (Braud et al. 2009).

Use of microbial inoculants for improving plant health in salinized soils holds promise for amelioration of salt stress and promotion of plant growth, along with improvement of disease resistance (Lugtenberg et al. 2013). Bacterial species that are closely associated with plant roots are of immense importance in alleviating the effects of stress and opening up new opportunities for development of technologies for salinity tolerance (Dodd and Pérez-Alfocea 2012). Earlier reports documented the usefulness of microbes in plant responses to various environmental extremes such as drought, soil and surface water salinity, and heavy metal stress. These kinds of approaches supports development of sustainable agricultural practices. Preinoculation of the germplasms of different crops such as pepper, tomato, and beans with PGPR strains tends to increase the plant biomass as well as the yield, reflecting enhanced tolerance of prevailing salinity.

4.2 Effects of Salinity on Plant Physiology

Optimal plant growth and agricultural yields are affected by salinity in many crops (Paul and Lade 2014). Various plants such as cereals, pulses, and other horticultural crops are greatly affected by prevailing salinity either in irrigation water or in the soil solution. Reports have documented decreased seed germination rates in response to salinity in wheat (Egamberdieva 2009), faba beans (Rabie et al. 2005), rice (Xu et al. 2011), maize, and soybean (Essa 2002). Reduced water uptake and improper mobilization of essential nutrients may be the reasons for the reduction in growth in response to salinity (Dolatabadian et al. 2011). Mobilization of some essential macronutrients in plants is affected by abundance of Na^+ and Cl^- ions in the rhizospheric region (Heidari and Jamshid 2010). Major physiological pathways such as protein synthesis, photosynthesis, and lipid metabolism are altered by the onset of salinity. The rate of carbon assimilation declines because prevailing salinity induces an osmotic imbalance and stomatal closure (Parida and Das 2005). The overall effects of osmotic stress in plants include a significant decrease in cell growth, a reduction in leaf area, chlorophyll degradation, and, most importantly, senescence (Paul and Lade 2014). The effects of salinity are multivariate, including hormonal and ionic imbalances, alterations in protein metabolism, and a decline in the enzyme activity that is responsible for nucleic acid metabolism. These morphological and physiological anomalies are attributable to osmotic imbalance and toxicity of salt ions. Apart from Na^+, Cl^- also disturbs the plant's regular physiology (e.g., photosynthesis). Excessive accumulation of these two ions in the rhizosphere induces competitive absorption of other mineral ions such as K^+, NO_3^-, and H_2PO_4 in the roots along with competition for translocation and for tissue-specific localization. Intake of salts by plants beyond their tolerance (storage capacity) leads to storage of ions in the intercellular spaces, which causes dehydration and death of plant tissues. Salinity affects the osmotic balance of root cells, which becomes lower than that in the surrounding rhizosphere solution, and this, in turn, influences water and nutrient uptake. Various primary effects of salinity tend to have various secondary consequences such as osmotic imbalance, oxidative stress, and ROS accumulation, leading to disintegration of membranes or other macromolecules. Implementation of both enzymatic and nonenzymatic antioxidant systems confers protection against stress-induced oxidative damage. In addition, the levels of expression of these antioxidant molecules can be used as indicators of the extent of oxidative stress in plants.

4.3 Impacts of Salinity on Rhizospheric Microbiodiversity

The microbial colony in the rhizosphere is an indispensable and dynamic element of soil composition, and it functions as an agent for recycling of nutrients as well as for transformation of organic matter to provide nutrients for optimum growth of the plants. Most microorganisms colonize the plant rhizosphere, rather than the surrounding soil, because of its favorable environment. The microbes in the rhizosphere soil have significant influences on the survival of plants in direct and indirect ways. As the microbial community modulates the nutrient availability to the plants, any alterations in the community as a result of environmental factors influence the nutrient supply to plants from the soil. Increased microbial populations in the rhizosphere have many effects, particularly in proper functioning of the ecosystem, pollutant degradation, and decontamination of contaminated environments (Wenzel 2009). The rhizospheric microbial community structure is affected by soil sodium chloride, influencing root exudates (Nelson and Mele 2007). The density of microbes such as *Pseudomonas* species has been reported to be significantly decreased by salinity in the rhizosphere of rice. A number of *Pseudomonas* species are habituated to saline soil; these include *P. stutzeri*, *P. fluorescens*, *P. aeruginosa*, *P. putida*, and *P. mendocina* (Egamberdieva and Lugtenberg 2014). The salt-tolerant nitrogen-fixing bacterium *Swaminathania salitolerans* has been obtained from rice associated with mangroves (Loganathan and Nair 2004). Apart from *Pseudomonas* species, various salt-resistant rhizobacteria such as *Serratia marcescens*, *P. aeruginosa*, and *Achromobacter xylosoxidans* have been reported to be present in the rhizosphere of rice. Similarly, various reports have suggested that the rhizosphere of salt-grown wheat is a potential PGPR reservoir (Egamberdiyeva and Islam 2008). It is evident from the consistent presence of *P. aeruginosa* in saline soils that this species is widespread in the rhizosphere soil of salt-grown plants. Microorganisms and plants compete for the available nutrients. Beneficial bacteria are attracted to the plants by secretions from roots in the rhizospheric zone. To survive in a saline environment, microbes have to expend extra energy for maintenance of the osmotic balance between both sides of their plasma membranes and also have to exclude Na^+ (Jiang et al. 2007). In saline conditions, plants use most of the available rhizospheric osmolytes, depriving microbes of those osmolytes and restricting bacterial growth. Salinity affects the composition of protein, lipopolysaccharides, and exopolysaccharides present on the bacterial surface. In addition, salinity affects the arrangement of membrane glucans that regulate signaling on the bacterial surface, affecting the mobility of bacteria (Ibekwe et al. 2010). An alteration in bacterial diversity is one of the biomarkers for determination of salinity effects on the quality of soil. Moreover, early detection of salinity in soil can be helpful for implementation of remedial action to improve crop survival and yield (Ibekwe et al. 2010). Halotolerant rhizobacteria implement an array of adaptive measures to maintain survival and growth under high salinity. Their adaptive mechanisms include (1) modification of the structural texture of the cell membrane and cell wall to minimize salt uptake; (2) efficient osmotic adjustment through regulation of intracellular ion concentrations by K^+/Na^+ ion transporters and Na^+/H^+ antiporters; (3) endogenous biosynthesis of trehalose, sucrose, glycerol, and glycosyl, which serve as compatible solutes; (4) biosynthesis of enzymes that can withstand high salinity; and (5) production of biofilms composed of exopolysaccharides (Sandhya et al. 2010b; Qin et al. 2016). Besides these mechanisms, other features of salt-tolerant rhizobacteria such as higher guanine–cytosine (GC) content, high levels of low hydrophobic proteins, less formation of helices, and higher levels of stabilized coil structures may enhance their halotolerance characteristics (Szymańska et al. 2016).

4.4 Salt Tolerance Mechanisms of Rhizospheric PGPRs

Microorganisms belonging to the domains of Bacteria and Archaea have the unique property of maintaining cellular integrity in response to changes in the osmolarity of the external soil solution. It was discovered that apart from the members of the Halobacteriaceae family, other species are also adaptable to high levels of NaCl (Oren 2002). Lately, it has been suggested that apart from archaea, the prokaryotic diversity in high-saline environments also includes different species of bacteria. This hypothesis has been further supported by isolation of a wide array of salt-tolerant bacteria at all latitudes. Salt-tolerant bacteria can grow on plant parts pickled in a salt solution; for example, *Staphylococcus xylosus* was found on plant parts soaked in a salt solution containing ~7.2% NaCl (Abou-Elela et al. 2010). Bacterial genera such as *Nesterenkonia*, *Stenotrophomonas*, and *Litoribacter* have been found in a saline area of northeast China (Shi et al. 2012), and *Mesorhizobium alhagi* has been found in the northwestern desert of China (Zhou et al. 2012a). High salinity can be withstood by avoidance of high cytoplasmic salt concentrations and water loss through specific cell wall construction and composition, which are known to be particular characteristics of archaea and cyanobacteria. The highly resistant lipid membranes of archaea, which are resistant to temperature and mechanical anomalies, make these species salt tolerant. In comparison with eukaryotes, the lipid membranes of archaea make them better suited to harsh environmental conditions for better survival. The cell walls of some members of Halobacteriales contain an S-layer made up of glycoprotein, which is arranged in a two-dimensional lattice. The membranes are made up of about 40–50% glycoprotein and are stabilized by the presence of NaCl. In a similar way, the cyanobacterial membrane system contains a two-membrane structure, i.e., an inner membrane and an outer membrane. The outer membrane encloses the periplasmic space, and the inner membrane surrounds the cytoplasmic content. Another membrane system is present in chloroplasts as thylakoid membranes, which accommodate the photosynthetic complexes and the electron transport chain (Ruppel et al. 2013). Apart from this, the coccoid *Natronococcus occultus* has been reported to have a thick cell wall, which requires a certain amount of salt for its structural integrity. The integrity of the wall of *Halococcus*, which is made up of poly(L-glutamine) glycoconjugate, differs greatly depending on the presence or absence of salt. In addition, several rhizobacterial species secrete salinity stress alleviator exopolysaccharides (Upadhyay et al. 2011). The role of these polysaccharides in reducing salt stress was demonstrated in *Nostoc* sp. The presence of salt in the growth media did not alter the composition of the wall in terms of the ratio of sugars in comparison with the control media (Yoshimura et al. 2012). The intracellular ionic concentration is maintained by various ion channels, resulting in osmotic adjustment. The level of Na^+ in the cell is kept low by a proton electrochemical gradient using Na^+/H^+ antiporters. The essential role of K^+ includes maintenance of membrane integrity and energetics, pH regulation, and, most importantly, regulation of gene expression (Hagemann 2011). Both Na^+ and K^+ have the same chemical and physical properties. Na^+ has the same-sized hydration shell as K^+; hence, it competes with K^+ uptake and causes K^+ deficiency in plants. In contrast, an optimum level of K^+ in the intracellular matrix suggests uninterrupted K^+ uptake from the soil solution (Hagemann 2011).

4.5 PGPRs Isolated from Different Halophytic Rhizospheres

Microorganisms that are associated with soil and a halophytic rhizosphere influence the growth and development of plants both beneficially and detrimentally. The growth of plants inoculated with PGPRs can be altered directly and/or indirectly. For soil microorganisms the plant rhizosphere is a favored ecological niche because of rich nutrient availability. The maximum activity of beneficial bacteria is mainly accomplished in the rhizosphere. This is because of the attractants and nutrients secreted in the form of different organic compounds from the roots, attracting an array of bacterial species (Szymańska et al. 2016). Beneficial effects on plants are exerted by various soil microorganisms, and sustainable crop production is procured by opposite effects on soilborne pests. Adverse conditions make microorganisms more tolerant and make them potential candidates for use in various environmental stress conditions along with salinity. Nevertheless, the diversity and growth of various bacterial species are affected by prevailing rhizospheric salinity (Egamberdiyeva and Islam 2008). Diverse plant–microbe interactions in rhizosphere soil greatly differ from species to species and also differ from those occurring in the surrounding soil. Because of rich nutrient availability, the rhizosphere is an appropriate survival niche for soil microorganisms (Ahmad et al. 2008). Studies on the genetics and morphology of rhizobacterial species have further clarified the communication mechanisms that function between the host root system and microorganisms (Tripathi et al. 2002). In lands that are greatly degraded by salinity, the microbial community is greatly influenced. However, halotolerant bacteria seem to grow abundantly in soils with a salt concentration of up to 30% and in lands that are devoid of salt (Khan et al. 2016). Therefore, they can thrive in rhizosphere soil of various halophytic species with a low water potential (Ruppel et al. 2013). More astonishingly, the plant growth–promoting properties of various PGPRs isolated from halophytic rhizospheres are maintained under high salinity. The salt-tolerant PGPR *Kushneria* species YCWA18, isolated from the eastern coast of China, has demonstrated optimal growth even in a salt solution containing 20% NaCl (Zhu et al. 2011). Various other PGPRs that have been isolated have salt-tolerant characteristics, shown by their growth in NaCl concentrations of up to 25%. These include *Arthrobacter*, *Halomonas*, *P. mendocina*, *Bacillus pumilus*, and *Nitrinicola lacisaponensis*. Along with maintenance of growth, these species have other growth-promoting features such as efficient phosphorus solubilization and production of indoleacetic acid (IAA), siderophores, and 1-aminocyclopropane-1-carboxylate (ACC) deaminase (Tiwari et al. 2011). A plant growth–promoting (PGP) trait is defined as the capability to provide P to the host in a P-limited environment. This phytohormone-like feature of PGPRs stimulates plant growth, providing sufficient Fe to the host and reducing the level of ACC deaminase, a precursor of ethylene production (Etesami and Beattie 2018). From earlier reports it is known that the nodules of leguminous species are more salt sensitive than the rhizobacteria present in the rhizosphere of legumes. Isolation of bacterial species from leguminous rhizospheres mainly include *Bacillus* spp., which are Gram-positive bacteria. Among rhizospheric PGPRs, *Pseudomonas* spp. in the rhizosphere of rice is most common (Rangarajan et al. 2002). These species includes *P. fluorescens*, *P. pseudoalcaligenes*, and *P. diminuta*. Fluorescent pseudomonads, *P. alcaligenes*, and *P. pseudoalcaligenes* have been found in both saline and nonsaline conditions. However, these species, isolated from the wild rice *S. salitolerans*, were capable of nitrogen fixation and increased P solubilization under salinity. Nonetheless, there is a lack of a definitive taxonomic categorization of PGPRs growing in saline soil (Loganathan and Nair 2004). Besides wide dominance of *Pseudomonas* and *Flavobacterium* spp. over *Arthrobacter* and *Bacillus* spp., *Bacillus* spp. are widespread in a variety of natural habitats. However, *Azotobacter* spp. have been isolated mostly from salt-affected areas. This suggests a diverse habitat range and efficient adaptability of *Bacillus*. Apart from the wild rice variety, various salt-tolerant bacteria have also been isolated from *Salicornia brachiata*. These include *Haererehalobacter*, *Zhihengliuella*, *Rhizobium radiobacter*, and *Brachybacterium saurashtrense*. These species were reported to have plant growth–promoting effects on *S. brachiata*, such as enhanced P solubilization, IAA production, and ACC deaminase synthesis under salinity (Jha et al. 2012). Apart from bacteria isolated in India, *Zhihengliuella* isolated from northwest China can withstand NaCl concentrations of up to 25% (Zhang et al. 2007). The rhizosphere of rice is reported to colonize various N_2-fixing endophytic diazotrophs such as *Pseudomonas* spp. (Jha et al. 2009). In addition, from the endorhizosphere of *Salicornia europea*, *P. pseudoalcaligenes* has been isolated (Ozawa et al. 2007). Of the rhizospheric bacteria that were isolated from the halophyte *Aster tripolium*, 36% were firmicutes, 9% were proteobacteria, and 55% were actinobacteria (Szymańska et al. 2016). The diversity of bacterial species in the roots of *A. tripolium* is limited to Gram-positive bacteria. Growth of Gram-positive bacteria in salt-dominated areas has previously been reported (Damodaran et al. 2013). There are primarily three genera of bacteria that dominate the rhizosphere soil of plants: *Streptomyces*, *Bacillus*, and *Pseudomonas* (Bouizgarne 2013). *Herbaspirillum*, *Azospirillum*, *P. alcaligenes*, *Gluconacetobacter*, *Serratia*, and many other species have also been identified in rhizosphere soil (Szymańska et al. 2016). Signature lipid biomarker (SLB) analysis is of great importance for characterizing diversity in the structure and function of rhizospheric microorganisms (Mrozik et al. 2014). PGPR strains isolated from *Suaeda fruticosa*, which show optimum growth even under a 300-mM NaCl concentration, showed potential PGP characteristics, and their traits were screened under high salinity (Goswami et al. 2014). Isolation of rhizospheric bacteria from the medicinal plant *Ajuga bracteosa* has also been reported (Kumar et al. 2012). The PGP trait is often altered by biosynthesis of chitinase, ion-chelating siderophores, and microbial antagonists (such as hydrogen cyanide (HCN)), and, as a result, the growth of the plant is affected (Zhou et al. 2012b). PGPRs such as *Bacillus* have been isolated from the rhizosphere soil of various halophytes such as *Atriplex leucoclada*, *Salicornia virginica*, *Haloxylon salicornicum*, and *Arthrobacter pascens* (Ullah and Asghari 2015). The rhizospheres of various halophytes have been reported to colonize a variety of salt-tolerant PGPR species. These halophytic species include *Salicornia bigelovii* (Rueda-Puente et al. 2010), *Halocnemum strobilaceum* (Al-Mailem et al. 2010), *S. brachiata* (Jha et al. 2012), *Sesuvium portulacastrum* (Bian et al. 2011; Anburaj et al. 2012), and *Acacia* spp. (Boukhatem et al. 2012). Apart from the diversity of the halophytic rhizospheres in which these PGPRs were found, they were also isolated from diverse habitats ranging from saline to alkali–saline and desert soils.

4.6 Growth-Stimulating Mechanisms of PGPRs Under Stress Conditions

The mechanisms by which PGPRs enhance the growth of plants mainly include growth hormone biosynthesis and improved uptake and mobilization of nutrients from soil to plant parts. Apart from these, PGPR-induced tolerance of salinity greatly influences the survival of plants under harsh saline conditions (Glick 2010; Sandhya et al. 2010a; Ma et al. 2011). However, different free-living soil bacteria belonging to particular genera may have different degrees of influence on plant growth. To alleviate stress effects on plants, rhizobacterial species can be used in different ways. PGPRs have multidimensional effects on plant growth, both direct and indirect. The indirect methods include protection from negative effects of plant pathogens through secretion of antipathogenic substances (Nadeem et al. 2013). The growth-promoting feature of PGPRs is implemented by synthesis of various secondary metabolites with significance potential for growth promotion or enhancement of nutrient uptake by plants. In addition, they enhance the growth of the host plant by enhancing nitrogen fixation and increasing the availability of P to the plants.

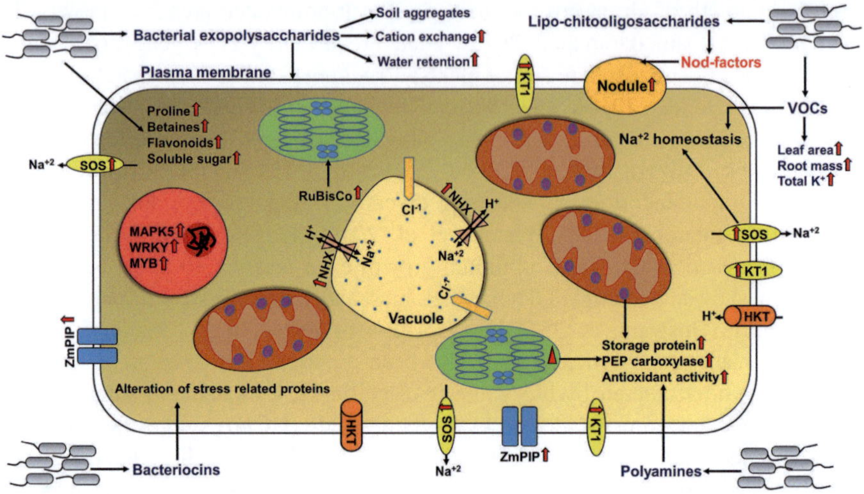

Fig. 4.1 Cellular mechanisms involved in alleviation of salt stress through interaction of plant growth–promoting rhizobacteria (*PGPRs*) and plants. Increases in the activity of ion transporters maintain the osmotic balance and minimize osmotic damage. Bacterial secretions such as exopolysaccharides increase cation exchange and water retention; polyamines increase the levels of storage protein and antioxidant activity; bacteriocins are responsible for alterations in stress-related proteins and lipo-chitooligosaccharides which increases nodulation by increasing the activity of Nod factors. An *up arrow* represents an increase in activity; a *down arrow* represents a decrease in activity

The halophytic rhizosphere contains a diverse ecological colonization of microorganisms that are salt tolerant and induce plant growth. PGPRs adopt several mechanisms to enhance growth in plants and protect them from various environmental stresses such as salinity and drought. The cellular mechanisms of PGPR–plant interactions in ameliorating the salinity effect are depicted in Fig. 4.1. These mechanisms include synthesis of ACC deaminase, enhancement of biological nitrogen fixation, phytohormone production, and exopolysaccharide synthesis. The abilities of PGPRs to enhance plant growth in both conventional and extreme habitats, and to facilitate plant adaptation to extreme environments, have been studied (Hayat et al. 2010). Rhizospheric bacteria promote plant growth in direct and indirect ways. The direct mechanisms include better nutrient fixation and solubilization, as well as enhanced phytohormone synthesis. An indirect mechanism is minimization of the deleterious effects of phytopathogens through biocontrol of their growth on the host plant (Mapelli et al. 2013). Among the different environmental stresses, salinity has the most deleterious effects on plant growth and crop yields (Mayak et al. 2004). Halophytic species are dominant in the dry regions of the world. Halophytes have their own ability to reclaim agricultural soil affected by high salinity (Ravindran et al. 2007). A number of works have lately enriched the existing understanding of host–microbe interactions and the beneficial roles of rhizospheric microbes in enhancing plant resistance to various abiotic stresses. These studies have

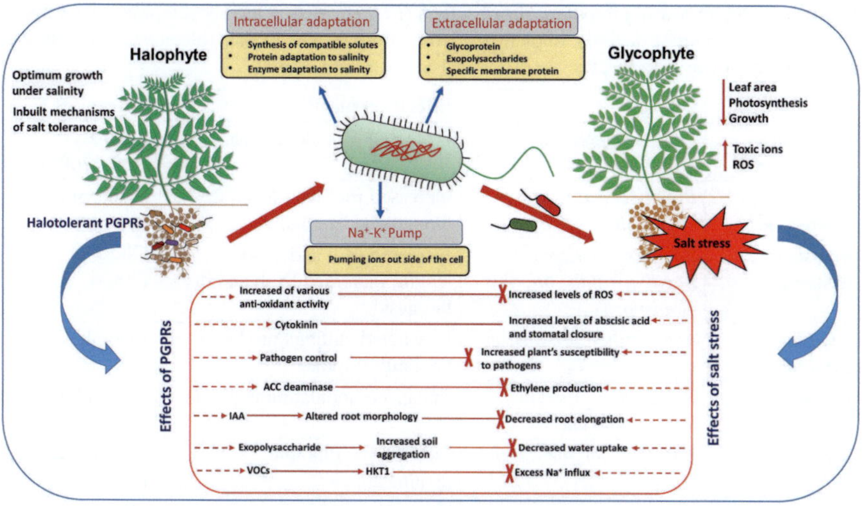

Fig. 4.2 Salt tolerance mechanisms of halotolerant plant growth–promoting rhizobacteria (*PGPRs*) and beneficial effects of PGPR inoculation in ameliorating detrimental effects of salinity

delivered substantial knowledge about (1) plant–microbe interactions, (2) mechanisms of plant growth under saline conditions, and (3) potential bacterial strains that can be used as biofertilizers in saline lands (Mapelli et al. 2013). The effects of salinity on glycophytes and the mechanisms of PGPR-mediated salt tolerance are depicted in Fig. 4.2. This type of tolerance that is provided by PGPRs is termed systemic tolerance (Yang et al. 2009). PGPRs that induce plant growth fall into three categories: biofertilizers, phytostimulators, and biopesticides. Biofertilizers enhance the availability of N and P to the host, phytostimulators induce biosynthesis of phytohormones, and biopesticides reduce the risks of pathogen attacks on the host. Various studies have illustrated the beneficial role of PGPRs in alleviating stress effects on various crop species (Ali et al. 2009; Kohler et al. 2009; Sandhya et al. 2010a). Enhanced growth through PGPR inoculation is achieved by extensive root formation stimulated by increased production of phytohormones (such as IAA and nitric acid) and increased biosynthesis of ACC deaminase (Dimkpa et al. 2009).

4.6.1 Growth Promotion by Enhanced ACC Deaminase Activity

Elevated biosynthesis of ACC deaminase reduces ethylene production, which confers tolerance to different environmental stresses (Lucy et al. 2004; Bal et al. 2013). Many reports have described the beneficial influence of PGPRs on growth in terms of enhanced ACC deaminase activity and improvement of salinity tolerance in host plants (Table 4.1). Ethylene, a plant stress hormone, is released as a stress response upon encounters with various environmental stresses. Salinity causes

Table 4.1 Plant growth–promoting rhizobacteria (*PGPRs*) with 1-aminocyclopropane-1-carboxylate (*ACC*) deaminase activity implemented for improvement of salt tolerance capabilities in various glycophytic crop species

Plant species	PGPRs	Effects on plants	References
Brassica campestris	*Methylobacterium fujisawaense*	Promoted root elongation	Madhaiyan et al. (2006)
Brassica campestris	*Brevibacterium epidermidis* RS15	Increased root length and dry biomass accumulation	Siddikee et al. (2010)
Brassica campestris	*Bacillus circulans*, *Bacillus firmus*, *Bacillus globisporus*	Increased root and shoot length; increased fresh and dry biomass	Ghosh et al. (2003)
Dianthus caryophyllus	*Azospirillum brasilense*	Inoculated cuttings produced the longest roots	Li et al. (2005)
Pisum sativum	*Rhizobium leguminosarum*	Enhanced nodulation in plants	Ma et al. (2003)
Zea mays	*Enterobacter sakazakii*	Increased agronomic parameters	Babalola et al. (2003)
Zea mays	*Pseudomonas*	Root elongation	Shaharoona et al. (2006)
Zea mays	*Pseudomonas syringae*, *Enterobacter aerogenes*, *Pseudomonas fluorescens*	High relative water; high chlorophyll; low proline content	Nadeem et al. (2007)
Oryza sativa	*Variovorax*, *Rhodococcus*	Increased root and shoot length; increased root fresh biomass	Bal et al. (2013)
Oryza sativa	*Enterobacter*	Increased growth parameters and activities of various antioxidant enzymes	Sarkar et al. (2017)
Lycopersicon esculentum	*Pseudomonas brassicacearum*	Increased root biomass	Belimov et al. (2007)
Lycopersicon esculentum	*Bacillus pumilus*, *Bacillus subtilis*	Increased vigor index under saline–sodic conditions	Damodaran et al. (2013)
Lycopersicon esculentum	*Achromobacter*	Improved rate of photosynthesis (in comparison with noninoculated plants) under saline conditions	Mayak et al. (2004)
Lactuca sativa	*Azospirillum*	Better germination and vegetative growth under salinity	Barassi et al. (2006)
Arachis hypogaea	*Pseudomonas fluorescens*	Greater growth (in comparison with noninoculated plants) under saline conditions	Saravanakumar and Samiyappan (2007)
Triticum aestivum	*Pseudomonas*, *Serratia*	Improved shoot and root growth and growth-contributing parameters	Zahir et al. (2009)

(continued)

Table 4.1 (continued)

Plant species	PGPRs	Effects on plants	References
Avena sativa	*Klebsiella*	Increased growth parameters; lower levels of various stress indicators such as MDA and proline	Sapre et al. (2018)
Avena sativa	*Acinetobacter*	Increased growth parameters and antioxidant enzyme activity	Xun et al. (2015)

various physiological changes in plant tissues by means of overproduction of ethylene through increased synthesis of ACC (Tank and Saraf 2010). PGPRs facilitate plant growth through enhanced biosynthesis of ACC deaminase under various abiotic stress conditions. Earlier studies described plant growth promotion under various abiotic stresses and pathogen attack, with PGPRs reducing the ethylene level through enhanced ACC deaminase activity. Numerous species of Gram-negative bacteria (Wang et al. 2000; Babalola et al. 2003), Gram-positive bacteria (Ghosh et al. 2003; Belimov et al. 2007), and rhizobia (Ma et al. 2003; Uchiumi et al. 2004) have been reported to show ACC deaminase activity. The PGP capabilities of numerous microbial species—such as *Methylobacterium fujisawaense*, *Burkholderia*, *Ralstonia solanacearum*, *Enterobacter*, *Agrobacterium genomovars*, and *Rhizobium*—have been studied (Hontzeas et al. 2004; Pandey et al. 2005; Blaha et al. 2006). ACC deaminase induces plant growth via reduced biosynthesis of ethylene through conversion of this ethylene precursor (ACC) to α-ketobutyrate and ammonia (Saleem et al. 2007). Further reduction of ethylene biosynthesis can be achieved by ACC deaminase competing with ACC oxidase, which has more affinity for ACC, resulting in lower endogenous synthesis of plant ethylene (Saleem et al. 2007). To achieve this, the level of ACC deaminase in the plant must be a minimum of 100 times higher than the level of ACC oxidase. Biosynthesis of IAA was occurred from the tryptophan and secreted by PGPR and get adsorbed by the plants. The newly produced IAA is added to the existing endogenous IAA pool and induces growth. Apart from this, IAA facilitates conversion of S-adenosylmethionine (SAM) into ACC by stimulating ACC synthetase activity (Saleem et al. 2007). The ACC deaminase activity of rhizobial microorganisms can stimulate ACC production and exudation from plant roots, which provides a unique nitrogen source to the microorganisms to stimulate root growth. Alterations in the ACC deaminase activity of various microorganisms under environmental stresses play a positive role in phytoremediation. The ACC deaminase activity of rhizobacteria induces growth of plants and mediates rhizodegradation through root exudates, biotransforming noxious elements. Apart from this, phytoremediation of heavy metals is achieved through the presence of rhizobacteria in the rhizosphere soil of various halophytes (Singh and Jha 2016; Singh et al. 2015). Rhizobacteria producing ACC deaminase enhance plant growth by implementation of different strategies such as alleviation of detrimental effects of various environmental stresses, delayed senescence, protection against phytopathogens, and induction of nodulation in legumes (Ali et al. 2012). PGPRs that inhabit saline environments and show ACC deaminase activity have been isolated and used for alleviation of negative effects of salinity on the growth of various crop plants. For example, the growth of maize shoots and roots was enhanced by 2.3-fold and 3.3-fold upon inoculation with *P. fluorescens* N3 and *P. putida* (PGPRs with ACC deaminase activity), respectively, in comparison with controls (Khan et al. 2016). Similarly, enhanced nodule formation in leguminous crops inoculated with ACC deaminase–producing rhizobacteria was also reported (Shaharoona et al. 2006). Moreover, it has been documented in wheat plants that the PGPR strain *Azospirillum brasilense* FP2 reduces ACC oxidase activity (Camilios-Neto et al. 2014). In addition, higher activity of ACC deaminase was reported in the PGPR strain *Pseudomonas maricaloris*, isolated from *Avicennia marina* (El-Tarabily and Youssef 2010). Upon inoculation with these strains, seedlings exhibited decreased level of ACC and higher growth rates under salinity stress (El-Tarabily and Youssef 2010). Reductions of 44%, 53%, and 57% in ethylene levels were observed in pepper plants after inoculation with *Zhihengliuella alba*, *Brevibacterium iodinum*, and *Bacillus licheniformis* (isolated from halophytes), respectively.

4.6.2 Growth Promotion by Enhanced Endogenous Phytohormone Production

Phytohormones are essential in regulating the stress responses of plants to various environmental stresses, and in maintaining normal plant physiology and conferring tolerance of salinity and other environmental stresses (Ryu and Cho 2015). The responses of plants to salinity induce a variety of alterations in molecular, biochemical, and physiological aspects. The responses vary according to the frequency and intensity of the salinity, the environmental conditions, the physical and chemical properties of the soil, and the age of the plants (Kumari et al. 2015). Application of exogenous phytohormones can be useful for amelioration of the adverse effect of salinity (Zahir et al. 2010). The hormonal status of plants is modulated by soil bacteria, which release exogenous phytohormones that contribute to enhanced salinity tolerance (Ilangumaran and Smith 2017). Exogenous phytohormones can alter the levels of endogenous phytohormones, providing a way to counter salt stress conditions (Ilangumaran and Smith 2017). Modulation of the plant hormonal balance is one of the strategies that PGPRs adapt to enhance plant growth (Tsukanova et al. 2017). Among phytohormones, most of the salt-tolerant PGPRs induce biosynthesis of IAA and, as a result, improve the growth of salinity-treated seedlings (Tiwari et al. 2011). PGPRs may also alter root-to-shoot hormonal signaling, improving crop salt tolerance (Yang et al. 2009). Higher production of IAA in plants, induced by salt-tolerant rhizobacteria, was shown to be responsible for increased root growth in coastal soils (Siddikee et al. 2011). The species from the roots of which the PGPRs were isolated included *Argentina* and *Prosopis strombulifera* (Sgroy et al. 2009; Tiwari et al. 2011; Piccoli et al. 2011; Bian et al.

2011). Salinity-induced levels of abscisic acid (ABA) were counteracted by the *Bacillus amyloliquefaciens* strain SQR9, which improved salt stress tolerance in maize by increasing chlorophyll and total soluble sugar content, with increased activity of various antioxidant enzyme components (Chen et al. 2016). Bacterial cytokinins can also be responsible for salt tolerance, but fewer studies have reported this. PGPR strains enhanced growth by 52% and alleviated salinity-induced dormancy in wheat (Egamberdieva 2009). Improvements in lettuce seedling growth under water stress conditions by cytokinin-producing *Bacillus subtilis* were reported by Arkhipova et al. (2007). Synthesis of excess ethylene is regulated by PGPRs producing ACC deaminase (Glick et al. 2007). Inoculation of *Solanum lycopersicum* with *P. putida* UW4 increased the shoot biomass under saline conditions (Yan et al. 2014). IAA accumulation on plants induces the activity of the ACC synthase gene, resulting in greater accumulation of ACC. Higher concentrations of ACC lead to greater production of ethylene. PGPRs with ACC deaminase activity reduce accumulation of ACC by breaking down excess ACC, which, in turn, reduces ethylene production under various environmental stresses. Several studies have documented the role of bacterial ABA in influencing the ABA status of plants. Modulation of the ABA signaling cascade of wheat has been reported upon inoculation with halotolerant *Dietzia natronolimnaea* STR1 via upregulation of different ABA-responsive genes (Bharti et al. 2016). Similarly, the ABA-producing bacterial strains *Burkholderia cepacia* SE4 and *Promicromonospora* sp. SE188 enhanced shoot biomass accumulation in *Cucumis sativus* under saline conditions (Kang et al. 2014). Cotton seedlings inoculated with *P. putida* Rs-198 had a lower ABA content and increased biomass accumulation under salt stress (Yao et al. 2010). *Arthrobacter protophormiae* SA3–inoculated wheat showed greater salinity tolerance, with increased levels of IAA and less accumulation of ABA and ACC.

4.6.3 Enhancement of Micronutrient Availability to Plants by Siderophore-Producing PGPRs

Siderophores are Fe-chelating compounds and have high affinity for iron. Siderophores make iron more accessible to plants by formation of an Fe-siderophore complex. In saline and sodic soils the level of iron available for uptake by plants is limited. The low availability of iron and other micronutrients needed by the plants in this type of soil results in iron deficiency (Abbas et al. 2015). The siderophore production ability of PGPRs increases the availability of micronutrients to plants (Navarro-Torre et al. 2017; Zhou et al. 2017). Bacterial siderophores are used by plants as an iron source. In addition, the bacterial siderophores prevent the spread of harmful microorganisms and their deleterious effects on plant growth (Shailendra Singh 2015). Earlier reports described the ability of PGPRs to synthesize siderophores (Kuffner et al. 2008). Along with auxin production, plant growth in metal-polluted soil is supported by siderophore-producing soil and rhizospheric bacteria. Bacterial stains such as *Pseudomonas* sp. *BE3dil*, *Janthinobacterium lividum*, *S. marcescens*, and *Flavobacterium frigidimaris* have been reported to have the ability to produce siderophores (Kuffner et al. 2008). Previous reports have suggested that plant growth can be enhanced by siderophore-producing PGPRs (Glick 2003).

4.7 Implementation of PGPRs in Development of Salt-Tolerant Crop Species

Growth improvements in crops grown in saline conditions by use of different salt-tolerant PGPRs have been reported in several studies (Mayak et al. 2004; Shukla et al. 2012; Qin et al. 2016) (Table 4.2). In the modern era, various environmental factors such as salinity have caused reductions in agricultural productivity in developing countries (Shukla et al. 2012). Salinity alters many basic physiological processes of plants, affecting their yields. Various factors exert different negative effects on growth and cause growth reductions in plants grown under high salinity—most importantly, imbalances between nutrient uptake and distribution. These further generate competitions between essential macronutrients such as K^+, Ca^{2+}, and NO_3^-, and prevailing Na^+ and Cl^-. To counter this, implementation of PGPR-based salinity tolerance to boost beneficial mechanisms in various agricultural crops is a revolutionary technique being used alongside traditional methods (Jha et al. 2012). As discussed earlier, several mechanisms are employed by halotolerant PGPRs to improve plant growth, such as (1) efficient regulation of antioxidant defenses for scavenging of ROS (Jha and Subramanian 2014; Qin et al. 2016); (2) improved fixation of atmospheric nitrogen, resulting in improved nutrition supply to the plants, better P solubilization, and increased production of siderophores; (3) restriction of toxic Na^+ and Cl^- within the tolerance limits of plants by selective absorption of ions and maintenance of a high K^+-to-Na^+ ratio (Shukla et al. 2012; Etesami 2018); and (4) secretion of extracellular polymeric substances (EPSs) forming a physical barrier, called a rhizosheath, which reduces Na^+ uptake by the roots and translocation to the shoots. Halotolerant PGPRs improve the soil structure by producing EPSs. These promote aggregation of the soil, which, in turn, increases the water retention capacity of the soil and increases nutrient availability to the plants. A functional characteristic of bacteria is that they contain functional groups such as hydroxyl, sulfhydryl, carboxyl, and phosphoryl groups, which bind to Na^+ and reduce its uptake and translocation (Nunkaew et al. 2014). Various halotolerant bacteria such as *Halomonas variabilis*, *Enterobacter*, *Paenibacillus*, *Planococcus rifietoensis*, and *Aeromonas hydrophila/caviae* produce EPSs and induce biofilm synthesis (Qurashi and Sabri 2012; Khan et al. 2016). Various studies have described the important role of PGPRs in alleviating salt stress in a variety of crops such as groundnut, rice, tomato, and red pepper (Shukla et al. 2012; Bal et al. 2013; Upadhyay and Singh 2015). In groundnut grown under high salinity, increased levels of chlorophyll, proline, and malondialdehyde, along with other physiological parameters, were

Table 4.2 Interaction of plant growth–promoting rhizobacteria (*PGPRs*) with various crop species and responses of the host plants

Crop	PGPR strain	Halophytic rhizosphere from which PGPRs were isolated	Plant response to PGPRs	References
Solanum lycopersicum	*Achromobacter piechaudii* ARV8	*Lycium shawii*	Increased fresh weight, dry weight, and water use efficiency of tomato through decreased ethylene production under stress	Mayak et al. (2004)
Solanum lycopersicum	*Azotobacter*, *Bacillus safensis*, *Bacillus subtilis*, *Rhizobium*	Halophyte grasses	Increased IAA, siderophores, and phosphate solubilization	Damodaran et al. (2013)
Zea mays	*Bacillus pumilus*, *Arthrobacter aurescens*, *Arthrobacter pascens*	*Haloxylon salicornicum*, *Lespedeza bicolor*, *Salicornia virginica*	Increased growth parameters; higher chlorophyll and sugar content; increased activity of antioxidant enzymes such as CAT and SOD	Ullah and Asghari (2015)
Eggplant	*Bacillus subtilis*	–	Increased plant auxin synthesis with changing regulation of the appropriate mechanisms induced by bacterial production of auxin and auxin precursors during root colonization	Bochow et al. (2001)
Vigna radiata	*Pseudomonas syringae*, *Pseudomonas fluorescens*	Salt-affected *Vigna radiata*	Increased growth parameters such as shoot and root fresh weight; improved water use efficiency	Ahmad et al. (2012)
Triticum aestivum	*Bacillus insolitus*, *Aeromonas hydrophila*	Salt-affected wheat	Improved growth parameters in inoculated samples (in comparison with uninoculated samples); higher Ca^{2+} levels in inoculated samples contributed to lower Na^{+} uptake	Ashraf et al. (2004)

(continued)

Table 4.2 (continued)

Crop	PGPR strain	Halophytic rhizosphere from which PGPRs were isolated	Plant response to PGPRs	References
Triticum aestivum	*Bacillus subtilis*, *Arthrobacter*	Saline area of the eastern Gangetic Plain	Inoculated samples showed lower activity of antioxidant enzymes such as APX, CAT, and GR, suggesting lower ROS production	Upadhyay et al. (2012)
Raphanus sativus	*Staphylococcus kloosii*, *Kocuria erythromyxa*	Plants naturally grown on high salty soils in the upper Çoruh Valley	Bacterial treatments under salinity conditions had positive effects on plant growth, with increased chlorophyll content	Hayat et al. (2010)
Capsicum annuum	*Brevibacterium iodinum*, *Bacillus licheniformis*	West coast soil of the Yellow Sea	Enhancement of growth and salt tolerance by regulation of stress ethylene synthesis	Siddikee et al. (2011)
Oats	*Acinetobacter*, *Klebsiella*	Salt-tolerant wheat	Enhancement of growth; less electrolyte leakage; lower MDA levels	Sapre et al. (2018)
Oryza sativa	*Enterobacter*	Salt-tolerant rice	Increased phosphate solubilization, IAA production, and siderosphere production	Kim et al. (2014)

APX ascorbate peroxidase, *CAT* catalase, *GR* glutathione reductase, *IAA* indoleacetic acid, *MDA* malondialdehyde, *ROS* reactive oxygen species, *SOD* superoxide dismutase

reported in response to inoculation with beneficial PGPRs (Shukla et al. 2012). Improved salt tolerance of oat plants inoculated with halotolerant PGPRs such as *Acinetobacter* has also been reported (Xun et al. 2015). In salinity tolerance studies of barley (Chang et al. 2014) and oats (Sapre et al. 2018), plants inoculated with various PGPRs showed higher tolerance of salinity than uninoculated control plants. The salinity tolerance of rice has been improved by inoculation with a halotolerant *Enterobacter* PGPR, which alleviated the negative effects of salinity and improved the growth of the plants (Sarkar et al. 2017). Increases in the germination rate, the seedling growth rate, the length and biomass of shoots and roots, and chlorophyll levels have been reported in rice inoculated with an *Enterobacter* PGPR. The enhanced growth of the plants inoculated with *Enterobacter* strain P23 may have been attributable to its PGP traits. Reduced activity of various antioxidant enzymes along with proficient growth of rice under saline conditions suggest that ROS generation is reduced upon inoculation with PGPRs (Sarkar et al. 2017). Similar findings have been documented in lettuce inoculated with *Rhizobium* sp. and *Serratia* sp. (Lee et al. 2005), in wheat inoculated with *Arthrobacter* sp. and *B. subtilis* (Upadhyay et al. 2012), and in rice inoculated with *B. pumilus* and *P. pseudoalcaligenes* (Jha and Subramanian 2014) under saline conditions. In addition to limiting ROS production, PGPRs increase biosynthesis of IAA, which, in turn, induces plant growth and, most importantly, root growth. Achievement of salinity tolerance in tomato through inoculation of the plants with *Enterobacter* sp. EJ01 was demonstrated by Kim et al. (2014). The salinity tolerance may have been attributable to transcriptional activation of various stress-responsive factors prior to the induction of salinity (Kim et al. 2014).

4.8 Conclusion and Future Perspectives

PGPRs isolated from a saline rhizosphere are able to mitigate the deleterious impacts of salinity and increase germination as well as the viability index, shoot and root length, dry biomass production, and, most importantly, the agricultural yield. PGPRs are used in organic farming practice as biofertilizers and biocontrol agents. Exploration of soil microbe diversity is essential to meet global food demands despite increasing deterioration of soil quality caused by increases in the prevalence and severity of salinity, which affect agricultural productivity. Plants have developed efficient strategies to enhance their tolerance of abiotic stresses such as high salinity and drought. PGPRs associated with the halophytic rhizosphere can mitigate the negative effects of prevailing salinity in a cost-effective manner where there is an urgent need for salt-tolerant crop varieties. Therefore, inoculation of agricultural crops with salt-tolerant PGPRs is an efficient strategy for sustainable agriculture in salt-affected areas. Research on PGPRs must be carried out to open up new avenues of opportunity to develop salinity-tolerant agricultural crops. Research is now being carried out to deepen our fundamental knowledge of microbe–halophyte interactions and to take that knowledge to another level for production of salt-tolerant crop varieties. Unraveling the molecular mechanisms of salt tolerance of PGPRs will allow us to engineer bacteria with enhanced ability to stimulate plant growth under saline conditions. Furthermore, the efficiency of PGPRs for stimulation of plant growth under salinity must be assessed over longer periods of time for relevant crop production. Apart from being useful for production of salt-tolerant crop varieties, PGPRs can be implemented in phytodesalinization of salt-affected agricultural lands, phytoremediation, and biological control of various types of pathogen attack. Strategies for rapid identification of efficient PGPR strains to be used as bioinoculants must be developed to improve growth of crops under saline conditions. With increasing exposure of agricultural lands to various environmental stresses, it is now essential to develop salt-tolerant crop varieties, and utilization of stress-tolerant PGPRs is paving the way toward fulfilling this goal.

Acknowledgements A.K.P. received a grant from the Science and Engineering Research Board (SERB) (grant number SB/SO/PS-14/2014), Department of Science and Technology (DST), Government of India (New Delhi, India), which is duly acknowledged. This manuscript has been assigned the Council of Scientific and Industrial Research–Central Salt and Marine Chemicals Research Institute (CSIR-CSMCRI) registration number PRIS 074/2018.

References

Abbas G, Saqib M, Akhtar J, Anwar M (2015) Interactive effects of salinity and iron deficiency on different rice genotypes. J Plant Nutr Soil Sci 178:306–311. https://doi.org/10.1002/jpln.201400358

Abou-Elela SI, Kamel MM, Fawzy ME (2010) Biological treatment of saline wastewater using a salt-tolerant microorganism. Desalination 250:1–5. https://doi.org/10.1016/j.desal.2009.03.022

Ahmad F, Ahmad I, Khan MS (2008) Screening of free-living rhizospheric bacteria for their multiple plant growth promoting activities. Microbiol Res 163:173–181. https://doi.org/10.1016/j.micres.2006.04.001

Ahmad M, Zahir ZA, Asghar HN, Arshad M (2012) The combined application of rhizobial strains and plant growth promoting rhizobacteria improves growth and productivity of mung bean (*Vigna radiata* L.) under salt-stressed conditions. Ann Microbiol 62:1321–1330. https://doi.org/10.1007/s13213-011-0380-9

Ali SZ, Sandhya V, Grover M et al (2009) *Pseudomonas* sp. strain AKM-P6 enhances tolerance of sorghum seedlings to elevated temperatures. Biol Fertil Soils 46:45–55. https://doi.org/10.1007/s00374-009-0404-9

Ali S, Charles TC, Glick BR (2012) Delay of flower senescence by bacterial endophytes expressing 1-aminocyclopropane-1-carboxylate deaminase. J Appl Microbiol 113:1139–1144. https://doi.org/10.1111/j.1365-2672.2012.05409.x

Al-Mailem DM, Sorkhoh NA, Marafie M et al (2010) Oil phytoremediation potential of hypersaline coasts of the Arabian Gulf using rhizosphere technology. Bioresour Technol 101:5786–5792. https://doi.org/10.1016/j.biortech.2010.02.082

Anburaj R, Nabeel MA, Sivakumar T, Kathiresan K (2012) The role of rhizobacteria in salinity effects on biochemical constituents of the halophyte *Sesuvium portulacastrum*. Russ J Plant Physiol 59:115–119. https://doi.org/10.1134/S1021443712010025

Arkhipova TN, Prinsen E, Veselov SU et al (2007) Cytokinin producing bacteria enhance plant growth in drying soil. Plant Soil 292:305–315. https://doi.org/10.1007/s11104-007-9233-5

Ashraf M, Hasnain S, Berge O, Mahmood T (2004) Inoculating wheat seedlings with exopolysaccharide-producing bacteria restricts sodium uptake and stimulates plant growth under salt stress. Biol Fertil Soils 40:157–162. https://doi.org/10.1007/s00374-004-0766-y

Babalola OO, Osir EO, Sanni AI et al (2003) Amplification of 1-amino-cyclopropane-1-carboxylic (ACC) deaminase from plant growth promoting rhizobacteria in *Striga*-infested soil. Afr J Biotechnol 2:157–160. https://doi.org/10.4314/ajb.v2i6.14791

Barassi CA, Ayrault G, Creus CM et al (2006) Seed inoculation with *Azospirillum mitigates* NaCl effects on lettuce. Sci Hortic (Amsterdam) 109:8–14. https://doi.org/10.1016/j.scienta.2006.02.025

Bal HB, Nayak L, Das S, Adhya TK (2013) Isolation of ACC deaminase producing PGPR from rice rhizosphere and evaluating their plant growth promoting activity under salt stress. Plant Soil 366:93–105. https://doi.org/10.1007/s11104-012-1402-5

Belimov AA, Dodd IC, Safronova VI et al (2007) *Pseudomonas brassicacearum* strain Am3 containing 1-aminocyclopropane-1-carboxylate deaminase can show both pathogenic and growth-promoting properties in its interaction with tomato. J Exp Bot 58:1485–1495. https://doi.org/10.1093/jxb/erm010

Bharti N, Pandey SS, Barnawal D et al (2016) Plant growth promoting rhizobacteria *Dietzia natronolimnaea* modulates the expression of stress responsive genes providing protection of wheat from salinity stress. Sci Rep 6:1–16. https://doi.org/10.1038/srep34768
Bian G, Zhang Y, Qin S et al (2011) Isolation and biodiversity of heavy metal tolerant endophytic bacteria from halotolerant plant species located in coastal shoal of Nantong. Wei Sheng Wu Xue Bao 51:1538–1547
Blaha D, Prigent-Combaret C, Mirza MS, Moënne-Loccoz Y (2006) Phylogeny of the 1-aminocyclopropane-1-carboxylic acid deaminase-encoding gene acdS in phytobeneficial and pathogenic proteobacteria and relation with strain biogeography. FEMS Microbiol Ecol 56:455–470. https://doi.org/10.1111/j.1574-6941.2006.00082.x
Bochow H, El-Sayed SF, Junge H et al (2001) Use of Bacillus subtilis as biocontrol agent. IV. Salt-stress tolerance induction by Bacillus subtilis FZB24 seed treatment in tropical vegetable field crops, and its mode of action. J Plant Dis Prot 108:21–30. https://doi.org/10.2307/43215378
Bouizgarne B (2013) Bacteria for plant growth promotion and disease management. In: Maheshwari DK (ed) Bacteria in agrobiology: disease management. Heidelberg: Springer, 15–47
Boukhatem ZF, Domergue O, Bekki A et al (2012) Symbiotic characterization and diversity of rhizobia associated with native and introduced acacias in arid and semi-arid regions in Algeria. FEMS Microbiol Ecol 80:534–547. https://doi.org/10.1111/j.1574-6941.2012.01315.x
Braud A, Jézéquel K, Bazot S, Lebeau T (2009) Enhanced phytoextraction of an agricultural Cr- and Pb-contaminated soil by bioaugmentation with siderophore-producing bacteria. Chemosphere 74:280–286. https://doi.org/10.1016/j.chemosphere.2008.09.013
Camilios-Neto D, Bonato P, Wassem R et al (2014) Dual RNA-seq transcriptional analysis of wheat roots colonized by *Azospirillum brasilense* reveals up-regulation of nutrient acquisition and cell cycle genes. BMC Genomics 15:1–13. https://doi.org/10.1186/1471-2164-15-378
Chang P, Gerhardt KE, Huang XD et al (2014) Plant growth–promoting bacteria facilitate the growth of barley and oats in salt-impacted soil: implications for phytoremediation of saline soils. Int J Phytoremediation 16:1133–1147. https://doi.org/10.1080/15226514.2013.821447
Chen L, Liu Y, Wu G et al (2016) Induced maize salt tolerance by rhizosphere inoculation of *Bacillus amyloliquefaciens* SQR9. Physiol Plant. https://doi.org/10.1111/ppl.12441
Damodaran T, Sah V, Rai RB et al (2013) Isolation of salt tolerant endophytic and rhizospheric bacteria by natural selection and screening for promising plant growth–promoting rhizobacteria (PGPR) and growth vigour in tomato under sodic environment. Afr J Microbiol Res 7:5082–5089. https://doi.org/10.5897/AJMR2013.6003
Dimkpa C, Weinand T, Asch F (2009) Plant–rhizobacteria interactions alleviate abiotic stress conditions. Plant Cell Environ 32:1682–1694. https://doi.org/10.1111/j.1365-3040.2009.02028.x
Dodd IC, Pérez-Alfocea F (2012) Microbial amelioration of crop salinity stress. J Exp Bot 63:3415–3428. https://doi.org/10.1093/jxb/ers033
Dolatabadian A, Modarres Sanavy SAM, Ghanati F (2011) Effect of salinity on growth, xylem structure and anatomical characteristics of soybean. Not Sci Biol 3:41–45
Egamberdieva D (2009) Alleviation of salt stress by plant growth regulators and IAA producing bacteria in wheat. Acta Physiol Plant 31:861–864. https://doi.org/10.1007/s11738-009-0297-0
Egamberdieva D, Lugtenberg B (2014) Use of plant growth–promoting rhizobacteria to alleviate salinity stress in plants. In: Miransari M (ed) Use of microbes for the alleviation of soil stresses, vol 1. Springer, New York, pp 73–96
Egamberdiyeva D, Islam KR (2008) Salt-tolerant rhizobacteria: plant growth promoting traits and physiological characterization within ecologically stressed environments. In: Ahmad I, Pichtel J, Hayat S (eds) (eds) Plant-bacteria interactions: strategies and techniques to promote plant growth. Wiley, Weinheim, pp 257–281
El-Tarabily KA, Youssef T (2010) Enhancement of morphological, anatomical and physiological characteristics of seedlings of the mangrove *Avicennia marina* inoculated with a native

phosphate-solubilizing isolate of *Oceanobacillus picturae* under greenhouse conditions. Plant Soil 332:147–162. https://doi.org/10.1007/s11104-010-0280-y

Essa TA (2002) Effect of salinity stress on growth and nutrient composition of three soybean (*Glycine max* L. Merrill) cultivars. J Agron Crop Sci 188:86–93. https://doi.org/10.1046/j.1439-037X.2002.00537.x

Etesami H (2018) Can interaction between silicon and plant growth promoting rhizobacteria benefit in alleviating abiotic and biotic stresses in crop plants? Agric Ecosyst Environ 253:98–112. https://doi.org/10.1016/j.agee.2017.11.007

Etesami H, Beattie GA (2018) Mining halophytes for plant growth–promoting halotolerant bacteria to enhance the salinity tolerance of non-halophytic crops. Front Microbiol 9:148. https://doi.org/10.3389/fmicb.2018.00148

Ghosh S, Penterman JN, Little RD et al (2003) Three newly isolated plant growth–promoting bacilli facilitate the seedling growth of canola, *Brassica campestris*. Plant Physiol Biochem 41:277–281. https://doi.org/10.1016/S0981-9428(03)00019-6

Glick BR (2003) Phytoremediation: synergistic use of plants and bacteria to clean up the environment. Biotechnol Adv 21:383–393. https://doi.org/10.1016/S0734-9750(03)00055-7

Glick BR (2010) Using soil bacteria to facilitate phytoremediation. Biotechnol Adv 28:367–374. https://doi.org/10.1016/j.biotechadv.2010.02.001

Glick BR, Cheng ÆZ, Czarny ÆJ (2007) Promotion of plant growth by ACC deaminase-producing soil bacteria. Eur J Plant Pathol 119:329–339. https://doi.org/10.1007/s10658-007-9162-4

Goswami D, Dhandhukia P, Patel P, Thakker JN (2014) Screening of PGPR from saline desert of Kutch: growth promotion in *Arachis hypogaea* by *Bacillus licheniformis* A2. Microbiol Res 169:66–75. https://doi.org/10.1016/j.micres.2013.07.004

Habib SH, Kausar H, Saud HM (2016) Plant growth–promoting rhizobacteria enhance salinity stress tolerance in okra through ROS-scavenging enzymes. Biomed Res Int 2016:6284547. https://doi.org/10.1155/2016/6284547

Hagemann M (2011) Molecular biology of cyanobacterial salt acclimation. FEMS Microbiol Rev 35:87–123. https://doi.org/10.1111/j.1574-6976.2010.00234.x

Hayat R, Ali S, Amara U et al (2010) Soil beneficial bacteria and their role in plant growth promotion: a review. Ann Microbiol 60:579–598. https://doi.org/10.1007/s13213-010-0117-1

Heidari M, Jamshid P (2010) Interaction between salinity and potassium on grain yield, carbohydrate content and nutrient uptake in pearl millet. J Agric Biol Sci 5:39–46

Hontzeas N, Zoidakis J, Glick BR, Abu-Omar MM (2004) Expression and characterization of 1-aminocyclopropane-1-carboxylate deaminase from the rhizobacterium *Pseudomonas putida* UW4: a key enzyme in bacterial plant growth promotion. Biochim Biophys Acta—Proteins Proteomics 1703:11–19. https://doi.org/10.1016/j.bbapap.2004.09.015

Ibekwe AM, Papiernik SK, Yang CH (2010) Influence of soil fumigation by methyl bromide and methyl iodide on rhizosphere and phyllosphere microbial community structure. J Environ Sci Heal—Part B Pestic Food Contam Agric Wastes 45:427–436. https://doi.org/10.1080/03601231003800131

Ilangumaran G, Smith DL (2017) Plant growth promoting rhizobacteria in amelioration of salinity stress: a systems biology perspective. Front Plant Sci 8:1–14. https://doi.org/10.3389/fpls.2017.01768

Jaleel CA, Riadh K, Gopi R et al (2009) Antioxidant defense responses: physiological plasticity in higher plants under abiotic constraints. Acta Physiol Plant 31:427–436. https://doi.org/10.1007/s11738-009-0275-6

Jha Y, Subramanian RB (2014) PGPR regulate caspase-like activity, programmed cell death, and antioxidant enzyme activity in paddy under salinity. Physiol Mol Biol Plants 20:201–207. https://doi.org/10.1007/s12298-014-0224-8

Jha B, Thakur MC, Gontia I et al (2009) Isolation, partial identification and application of diazotrophic rhizobacteria from traditional Indian rice cultivars. Eur J Soil Biol 45:62–72. https://doi.org/10.1016/j.ejsobi.2008.06.007

Jha B, Gontia I, Hartmann A (2012) The roots of the halophyte *Salicornia brachiata* are a source of new halotolerant diazotrophic bacteria with plant growth–promoting potential. Plant Soil 356:265–277. https://doi.org/10.1007/s11104-011-0877-9

Jiang H, Dong H, Yu B et al (2007) Microbial response to salinity change in Lake Chaka, a hypersaline lake on Tibetan plateau. Environ Microbiol 9:2603–2621. https://doi.org/10.1111/j.1462-2920.2007.01377.x

Kang SM, Khan AL, Waqas M et al (2014) Plant growth–promoting rhizobacteria reduce adverse effects of salinity and osmotic stress by regulating phytohormones and antioxidants in *Cucumis sativus*. J Plant Interact 9:673–682. https://doi.org/10.1080/17429145.2014.894587

Khan M, Böer B, Öztürk M et al (2016) Sabkha ecosystems. Springer, Cham

Kim K, Jang Y-J, Lee S-M et al (2014) Alleviation of salt stress by *Enterobacter* sp. EJ01 in tomato and *Arabidopsis* is accompanied by up-regulation of conserved salinity responsive factors in plants. Mol Cells 37:109–117. https://doi.org/10.14348/molcells.2014.2239

Kohler J, Hernández JA, Caravaca F, Roldán A (2009) Induction of antioxidant enzymes is involved in the greater effectiveness of a PGPR versus AM fungi with respect to increasing the tolerance of lettuce to severe salt stress. Environ Exp Bot 65:245–252. https://doi.org/10.1016/j.envexpbot.2008.09.008

Kuffner M, Puschenreiter M, Wieshammer G et al (2008) Rhizosphere bacteria affect growth and metal uptake of heavy metal accumulating willows. Plant Soil 304:35–44. https://doi.org/10.1007/s11104-007-9517-9

Kumar G, Kanaujia N, Bafana A (2012) Functional and phylogenetic diversity of root-associated bacteria of *Ajuga bracteosa* in Kangra Valley. Microbiol Res 167:220–225. https://doi.org/10.1016/j.micres.2011.09.001

Kumari A, Das P, Parida AK, Agarwal PK (2015) Proteomics, metabolomics, and ionomics perspectives of salinity tolerance in halophytes. Front Plant Sci 6:1–20. https://doi.org/10.3389/fpls.2015.00537

Lee KD, Han HS, Lee KD (2005) Plant growth promoting rhizobacteria effect on antioxidant status, photosynthesis, mineral uptake and growth of lettuce under soil salinity. Res J Agric Biol Sci 1:210–215

Li Q, Saleh-Lakha S, Glick BR (2005) The effect of native and ACC deaminase-containing *Azospirillum brasilense* Cd1843 on the rooting of carnation cuttings. Can J Microbiol 51:511–514. https://doi.org/10.1139/w05-027

Loganathan P, Nair S (2004) *Swaminathania salitolerans* gen. nov., sp. nov., a salt-tolerant, nitrogen-fixing and phosphate-solubilizing bacterium from wild rice (*Porteresia coarctata* Tateoka). Int J Syst Evol Microbiol 54:1185–1190. https://doi.org/10.1099/ijs.0.02817-0

Lucy M, Reed E, Glick BR (2004) Applications of free living plant growth–promoting rhizobacteria. Antonie Van Leeuwenhoek 86:1–25. https://doi.org/10.1023/B:ANTO.0000024903.10757.6e

Lugtenberg BJJ, Malfanova N, Kamilova F, Berg G (2013) Plant growth promotion by microbes. In: de Bruijn FJ (ed) (ed) Molecular microbial ecology of the rhizosphere, 1st edn. Wiley Blackwell, Singapore, pp 561–574

Ma W, Sebestianova SB, Sebestian J et al (2003) Prevalence of 1-aminocyclopropane-1-carboxylate deaminase in *Rhizobium* spp. Antonie van Leeuwenhoek, Int J Gen Mol Microbiol 83:285–291. https://doi.org/10.1023/A:1023360919140

Ma Y, Prasad MNV, Rajkumar M, Freitas H (2011) Plant growth promoting rhizobacteria and endophytes accelerate phytoremediation of metalliferous soils. Biotechnol Adv 29:248–258. https://doi.org/10.1016/j.biotechadv.2010.12.001

Madhaiyan M, Poonguzhali S, Ryu J, Sa T (2006) Regulation of ethylene levels in canola (B*rassica campestris*) by 1-aminocyclopropane-1-carboxylate deaminase-containing *Methylobacterium fujisawaense*. Planta 224:268–278. https://doi.org/10.1007/s00425-005-0211-y

Manchanda G, Garg N (2008) Salinity and its effects on the functional biology of legumes. Acta Physiol Plant 30:595–618

Mapelli F, Marasco R, Rolli E et al (2013) Potential for plant growth promotion of rhizobacteria associated with *Salicornia* growing in Tunisian hypersaline soils. Biomed Res Int 2013:248078. https://doi.org/10.1155/2013/248078

Mayak S, Tirosh T, Glick BR (2004) Plant growth–promoting bacteria confer resistance in tomato plants to salt stress. Plant Physiol Biochem 42:565–572. https://doi.org/10.1016/j.plaphy.2004.05.009

Mrozik A, Nowak A, Piotrowska-Seget Z (2014) Microbial diversity in waters, sediments and microbial mats evaluated using fatty acid-based methods. Int J Environ Sci Technol 11:1487–1496. https://doi.org/10.1007/s13762-013-0449-z

Munns R, Tester M (2008) Mechanisms of salinity tolerance. Annu Rev Plant Biol 59:651–681. https://doi.org/10.1146/annurev.arplant.59.032607.092911

Nadeem SM, Zahir ZA, Naveed M, Arshad M (2007) Preliminary investigations on inducing salt tolerance in maize through inoculation with rhizobacteria containing ACC deaminase activity. Can J Microbiol 53:1141–1149. https://doi.org/10.1139/W07-081

Nadeem SM, Zahir ZA, Naveed M, Nawaz S (2013) Mitigation of salinity-induced negative impact on the growth and yield of wheat by plant growth–promoting rhizobacteria in naturally saline conditions. Ann Microbiol 63:225–232. https://doi.org/10.1007/s13213-012-0465-0

Navarro-Torre S, Barcia-Piedras JM, Mateos-Naranjo E et al (2017) Assessing the role of endophytic bacteria in the halophyte *Arthrocnemum macrostachyum* salt tolerance. Plant Biol 19:249–256. https://doi.org/10.1111/plb.12521

Nelson DR, Mele PM (2007) Subtle changes in rhizosphere microbial community structure in response to increased boron and sodium chloride concentrations. Soil Biol Biochem 39:340–351. https://doi.org/10.1016/j.soilbio.2006.08.004

Nunkaew T, Kantachote D, Nitoda T et al (2014) Characterization of exopolymeric substances from selected *Rhodopseudomonas palustris* strains and their ability to adsorb sodium ions. Carbohydr Polym 115:334–341. https://doi.org/10.1016/j.carbpol.2014.08.099

Oren A (2002) Molecular ecology of extremely halophilic archaea and bacteria. FEMS Microbiol Ecol 39:1–7. https://doi.org/10.1071/FP12355

Ozawa T, Wu J, Fujii S (2007) Effect of inoculation with a strain of *Pseudomonas pseudoalcaligenes* isolated from the endorhizosphere of *Salicornia europea* on salt tolerance of the glasswort. Soil Sci Plant Nutr 53:12–16. https://doi.org/10.1111/j.1747-0765.2007.00098.x

Pandey P, Kang SC, Maheshwari DK (2005) Isolation of endophytic plant growth promoting *Burkholderia* sp. MSSP from root nodules of *Mimosa pudica*. Curr Sci 89:177–180

Parida AK, Das AB (2005) Salt tolerance and salinity effects on plants: a review. Ecotoxicol Environ Saf 60:324–349. https://doi.org/10.1016/j.ecoenv.2004.06.010

Paul D, Lade H (2014) Plant-growth-promoting rhizobacteria to improve crop growth in saline soils: a review. Agron Sustain Dev 34:737–752. https://doi.org/10.1007/s13593-014-0233-6

Piccoli P, Travaglia C, Cohen A et al (2011) An endophytic bacterium isolated from roots of the halophyte *Prosopis strombulifera* produces ABA, IAA, gibberellins A1 and A3 and jasmonic acid in chemically-defined culture medium. Plant Growth Regul 64:207–210. https://doi.org/10.1007/s10725-010-9536-z

Qin Y, Druzhinina IS, Pan X, Yuan Z (2016) Microbially mediated plant salt tolerance and microbiome-based solutions for saline agriculture. Biotechnol Adv 34:1245–1259. https://doi.org/10.1016/j.biotechadv.2016.08.005

Qurashi AW, Sabri AN (2012) Bacterial exopolysaccharide and biofilm formation stimulate chickpea growth and soil aggregation under salt stress. Braz J Microbiol 43:1183–1191. https://doi.org/10.1590/S1517-83822012000300046

Rabie GH, Aboul-Nasr MB, Al-Humiany A (2005) Increased salinity tolerance of cowpea plants by dual inoculation of an arbuscular mycorrhizal fungus *Glomus clarum* and a nitrogen-fixer *Azospirillum brasilense*. Mycobiology 33:51–60. https://doi.org/10.4489/myco.2005.33.1.051

Rangarajan S, Saleena LM, Nair S (2002) Diversity of *Pseudomonas* spp. isolated from rice rhizosphere populations grown along a salinity gradient. Microb Ecol 43:280–289. https://doi.org/10.1007/s00248-002-2004-1

Ravindran KC, Venkatesan K, Balakrishnan V et al (2007) Restoration of saline land by halophytes for Indian soils. Soil Biol Biochem 39:2661–2664. https://doi.org/10.1016/j.soilbio.2007.02.005

Rodriguez R, Redman R (2008) More than 400 million years of evolution and some plants still can't make it on their own: plant stress tolerance via fungal symbiosis. J Exp Bot 59:1109–1114. https://doi.org/10.1093/jxb/erm342

Rueda-Puente EO, Castellanos-Cervantes T, Díaz De León-Álvarez JL, Preciado P, Almaguer-Vargas G (2010) Bacterial community of rhizosphere associated to the annual halophyte *Salicornia bigelovii* (Torr.). Terra Latinoam 28:345–353

Ruppel S, Franken P, Witzel K (2013) Properties of the halophyte microbiome and their implications for plant salt tolerance. Funct Plant Biol 40:940–951. https://doi.org/10.1071/FP12355

Ryu H, Cho Y-G (2015) Plant hormones in salt stress tolerance. J Plant Biol 58:147–155. https://doi.org/10.1007/s12374-015-0103-z

Saleem M, Arshad M, Hussain S, Bhatti AS (2007) Perspective of plant growth promoting rhizobacteria (PGPR) containing ACC deaminase in stress agriculture. J Ind Microbiol Biotechnol 34:635–648. https://doi.org/10.1007/s10295-007-0240-6

Sandhya V, Ali SZ, Grover M et al (2010a) Effect of plant growth promoting *Pseudomonas* spp. on compatible solutes, antioxidant status and plant growth of maize under drought stress. Plant Growth Regul 62:21–30. https://doi.org/10.1007/s10725-010-9479-4

Sandhya V, Ali SZ, Venkateswarlu B et al (2010b) Effect of osmotic stress on plant growth promoting *Pseudomonas* spp. Arch Microbiol 192:867–876. https://doi.org/10.1007/s00203-010-0613-5

Sapre S, Gontia-Mishra I, Tiwari S (2018) *Klebsiella* sp. confers enhanced tolerance to salinity and plant growth promotion in oat seedlings (*Avena sativa*). Microbiol Res 206:25–32. https://doi.org/10.1016/j.micres.2017.09.009

Saravanakumar D, Samiyappan R (2007) ACC deaminase from *Pseudomonas fluorescens* mediated saline resistance in groundnut (*Arachis hypogaea*) plants. J Appl Microbiol 102:1283–1292. https://doi.org/10.1111/j.1365-2672.2006.03179.x

Sarkar A, Ghosh PK, Pramanik K et al (2017) A halotolerant *Enterobacter* sp. displaying ACC deaminase activity promotes rice seedling growth under salt stress. Res Microbiol. https://doi.org/10.1016/j.resmic.2017.08.005

Sgroy V, Cassán F, Masciarelli O et al (2009) Isolation and characterization of endophytic plant growth–promoting (PGPB) or stress homeostasis–regulating (PSHB) bacteria associated to the halophyte *Prosopis strombulifera*. Appl Microbiol Biotechnol 85:371–381. https://doi.org/10.1007/s00253-009-2116-3

Shaharoona B, Arshad M, Zahir ZA, Khalid A (2006) Performance of *Pseudomonas* spp. containing ACC-deaminase for improving growth and yield of maize (*Zea mays* L.) in the presence of nitrogenous fertilizer. Soil Biol Biochem 38:2971–2975. https://doi.org/10.1016/j.soilbio.2006.03.024

Shailendra Singh GG (2015) Plant growth promoting rhizobacteria (PGPR): current and future prospects for development of sustainable agriculture. J Microb Biochem Technol 07:96–102. https://doi.org/10.4172/1948-5948.1000188

Shanker AK, Venkateswarlu B (2011) Abiotic stress in plants—mechanisms and adaptations, 1st edn. InTechOpen, Rijeka

Shi W, Takano T, Liu S (2012) Isolation and characterization of novel bacterial taxa from extreme alkali–saline soil. World J Microbiol Biotechnol 28:2147–2157. https://doi.org/10.1007/s11274-012-1020-7

Shukla PS, Agarwal PK, Jha B (2012) Improved salinity tolerance of *Arachis hypogaea* (L.) by the interaction of halotolerant plant-growth-promoting rhizobacteria. J Plant Growth Regul 31:195–206. https://doi.org/10.1007/s00344-011-9231-y

Siddikee MA, Glick BR, Chauhan PS et al (2011) Enhancement of growth and salt tolerance of red pepper seedlings (*Capsicum annuum* L.) by regulating stress ethylene synthesis with halotolerant bacteria containing 1-aminocyclopropane-1-carboxylic acid deaminase activity. Plant Physiol Biochem 49:427–434. https://doi.org/10.1016/j.plaphy.2011.01.015

Siddikee MA, Chauhan PS, Anandham R et al (2010) Isolation, characterization, and use for plant growth promotion under salt stress, of ACC deaminase-producing halotolerant bacteria derived from coastal soil. J Microbiol Biotechnol 20:1577–1584. https://doi.org/10.4014/jmb.1007.07011

Singh RP, Jha PN (2016) A halotolerant bacterium *Bacillus licheniformis* HSW-16 augments induced systemic tolerance to salt stress in wheat plant (*Triticum aestivum*). Front Plant Sci 7:1–18. https://doi.org/10.3389/fpls.2016.01890

Singh RP, Shelke GM, Kumar A, Jha PN (2015) Biochemistry and genetics of ACC deaminase: a weapon to "stress ethylene" produced in plants. Front Microbiol 6:1–14. https://doi.org/10.3389/fmicb.2015.00937

Szymańska S, Płociniczak T, Piotrowska-Seget Z et al (2016) Metabolic potential and community structure of endophytic and rhizosphere bacteria associated with the roots of the halophyte *Aster tripolium* L. Microbiol Res 182:68–79. https://doi.org/10.1016/j.micres.2015.09.007

Tank N, Saraf M (2010) Salinity-resistant plant growth promoting rhizobacteria ameliorates sodium chloride stress on tomato plants. J Plant Interact 5:51–58. https://doi.org/10.1080/17429140903125848

Tiwari S, Singh P, Tiwari R et al (2011) Salt-tolerant rhizobacteria–mediated induced tolerance in wheat (*Triticum aestivum*) and chemical diversity in rhizosphere enhance plant growth. Biol Fertil Soils 47:907–916. https://doi.org/10.1007/s00374-011-0598-5

Tripathi AK, Verma SC, Ron EZ (2002) Molecular characterization of a salt-tolerant bacterial community in the rice rhizosphere. Res Microbiol 153:579–584. https://doi.org/10.1016/S0923-2508(02)01371-2

Tsukanova KA, Chebotar V, Meyer JJM, Bibikova TN (2017) Effect of plant growth–promoting rhizobacteria on plant hormone homeostasis. S Afr J Bot 113:91–102. https://doi.org/10.1016/j.sajb.2017.07.007

Uchiumi T, Ohwada T, Itakura M et al (2004) Expression islands clustered on the symbiosis island of the *Mesorhizobium loti* genome. J Bacteriol 186:2439–2448. https://doi.org/10.1128/JB.186.8.2439-2448.2004

Ullah S, Asghari B (2015) Isolation of plant-growth-promoting rhizobacteria from rhizospheric soil of halophytes and its impact on maize (*Zea mays* L.) under induced soil salinity. Can J Microbiol 61:1–34. https://doi.org/10.1139/cjm-2014-0668

Upadhyay SK, Singh DP (2015) Effect of salt-tolerant plant growth–promoting rhizobacteria on wheat plants and soil health in a saline environment. Plant Biol 17:288–293. https://doi.org/10.1111/plb.12173

Upadhyay SK, Singh JS, Singh DP (2011) Exopolysaccharide-producing plant growth–promoting rhizobacteria under salinity condition. Pedosphere 21:214–222. https://doi.org/10.1016/S1002-0160(11)60120-3

Upadhyay SK, Singh JS, Saxena AK, Singh DP (2012) Impact of PGPR inoculation on growth and antioxidant status of wheat under saline conditions. Plant Biol 14:605–611. https://doi.org/10.1111/j.1438-8677.2011.00533.x

Wang C, Knill E, Glick BR, Défago G (2000) Effect of transferring 1-aminocyclopropane-1-carboxylic acid (ACC) deaminase genes into *Pseudomonas fluorescens* strain CHA0 and its *gac*A derivative CHA96 on their growth-promoting and disease-suppressive capacities. Can J Microbiol 46:898–907. https://doi.org/10.1139/w00-071

Wenzel WW (2009) Rhizosphere processes and management in plant-assisted bioremediation (phytoremediation) of soils. Plant Soil 321:385–408. https://doi.org/10.1007/s11104-008-9686-1

Xu GY, Rocha PSCF, Wang ML et al (2011) A novel rice calmodulin-like gene, OsMSR2, enhances drought and salt tolerance and increases ABA sensitivity in *Arabidopsis*. Planta 234:47–59. https://doi.org/10.1007/s00425-011-1386-z

Xun F, Xie B, Liu S, Guo C (2015) Effect of plant growth–promoting bacteria (PGPR) and arbuscular mycorrhizal fungi (AMF) inoculation on oats in saline–alkali soil contaminated by petroleum to enhance phytoremediation. Environ Sci Pollut Res 22:598–608. https://doi.org/10.1007/s11356-014-3396-4

Yan J, Smith MD, Glick BR, Liang Y (2014) Effects of ACC deaminase containing rhizobacteria on plant growth and expression of Toc GTPases in tomato (*Solanum lycopersicum*) under salt stress. Botany 92:775–781. https://doi.org/10.1139/cjb-2014-0038

Yang J, Kloepper JW, Ryu CM (2009) Rhizosphere bacteria help plants tolerate abiotic stress. Trends Plant Sci 14:1–4. https://doi.org/10.1016/j.tplants.2008.10.004

Yao L, Wu Z, Zheng Y et al (2010) Growth promotion and protection against salt stress by *Pseudomonas putida* Rs-198 on cotton. Eur J Soil Biol 46:49–54. https://doi.org/10.1016/j.ejsobi.2009.11.002

Yoshimura H, Kotake T, Aohara T et al (2012) The role of extracellular polysaccharides produced by the terrestrial cyanobacterium *Nostoc* sp. strain HK-01 in NaCl tolerance. J Appl Phycol 24:237–243. https://doi.org/10.1007/s10811-011-9672-5

Zahir ZA, Ghani U, Naveed M et al (2009) Comparative effectiveness of *Pseudomonas* and *Serratia* sp. containing ACC-deaminase for improving growth and yield of wheat (*Triticum aestivum* L.) under salt-stressed conditions. Arch Microbiol 191:415–424. https://doi.org/10.1007/s00203-009-0466-y

Zahir ZA, Shah MK, Naveed M, Akhter MJ (2010) Substrate-dependent auxin production by *Rhizobium phaseoli* improves the growth and yield of *Vigna radiata* L. under salt stress conditions. J Microbiol Biotechnol 20:1288–1294. https://doi.org/10.4014/jmb.1002.02010

Zhang Y-Q, Schumann P, Yu L-Y et al (2007) *Zhihengliuella halotolerans* gen. nov., sp. nov., a novel member of the family Micrococcaceae. Int J Syst Evol Microbiol 57:1018–1023. https://doi.org/10.1099/ijs.0.64528-0

Zhou M, Chen W, Chen H, Wei G (2012a) Draft genome sequence of *Mesorhizobium alhagi* CCNWXJ12-2 T, a novel salt-resistant species isolated from the desert of northwestern China. J Bacteriol 194:1261–1262. https://doi.org/10.1128/JB.06635-11

Zhou T, Chen D, Li C et al (2012b) Isolation and characterization of *Pseudomonas brassicacearum* J12 as an antagonist against *Ralstonia solanacearum* and identification of its antimicrobial components. Microbiol Res 167:388–394. https://doi.org/10.1016/j.micres.2012.01.003

Zhou N, Zhao S, Tian CY (2017) Effect of halotolerant rhizobacteria isolated from halophytes on the growth of sugar beet (*Beta vulgaris* L.) under salt stress. FEMS Microbiol Lett 364:1–8. https://doi.org/10.1093/femsle/fnx091

Zhu F, Qu L, Hong X, Sun X (2011) Isolation and characterization of a phosphate-solubilizing halophilic bacterium *Kushneria* sp. YCWA18 from Daqiao saltern on the coast of yellow sea of China. Evid-Based Complement Altern Med 2011:615032. https://doi.org/10.1155/2011/615032

5 Halotolerant Plant Growth-Promoting Fungi and Bacteria as an Alternative Strategy for Improving Nutrient Availability to Salinity-Stressed Crop Plants

Hassan Etesami and Hossein Ali Alikhani

Abstract

Nowadays, it is imperative to address the management of crop production in difficult environmental conditions in order to achieve the maximum potential of plant growth and yield for providing enough food. Increasing demand for plant products has been coupled with decreasing cultivated land due to the limitation of water and soil resources. Salinity is one of the most important abiotic stresses that both limit the production of agricultural products in arid and semiarid areas and decrease arable land across the world. Plant nutrition imbalances due to excessive absorption of sodium (Na^+), and chloride (Cl^-) ions also reduce the absorption of macronutrients such as potassium (K), calcium (Ca), magnesium (Mg), nitrogen (N), and phosphorus (P) and micronutrients such as iron (Fe), zinc (Zn), copper (Cu), manganese (Mn), and boron (B). To satisfy crop nutritional requirements under salinity stress, micro- and macronutrients are usually added to soil as chemical fertilizers; however synthesis of these fertilizers is highly energy-intensive processes and has long-term impacts on the environment in terms of eutrophication, soil fertility depletion, and carbon footprint. Such environmental concerns have led to the search for sustainable way of providing crops with nutrients. In this regard, plant growth-promoting salinity-tolerant microorganisms have been seen as best eco-friendly means for nutrition of salinity-stressed crop plants. These microorganisms increase the nutrient bioavailability through N_2 fixation and mobilization of key nutrients (P, K, and micronutrients) to the crop plants. Use of salinity-tolerant microorganisms is also known as an alternative, innovative, environmental friendly option to reduce the use of costly and

H. Etesami (✉) · H. A. Alikhani
Department of Soil Science, University College of Agriculture and Natural Resources, University of Tehran, Tehran, Iran
e-mail: hassanetesami@ut.ac.ir

M. Kumar et al. (eds.), *Saline Soil-based Agriculture by Halotolerant Microorganisms*, https://doi.org/10.1007/978-981-13-8335-9_5

non-environmental friendly chemical fertilizers and can represent a promising approach to increase nutrient bioavailability and facilitate the development of saline soil-based agriculture. In this chapter, current knowledge on the mechanisms used by plant growth-promoting rhizobacteria (PGPR) and arbuscular mycorrhizal (AM) fungi to influence soil nutrient bioavailability (N, P, K, and micronutrients) under salinity stress is discussed.

Keywords
Arbuscular mycorrhizal fungi · PGPR · Salinity stresses · Agriculture crops · Plant-microbe interactions · Saline soil-based agriculture

5.1 Introduction

Given the growing population of the world with an annual growth rate of 80 million per year and the food shortage crisis of the present century, it is necessary to look for a proper way to meet the food needs of the community in an adequate and desirable quality because one of the indicators of sustainable development is the issue of food security. In order to meet the food needs, given the declining trend in agricultural land that is mostly due to soil salinization, there is no solution other than the strategy of increasing production per unit area (maximum production) (Etesami and Beattie 2018). Due to the ever-increasing demand for food and the lack of salt non-affected soils, we should inevitably use the potential of salinity-stressed soils to produce important crop plants. Salinity of water and soil is among the most important factors limiting the growth of crop plants and production of agricultural products. The total area affected by salt in the world is constantly rising. According to the estimates, around 30–50% of the world's land is under salt stress, and in Iran, about 50% of the cultivated land is facing a salinity problem. In saline conditions, sodium ion (Na^+) concentration is usually higher than the concentration of micro- and macronutrients, and this leads to nutritional imbalances in plants under salt stress in a variety of ways. In general, the presence of excess salts in soil or irrigation water confronts the plant with salt stress. Salinity stress, irrespective of the mechanism of salinization, occurs due to excessive accumulation of cations and anions in the soil solution. These salts consist primarily of ions such as Na^+ and chlorine (Cl^-) and then bicarbonate (HCO_3^-), sulfate (SO_4^{2-}), calcium (Ca^{2+}), magnesium (Mg^{2+}), borate (BO_3^{-3}), and rarely nitrate (NO_3^-) (Sha Valli Khan et al. 2014). The presence of these cations and anions in the soil prevents the absorption of other essential nutrients such as nitrogen (N), phosphorus (P), potassium (K), iron (Fe), cupper (Cu), zinc (Zn), etc. (Giri et al. 2007; Munns and Tester 2008; Tester and Davenport 2003). To increase the availability of nutrients for salinity-stressed plants, large amounts of chemical fertilizers are required on a regular basis. In areas where there is a problem of water and soil salinity, salinity causes a 10–60% reduction in the yield of crops (FAO 2005). In these areas, farmers are trying to compensate for the damage caused by salinity by using more inputs such as fertilizer, seeds, and water, which sometimes accounts for up to 65% of the income of the farmers. But, in addition to increasing production costs, the excessive use of these fertilizers causes a lot of environmental problems including increasing pollution of water and soil resources (greenhouse gaseous emissions) as well as water eutrophication, imbalance of nutrients in the soil, decrease in the yield of agricultural products, and endangering the health of humans and other living organisms (Adesemoye and Kloepper 2009). Moreover, the use of chemical fertilizers in salt-affected soils is not a right choice and, if used, increases the salinity of the soil. Most crop plants are sensitive to salinity and cannot survive under high salinity concentrations (Siddiqui et al. 2009). In other words, such plants need microorganisms to grow in saline soils (Etesami and Beattie 2018). For this reason, in recent years, the use of plant growth-promoting microorganisms (PGPMs) as bio-fertilizers or biocontrol agents in agriculture has been considered by many researchers, and the increased growth of different crops by these microorganisms has been proven in greenhouse and field experiments. Use of PGPMs is known as an alternative, innovative, environmental friendly option to reduce the use of costly and non-environmental friendly chemical fertilizers (Rashid et al. 2016). PGPMs increase plant growth and yield through various mechanisms (Etesami and Maheshwari 2018).

Rhizosphere free living bacteria that directly and indirectly improve plant growth and health are named as plant growth-promoting rhizobacteria (PGPR). In indirect mechanisms, the PGPR indirectly modify the harmful effects of plant pathogens by using specific mechanisms, thereby increasing plant growth. But in direct mechanisms, these bacteria, by fixing nitrogen, producing the metabolites effective at plant growth, such as plant hormones indole-3-acetic acid (IAA), cytokinins (CK), and gibberellins (GAs), increasing the solubilization of insoluble compounds such as P and K through the production of organic and inorganic acids, producing siderophores (increase in the availability of micronutrients such as Fe, Zn, etc.), and reducing the effects of stress ethylene by the production of enzyme ACC (1-aminocyclopropane-1-carboxylate) deaminase, help plants to grow more efficiently (Etesami 2018; Etesami and Maheshwari 2018; Gamalero and Glick 2011; Glick 2014). Various genera of bacteria such as *Pseudomonas*, *Azospirillum*, *Azotobacter*, *Klebsiella*, *Enterobacter*, *Alcaligenes*, *Arthrobacter*, *Burkholderia*, *Bacillus*, *Serratia*, *Erwinia*, *Flavobacterium*, and *Rhizobium* have been reported as PGPR (Egamberdiyeva 2005; Glick 2014).

Another group of microorganisms that are closely associated with plants are fungi. Fungi, like bacteria, can colonize the rhizosphere, the rhizoplane, and in many cases within root tissues (endophyte). Arbuscular mycorrhizal (AM) fungi are among the most abundant fungi in agricultural soils. Mycorrhizal fungi are the fungi that form a symbiotic association with the roots of 83% of dicotyledonous plants and 79% of monocotyledons (Peterson et al. 2004). The role of mycorrhizal fungi in increasing the absorption of nutrients is important only in ecosystems that have nutrient deficiency (Cardoso and Kuyper 2006). These fungi can also confer plant beneficial effects through several mechanisms (Owen et al. 2015; Rawat and Tewari 2011; Waller et al. 2005).

In addition to negatively influencing plant growth, salinity also negatively affects microorganisms including bacteria (Greaves 1922) and AM fungi (e.g., negative effect on colonization capacity, spore germination, growth of fungal hyphae, and the formation and function of mycorrhizal symbiosis) (Giri et al. 2007; McMillen et al. 1998; Porcel et al. 2012; Sheng et al. 2008). Moreover, there are reports that activity of plant growth-promoting bacteria (PGPB), such as their ability to solubilize phosphate (Sharma et al. 2013) or produce enzyme ACC deaminase (Upadhyay et al. 2009), is affected by environmental conditions such as salinity stresses (Sánchez-Porro et al. 2009). Therefore, the microorganisms can increase levels of nutrients for plants under conditions of environmental stresses that are themselves resistant to these stresses (Etesami and Beattie 2018; Etesami and Maheshwari 2018). It has been known that microorganisms isolated from saline environments or halophytic plants have been able to well maintain their plant growth-promoting (PGP) traits under salinity stress (Etesami and Beattie 2018). These microorganisms usually have strategies to maintain their survival in saline environments (Vreeland 1987).

Halophiles are salt-loving organisms that grow well in relatively high salt concentrations in saline and hypersaline environments. Halophiles are present in all major groups of microorganisms (archaea, bacteria, and eukarya), and their salt differentiation is based on their salt tolerance level and salt needs (Ma et al. 2010). Halotolerant or halophilic bacteria form a diverse set of microorganisms that physiologically belong to different genera. Salt-resistant microorganisms usually have strategies to maintain their survival in saline environments (Vreeland 1987). Salinity-resistant bacteria through various mechanisms such as sodium excretion of cells; production of extracellular polysaccharides (EPS); intracellular accumulation of compatible soluble compounds, such as proline, trehalose, glycinebetaine (GB), sucrose, and glycerol; adaptation of proteins and enzymes to high concentrations of soluble ions; and the accumulation of K can withstand high levels of salt (Ruppel et al. 2013). Such halotolerant PGPR have been isolated both from halophytes (Etesami and Beattie 2018) and from salt-affected soils (Hingole and Pathak 2016; Orhan and Gulluce 2015). These bacteria could also increase the resistance of salinity-sensitive crop plants to salinity and improve the growth and yield of these plants under salinity stress (Etesami and Beattie 2018). For example, salinity-resistant bacteria with the ability to N_2 fixation or phosphate solubilization have potential for growth and yield of crop plants in saline soils as bio-fertilizers (Delgado-García et al. 2013). Also, these bacteria can play a special role in counteracting environmental stresses such as salinity and drought by producing bio-surfactants and EPS (Margesin and Schinner 2001).

AM fungi are also widely found in natural ecosystems and in salt-affected soils (Sengupta and Chaudhuri 2002) and are considered as suitable candidates for alleviating salinity stress in salinity-stressed plants (Garg and Chandel 2011). Mycorrhizal fungi such as *Glomus* sp. and *G. mosseae* also occur naturally in saline environments (Evelin et al. 2009). There are reports that salinity has not been able to prevent AMF sporulation and colonization of AM fungi (Aliasgharzadeh et al. 2001). Previous studies showed that AMF-inoculated plants grew better under salinity stress compared to non-inoculated plants (Feng et al. 2002; Giri et al. 2007; Hajiboland et al. 2010; Porcel et al. 2012; Sannazzaro et al. 2007; Zuccarini and Okurowska 2008). These fungi were able to improve plant growth, vigor, and performance and protect plant against the detrimental effects of salinity (Evelin et al. 2009; Garg and Chandel 2011; Giri et al. 2007; Kaya et al. 2009a; Kumar et al. 2010;

Porras-Soriano et al. 2009). In general, AM fungi can increase the resistance of AM plant to salinity by many mechanisms (including a combination of nutritional, biochemical, and physiological effects) (Evelin et al. 2009) such as:

1. Accumulating osmoregulators (e.g., proline, glycine betaine, etc.) (Garg and Manchanda 2009). These fungi can also synthesize trehalose in the extraradical mycelium, which serves as the main storage carbohydrate and also as an abiotic stress protectant (Garg and Chandel 2011; Ocón et al. 2007).
2. Improving host mineral micro- and macronutrient uptake (e.g., P, Zn, and Fe) (Asghari 2012; Miransari and Smith 2008) and ion balance (Giri et al. 2007; Wu et al. 2010).
3. Enhancing photosynthetic rate and water-use efficiency (WUE) (Colla et al. 2008; Wu et al. 2010).
4. Improving physiological processes such as water absorption capacity of salinity-stressed plants (e.g., by enhancing root hydraulic conductivity, WUE, higher K^+/Na^+ ratios, and compartmentalization of sodium within some plant tissues and favorably adjusting the osmotic balance and composition of carbohydrates) (Al-Karaki 2006; Giri et al. 2003; Ruiz-Lozano 2003).
5. Maintaining membrane integrity that facilitates compartmentalization within vacuoles and selective ion uptake (Rinaldelli and Mancuso 1996).
6. Lowering Na concentrations in salinity-stressed plants by dilution effects due to growth enhancement (Al-Karaki 2006).
7. Positively influencing the composition of mineral nutrients of plants grown under salt stress conditions (Al-Karaki and Clark 1998) by increasing and/or selective uptake of nutrients (Porcel et al. 2012). The extensive mycelial system of fungi can absorb nutrients such as N, P, K, Zn, Cu, and Fe. The nutrients absorbed by AM fungi are not only for these fungi, but only some of them are transmitted to host plant (Gosling et al. 2006).

Extensive research has been carried out on occurrence and functional diversity of agriculturally important microbes in stressed environments as reviewed by several authors (Grover et al. 2011; Paul 2013; Paul and Lade 2014; Venkateswarlu et al. 2008; Yang et al. 2009). Beneficial bacteria and fungi could improve plant performance under stress environments and, consequently, enhance yield both directly and indirectly (Banik et al. 2006; Barassi et al. 2006; Chakraborty et al. 2015; Choudhary 2012; Choudhary et al. 2015; Dahmardeh et al. 2009; Damodaran et al. 2014; Davies et al. 2011; del Amor Francisco and Cuadra-Crespo 2012; Dimkpa et al. 2009; Dolkar et al. 2018; Egamberdieva et al. 2008; Etesami and Alikhani 2016b; Fu et al. 2010; Gray and Smith 2005; Hamilton et al. 2016; Kaymak et al. 2009; Khan et al. 2012; Mayak et al. 2004b; Milošević et al. 2012; Nadeem et al. 2007; Paul 2012; Paul and Nair 2008; Ramadoss et al. 2013; Rojas-Tapias et al. 2012; Shrivastava and Kumar 2015; Soleimani et al. 2011; Tiwari et al. 2011; Upadhyay et al. 2009; Yang et al. 2009; Yao et al. 2010; Yildirim and Taylor 2005). The main objective of the present chapter is to highlight and discuss current knowledge on the mechanisms used by PGPR and AM fungi to influence soil nutrient (N, P, K, and Fe) bioavailability under salinity stress.

5.2 Salinity Stress

Among various environmental stresses, soil salinity, either resulting from natural or man-made processes, is one of the main concerns for agriculture because it converts arable land to nonarable land and affects crop production and agricultural sustainability especially in arid and semiarid regions of the world (Flowers 2004; Munns 2005). Soil salinization reduces 1–2% of the area that can be used for agriculture each year. Approximately 7% of the land on the planet and 20% of the total arable land are adversely affected by salinity. Soil salinity affects extensive areas of land in both developed and developing countries. The agricultural intensification, together with unfavorable natural conditions, has accelerated soil salinity in many parts of the world. A soil is considered to suffer from salinity if the electrical conductivity (EC) of its saturation is above 4 dS m^{-1} (America 2001), which is equivalent to approximately 40 mM NaCl (Munns and Tester 2008). Cultivated soils worldwide are becoming more saline from marginal irrigation water, excessive fertilization, and desertification processes (Munns and Tester 2008). The Food and Agriculture Organization (FAO) reported that more than 1 billion ha of land throughout the world were affected by salinity (Ahmad 2013; FAO 2008). Because of global climate change, the area of land affected by salinity is increasing day by day (Shrivastava and Kumar 2015). It has been estimated that more than 50% of the arable land would be salinized by the year 2050 (Jamil et al. 2011).

Salinity limits the productivity of agricultural crops in varying degrees with adverse effects on germination, plant vigor, and crop yield such as cereals (rice, wheat, and maize), forages (clover) or horticultural crops (potatoes and tomatoes) (Munns and Tester 2008). These crops are relatively susceptible to excessive concentration of salts, either dissolved in irrigation water or present in soil (rhizosphere) solution. At plant level, response to salinity stress is very complicated because it reflects the combination of the effects of drought stress and the special effect of ions. In order to understand the nature of salt stress, a definition must first be provided. Different definitions have been made for salinity so that, according to Shannon and Grieve (1998), salinity is the excessive concentration of soluble salts and mineral elements in soil solution that result in salt accumulation in the root area (rhizosphere), which makes water uptake by plant difficult. Soil salinization is a process in which the accumulation of soluble salts in the surface layer of the soil increases, and, as a result, the surface layer loses its potential as a growth medium for plants. In arid and semiarid soils, the problem is excess salts, which are mainly due to the lack of rainfall and high evaporation. Soil salinity is also due to other causes such as excessive use of chemical fertilizers, improper water quality for irrigation, deforestation, or compact cropping. Salinity also results in low osmotic potential of soil solution (osmotic effect) (Munns 2005; Munns and Tester 2008), and as a result, plants suffer from water shortages, resulting in excessive production of reactive oxygen species (ROS) such as peroxide anions, hydroxyl radical, hydrogen peroxide, and single oxygen, each of which can disrupt normal plant metabolism and lead to the destruction of plasma membranes and internal membrane systems (Parida and Das 2005). It has been widely reported that salinity limits plant growth and development, which leads to changes in physiology, morphology, and biochemical properties of plants. In general, salinity-induced stress reduces the yield of many crop plants because salt inhibits plant photosynthesis, protein synthesis, lipid metabolism, plant biomass, green pigments, stomatal conductance, and transpiration rate and lead to change in activity of enzymes such as superoxide dismutase (SOD), peroxidase (POD), ascorbate peroxidase (APX) and catalase (CAT), and osmolyte accumulation (such as proline, glycine betaine, total free amino acids, soluble sugars, and so on). In salinity stress conditions, seed germination stage, as one of the most important plant life stages, which plays an important role in plant establishment, is negatively affected by salinity. Salinity also leads to an imbalance of phytohormones (in particular IAA and ethylene) and thus decreases plant growth (Kaya et al. 2009b). High concentrations of soluble salts and some ions (Na^+ and Cl^-), due to causing specific ion effects (mainly Na^+ and Cl^-) (specific ion effect), which result in nutritional imbalances or ion cytotoxicity (Munns 2005; Munns and Tester 2008), undesirably also affect plant growth by limiting the absorption and transfer of essential nutrients to plant roots (Shrivastava and Kumar 2015), resulting in deficiency symptoms, and reduce the availability of nutrients (N, P, K, Ca, Mg, Fe, Cu, Zn, etc.) in soil (Barea et al. 2005; Moradi et al. 2011). Almost all micro- and macronutrient contents decrease in the roots and shoots of the plant with increasing NaCl concentration in the growth medium (Paul and Lade 2014).

In general, one of the most important consequences of salinity is the nutritional imbalances or ion cytotoxicity. The imbalance of nutrients, in turn, leads to a decline in the growth and development of plants. Plant nutrient management is the most practical and easiest way to overcome stress. The nutritional status of the plant mainly affects their ability to adapt to adverse environmental conditions and in particular non-biological stress factors. Several studies have shown that plants exposed to environmental stresses need the higher amount of nutrients to adapt to stress and reduce the negative effects of these stresses. In addition to having toxic effects on plant cells, soil nutrient imbalances can affect the plant's ability to survive or absorb water (Etesami 2018; Etesami and Maheshwari 2018). However, it has been found that salinity-resistant microorganisms can grow well in salt and high osmotic pressure conditions and partially meet the nutritional needs of the plant under salinity stress (Fig. 5.1 and Table 5.1) (Etesami and Beattie 2018).

5.3 PGPR and AM Fungi-Mediated Increase of N Availability Under Salinity Stress

Nitrogen (N) is the most vital nutrient for plant growth and productivity. Although there is about 78% N_2 in the atmosphere, soil N is mostly in organic forms and unavailable for plants. Deficiency of this nutrient in the soil results in a significant reduction in plant growth and yield. As mentioned above, one of the adverse effects of soil salinity on plants is to alter the uptake of nutrients by plants (decrease in uptake of nutrients with increase in soil salinity) (Cheng-Song et al. 2010; Rawal and Kuligod 2014). Studies indicate that salinity reduces N uptake/accumulation (Feigin

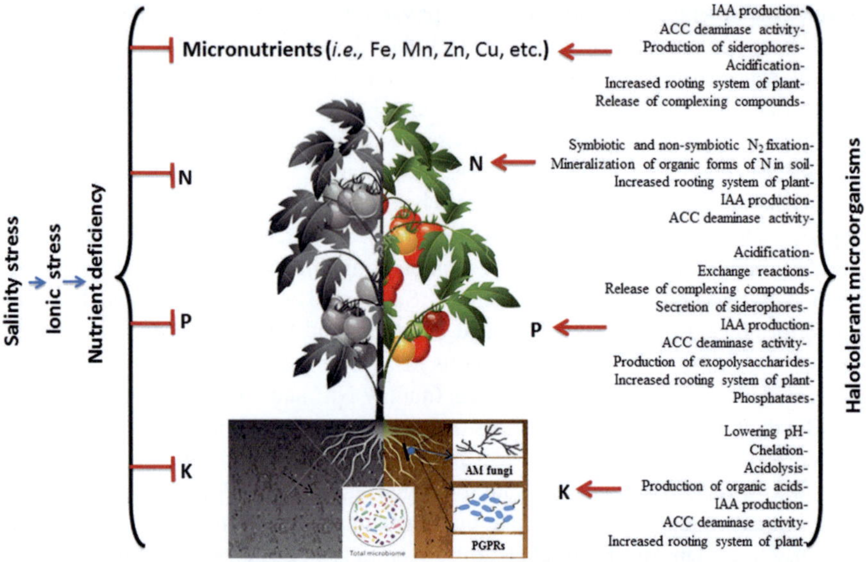

Fig. 5.1 An overview on the role of PGPR and mycorrhizal fungi to promote nutrient acquisition under salinity stress

1985). One of the ways to alleviate, to some extent, the detrimental effects of salinity and help to improve the yield of crops is to use chemical fertilizers. According to Rawal and Kuligod (2014), under salinity conditions, application of higher doses of N resulted in increased yield. In general, in most soils of saline and drought areas, the available N shortage for plants is present, and this shortage cannot be compensated for by using only chemical fertilizers at global level. On the other hand, the use of biological nitrogen fixation (BNF) process is more efficient and more favorable than chemical fertilizers (Shamseldin and Werner 2005). The atmospheric N_2 is converted into plant-utilizable forms by BNF which changes N_2 to ammonia (NH_4^+) by nitrogen-fixing PGPR using a complex enzyme system known as nitrogenase (Kim and Rees 1994). The importance of this process is saving on the use of chemical fertilizers, providing N in line with the plant's requirements and, consequently, increasing plant yield. Also, the excessive use of bio-fertilizers does not create environmental problems unlike chemical fertilizers. On the other hand, following the planting of leguminous plants, the soil is enriched in terms of N absorption, and the remaining effects of leguminous plants will be useful for subsequent planting (Shamseldin and Werner 2005). Today, in planning for sustainable agricultural systems, the use of this type of symbiosis is considered a basic necessity.

It is known that most legume plants are sensitive or relatively resistant to salinity (Dulormne et al. 2010; Garg and Chandel 2011; Jebara et al. 2010; Khadri et al. 2006; López and Lluch 2008). These plants are dependent on nitrogen fixation for their optimal growth (Chalk et al. 2010; Jebara et al. 2010). Salinity, in addition to its

Table 5.1 Potential application of PGPR and mycorrhizal fungi to improve nutrient availability to salinity-stressed plants

Nutrient	Experimental plant(s)	Microorganism(s)	Amelioration effect of microorganism on salinity-stressed plant	References
N	*Cajanus cajan* (L.)	*Glomus mosseae*	The symbiotic association with AM led to significant improvement in plant dry mass and nitrogen-fixing potential of nodules under salt stress. AM plants had lower trehalase activity under saline and non-saline conditions	Garg and Chandel (2011)
P, K, Zn, and Cu	*Acacia nilotica*	*Glomus fasciculatum*	Mycorrhizal plants maintained greater root and shoot biomass at all salinity levels compared to non-mycorrhizal plants. AM-inoculated plants had higher P, Zn, and Cu concentrations than un-inoculated plants	Giri et al. (2007)
			Mycorrhizal plants accumulated a higher concentration of K at all salinity levels. Mycorrhiza-inoculated plants had lower concentration of Na in shoot tissue. Mycorrhizal fungus alleviated deleterious effects of saline soils on plant growth that could be primarily related to improved P nutrition. The improved K^+/Na^+ ratios in root and shoot tissues of mycorrhizal plants may help in protecting disruption of K-mediated enzymatic processes under salt stress conditions	
P, K, and Zn	Soybean (*Glycine max* L.)	*Glomus etunicatum*	Soybean plants inoculated with the AM fungus had significantly higher fresh and dry	Sharifi et al. (2007)

(continued)

Table 5.1 (continued)

Nutrient	Experimental plant(s)	Microorganism(s)	Amelioration effect of microorganism on salinity-stressed plant	References
			weight, root proline, P, K and Zn compared to non-AM plants	
P, N, and Mg	*Sesbania aegyptiaca* and *S. grandiflora*	*Glomus macrocarpum*	Mycorrhizal seedlings had significantly higher root and shoot dry biomass production than non-mycorrhizal seedlings grown in saline soil. The content of chlorophyll was greater in the leaves of mycorrhiza-inoculated seedlings as compared to un-inoculated seedlings. The number of nodules was significantly higher in mycorrhizal plants than non-mycorrhizal plants. Mycorrhizal seedling tissue had significantly increased concentrations of P, N, and Mg but lower Na concentration than non-mycorrhizal seedlings	Giri and Mukerji (2004)
N	*Cajanus cajan* (L.) Mill sp.	*Glomus mosseae*	AM significantly improved nodulation, leghemoglobin content, and nitrogenase activity under salt stress. Activities of superoxide dismutase (SOD), catalase (CAT) and peroxidase (POX), and glutathione reductase (GR) increased markedly in mycorrhizal-stressed plants	Garg and Manchanda (2008)
P	*Phyllanthus amarus*	*Acinetobacter* sp. and *Bacillus* sp.	Both bacteria alone or in combination promoted a higher vigor index, germination (%), plant biomass, P	Joe et al. (2016)

(continued)

Table 5.1 (continued)

Nutrient	Experimental plant(s)	Microorganism(s)	Amelioration effect of microorganism on salinity-stressed plant	References
			content, plant phenolic content, radical scavenging, and antioxidative activity, compared to un-inoculated control	
N, K, and P	Wheat (*Triticum aestivum* L.)	*Bacillus pumilus*, *B. aquimaris*, *B. arsinicus*, *Arthrobacter* sp., *B. cereus*, *P. mendocina*, and *B. subtilis*	Wheat root dry weight and shoot biomass, proline and total soluble sugar accumulation, and percentage of N, K, and P in leaves of wheat increased significantly after inoculation with PGPR, as compared to un-inoculated plants. These PGPR showed maximum reduction of Na content in wheat leaves of about 23% at both 60 and 90 days after sowing and produced the best yield of around 17.8% more than the control	Upadhyay and Singh (2015)
N and P	*Salicornia bigelovii*	*Azospirillum halopraeferens*, *Azospirillum brasilense*, *Vibrio aestuarianus*, *Vibrio proteolyticus*, *Bacillus licheniformis*, and *Phyllobacterium* sp.	At the end of the growing season, the N and protein content of seeds significantly increased. The P content in foliage increased significantly in plants treated with bacterial strains	Bashan et al. (2000)
P	*Avicennia marina*	*Oceanobacillus picturae*	*O. picturae* significantly increased available P, decreased pH, positively enhanced nutrient uptake parameters in roots and shoots, increased stem circumference, number of xylem vessels, mean xylem diameter, and the hydraulically weighted xylem vessel diameter, compared with non-inoculated plants	El-Tarabily and Youssef (2010)

(continued)

Table 5.1 (continued)

Nutrient	Experimental plant(s)	Microorganism(s)	Amelioration effect of microorganism on salinity-stressed plant	References
N, P, K, and Ca	Groundnut (*Arachis hypogaea* L.)	*Brachybacterium saurashtrense* (JG-06), *Brevibacterium casei* (JG-08), and *Haererohalobacter* (JG-11)	Plant length, root length, shoot dry weight, root dry weight, and total biomass were significantly higher in inoculated plants compared to un-inoculated plants. The PGPR-inoculated plants were quite healthy and hydrated, whereas the un-inoculated plant leaves were desiccated in the presence of 100 mM NaCl. The inoculated plants also had a higher K^+/Na^+ ratio and higher Ca^{2+}, P, and N content	Shukla et al. (2012)
N, P, and K	Maize (*Zeya mays* L.)	*Pseudomonas* and *Enterobacter* spp.	Under stress conditions, more N, P, and K uptake and high K^+/Na^+ ratios were recorded in inoculated plants compared with the control	Nadeem et al. (2009)
N	Peanut	*Klebsiella*, *Pseudomonas*, *Agrobacterium*, and *Ochrobactrum*	A significant increase in total nitrogen (N) content (up to 76%) was observed over the non-inoculated control. Under salt stress, inoculated peanut seedlings maintained ion homeostasis, accumulated less reactive oxygen species (ROS), and showed enhanced growth compared to non-inoculated seedlings	Sharma et al. (2016)
K and Ca	Wheat	*Bacillus licheniformis* HSW-16	Inoculation of *B. licheniformis* HSW-16 protected wheat plants from growth inhibition	Singh and Jha (2016)

(continued)

Table 5.1 (continued)

Nutrient	Experimental plant(s)	Microorganism(s)	Amelioration effect of microorganism on salinity-stressed plant	References
			caused by NaCl and increased plant growth (6–38%) in terms of root length, shoot length, fresh weight, and dry weight. Ionic analysis of plant samples showed that the bacterial inoculation decreased the accumulation of Na^+ content (51%) and increased K^+ (68%) and Ca^{2+} content (32%) in plants at different concentration of NaCl	
N, K, P, Ca, Mg, S, Mn, Cu, and Fe	Strawberry	*Bacillus subtilis* EY2, *B. atrophaeus* EY6, *B. sphaericus* EY30, *Staphylococcus kloosii* EY37, and *Kocuria erythromyxa* EY43	PGPR inoculations significantly increased the growth, chlorophyll content, nutrient element content, and yield of strawberry plants. The highest N, K, P, Ca, Mg, S, Mn, Cu, and Fe concentrations were observed in the inoculated plants	Karlidag et al. (2013)
K, Ca, Mg, and P	Pea (*Pisum sativum* L. cv. Alderman)	*Variovorax paradoxus* 5C-2	*V. paradoxus* 5C-2 increased K, Ca, Mg, and P uptake and root to shoot K flow but decreased Na flow and increased Na deposition in roots. Thus, shoot K^+: Na^+ ratio increased following *V. paradoxus* 5C-2 inoculation	Wang et al. (2016)
N, P, and K	Wheat	*Pseudomonas putida*, *Enterobacter cloacae*, *Serratia ficaria*, and *Pseudomonas fluorescens*	Inoculation with the PGPR resulted a significant increase in the growth and yield parameters of wheat. The inoculated plants also improved the nutrient status of the wheat plants. The	Nadeem et al. (2013)

(continued)

Table 5.1 (continued)

Nutrient	Experimental plant(s)	Microorganism(s)	Amelioration effect of microorganism on salinity-stressed plant	References
			inoculated plants had low Na and high N, P, and K contents	
N, P, K, Ca, and Fe	Cotton	*Raoultella planticola*	*Raoultella planticola* reduced the quantities of ethylene and abscisic acid in cotton seedlings and increased indole-3-acetic acid content in cotton seedlings under salinity stress. The accumulation of N, P, K, Ca, and Fe in the cotton plants was increased significantly in *Raoultella planticola* treatment, whereas the uptake of Na in cotton seedlings decreased	Wu et al. (2012)
K, Ca, and Mg	Maize (*Zea mays* L.)	*Azotobacter sp. C5*	Chlorophyll, Mg, K, Ca, Mg, total polyphenol, and proline of the inoculated plant increased	Rojas-Tapias et al. (2012)

References: Bashan et al. (2000), El-Tarabily and Youssef (2010), Garg and Chandel (2011), Garg and Manchanda (2008), Giri et al. (2007), Giri and Mukerji (2004), Joe et al. (2016), Karlidag et al. (2013), Nadeem et al. (2009, 2013), Rojas-Tapias et al. (2012), Sharifi et al. (2007), Sharma et al. (2016), Shukla et al. (2012), Singh and Jha (2016), Upadhyay and Singh (2015), Wang et al. (2016), and Wu et al. (2012)

negative effect on plant growth, has a negative effect on the nitrogen fixation process, such as nodule growth (Abdelmoumen and El Idrissi 2009), symbiotic activity (Dulormne et al. 2010; Jebara et al. 2010), activity of the nitrogenase enzyme (Jebara et al. 2010), nodule respiration (Dulormne et al. 2010; López and Lluch 2008), nodule establishment (Aydi et al. 2008), leghemoglobin, acetylene-reduction activity, and nitrogen content of legume plants (Garg and Chandel 2011), for example, in soybean (*Glycine max*) (van Hoorn et al. 2001), mung bean (*Vigna mungo* L. Hepper) (Mensah and Ihenyen 2009), common bean (*Phaseolus vulgaris*), faba bean (*Vicia faba*) (Rabie and Almadini 2005), and goat's rue (*Galega officinalis*) (Egamberdieva et al. 2013; Paul and Lade 2014).

In addition to the effect of salinity on the nitrogen fixation process, salinity also affects the establishment and activity of the legume-*Rhizobium*, which is susceptible to salinity (Jebara et al. 2010). It is known that environmental stresses including salinity stress have a negative effect on the number of rhizobial isolates, BNF capacity (Duzan et al. 2004), and the ability to produce some bacterial metabolites

(Sánchez-Porro et al. 2009; Upadhyay et al. 2009; Yoon et al. 2001). There are different rhizobial strains in soil that their effect on their host is not the same in terms of BNF and dealing with environmental stresses (Rehman and Nautiyal 2002). In order to obtain the best bio-fertilizers, the exact identification of the isolates of each region and their biological ecology, the isolation of the specific rhizobial isolates of each plant, and the adaptation of these rhizobial isolates to different climatic conditions are necessary. Therefore, selection and breeding of genotypes and halotolerant rhizobial strains that can grow under stress conditions might be an effective tool at resolving the soil reclamation practices.

The combined use of AM fungi with rhizobial bacteria has also been suggested as an effective strategy to increase plant growth and nitrogen fixation under salinity stress (Chalk et al. 2006; Franzini et al. 2010; Garg and Chandel 2011). AM fungi could neutralize harmful effects of salinity on nodulation and nitrogen fixation and thereby alleviate the salinity-induced premature nodule senescence (Garg and Chandel 2011; Garg and Manchanda 2008, 2009; Goss and De Varennes 2002; Patreze and Cordeiro 2004) by different ways such as (i) increasing the fitness of host plants by enhancing shoot and root biomass (Garg and Chandel 2011); (ii) enhancing host plant P nutrition and other nutrients with low mobility, such as Fe, Cu, and Zn (Kaya et al. 2009a; Miransari and Smith 2008); (iii) decreasing uptake of Na by salinity-stressed plant (Al-Karaki 2006; Chakraborty et al. 2008; Giri et al. 2007); and (iv) stimulating the salinity-induced accumulation of trehalose, a carbohydrate in osmoprotection and osmotic adjustment against salinity stress, by inhibiting trehalase activity in AM plants and increasing total soluble sugars (Garg and Chandel 2011).

One of the responses of salinity-stressed plants (osmotic adjustment) against salt stress is various organic compounds, collectively known as compatible solutes or osmolytes (Contreras-Cornejo et al. 2009; Cortina and Culiáñez-Macià 2005). These compounds play an important role as an abiotic stress protectant in a wide variety of organisms (Contreras-Cornejo et al. 2009; Elbein et al. 2003; Fernandez et al. 2010; Garg and Chandel 2011). Trehalose (α-D-glucopyranosyl-1,1-α-D-glucopyranoside) is one of these osmolytes that has been detected in symbiotic nitrogen-fixing bacteria such as *Rhizobium* (the accumulation in bacteroids and in nodules of leguminous plants) (Müller et al. 2001). This compatible solute plays an important role in the maintenance of efficient nitrogen fixation, protection of bacterial nitrogenase activity, and whole plant tolerance under drought and salinity stresses (Farías-Rodríguez et al. 1998; Garg and Chandel 2011; López et al. 2008; Zacarías et al. 2004). The ability of AM fungus has been shown to increase plant nitrogen content under saline conditions. For an example, in a study, Giri and Mukerji (2004) showed that AM-inoculated plants had significantly greater concentration of N and the greater number of nodules than non-mycorrhizal plants under salinity stress. Increased N concentration under saline conditions may help to decrease Na uptake, which may be indirectly related to maintaining the chlorophyll content of the plant (Giri and Mukerji 2004). Improved nodulation and N fixation in mycorrhizal plants may be because of alleviation from P stress and possibly to absorption of some essential micronutrients, which leads both to enhanced growth of plants and has an indirect effect on the N-fixing system (Founoune et al. 2002).

Environmental conditions such as stresses affect the production of extracellular polysaccharides (EPS) of rhizobia strains, so that these strains produce more EPS to withstand adverse environmental conditions (Ashraf et al. 2004; Delavechia et al. 2003). It has been reported that salinity- and drought-tolerant rhizobial strains produce more EPS content and, by increasing the amount of EPS, tolerate unsatisfactory conditions and continue to grow (Ashraf et al. 2004; Delavechia et al. 2003). On the other hand, the survival of rhizobial strains in soil depends on the ability of these strains to bind to the host plant and to create a symbiotic relationship. Among these, EPS are of particular importance in the specialization of rhizobia in identifying these strains by the host plants (Werner 1992). The EPS of the rhizobial strains play an important role in the process of recognizing legume-rhizobium as well as protecting the bacterial cell against environmental stresses (Werner 1992). Therefore, EPS-producing salinity-resistant rhizobial strains can not only tolerate the conditions of salinity stress but can also act more successful in creating a symbiotic relationship with the host plant than salinity-sensitive rhizobial strains. In several experiments conducted in salinity and drought regions, it was observed that the salinity- and drought-resistant rhizobial isolates isolated from salt-affected soils were of significant efficiency (Shamseldin and Werner 2004, 2005).

Since essential plant nutrients are taken up from the soil by roots (Mills et al. 1996), good root growth is considered a prerequisite for enhanced plant development. The morphological modification of roots plays an important role in the mechanisms of stress response (Carmen and Roberto 2011). Microorganisms increase root system of plants by production of phytohormone IAA and ACC deaminase. In addition to the production of IAA, gibberellins and other growth regulators produced by PGPR can support the increase of root length, root surface area, and number of root tips, leading to enhanced uptake of nutrients, thereby improving plant health under stress conditions (Egamberdieva and Kucharova 2009; Etesami and Maheshwari 2018). Rhizobacteria-mediated root proliferation has been well proved, and it also works in stressed soils (Diby et al. 2005a). Inoculation of various plant species with PGPR has also reported to lead to enhanced formation of lateral roots and root hairs that can result in enhanced tolerance to abiotic stress (Etesami and Maheshwari 2018; Paul and Lade 2014). Promotion of root growth results in a larger root surface and can, therefore, have positive effects on water acquisition and nutrient (i.e., N) uptake (Diby et al. 2005b; Paul and Sarma 2006) that is expected to alleviate the stress effects in the plant (Chakraborty et al. 2006; Hamdia et al. 2004; Long et al. 2008; Paul and Sarma 2006). Microbial IAA is involved in many processes of nodule formation by rhizobia in legume plants, such as founder cell specification, nodule initiation and differentiation (IAA accumulation), nodule numbers, vascular bundle formation, and cell division and differentiation (Etesami et al. 2015b; Etesami and Beattie 2017). These three later events are more necessary for nodule formation (Glick 2012; Theunis 2005). In addition, due to having the function in increased root system, IAA-producing bacteria can provide more number of active sites and access to colonization for N_2-fixing bacteria or PGPR.

Previous studies indicate that salinity can either decrease (300 mM NaCl) (Dunlap and Binzel 1996) or increase (100 mM NaCl) (Albacete et al. 2008) root IAA accumulation in plant, which might affect cell elongation and root growth. It has been known that the bacteria isolated from saline environments have been able to produce IAA in the presence of salinity (Etesami and Beattie 2018). In a study, Sadeghi et al. (2012) demonstrated that a *Streptomyces* isolate increased plant growth in wheat and produced IAA in the presence of salt. The ability to modify plant stress levels by providing IAA, a molecule involved in the development of lateral roots, has also been reported for halotolerant bacteria isolated from saline habitats and halophytes (Bian et al. 2011; Marasco et al. 2012; Naz et al. 2009; Piccoli et al. 2011; Sgroy et al. 2009; Siddikee et al. 2010; Tiwari et al. 2011). *Arthrobacter* sp., *Nitrinicolalacis aponensis*, *Brachybacterium* sp., *B. saurashtrense*, *Brevibacterium casei*, *Br. halotolerans*, *Haererohalobacter* sp., *Klebsiella* sp., *Pseudomonas* sp., *P. stutzeri*, *P. pseudoalcaligenes*, *P. putida*, *P. mendocina*, *Agrobacterium tumefaciens*, *Ochrobactrum anthropi*, *Bacillus* sp., *B. cereus*, *B. pumilus*, *B. simplex*, *B. mycoides*, *B. aquimaris*, *B. licheniformis*, *B. mycoides*, *B. subtilis*, *Sporosarcina aquimarina*, *Serratia marcescens*, *S. plymuthica*, *S. plymuthica*, *Thalassospira permensis*, *Acinetobacter*, *Microbacterium* sp., *M. oxydans*, *Streptomyces* sp., *S. avidinii*, *S. europaeiscabiei*, *S. exfoliatus*, *S. umbrinus*, *S. griseoplanus*, *Rhodococcus* sp., *R. equi*, *R. erythropolis*, *Micrococcus luteus*, *Marinococcus halophilus*, *Nocardia* sp., *Halomonas* sp., *H. elongata*, *H. eurihalina*, *H. sinaiensis*, *H. halmophila*, *H. ilicicola*, *H. indalina*, *H. variabilis*, *H. xinjiangensis*, *H. taeheungii*, *Halobacillus trueperi*, *Nesterenkonia halobia*, *Oceanobacillus picturae*, *Kushneria*, *Virgibacillus olivae*, *Lysinibacillus fusiformis*, *Achromobacter xylosoxidans*, *Mesorhizobium* sp., *Zhihengliuella* sp., *Vibrio alginolyticus*, and *Cronobacter sakazakii* are some examples of IAA-producing salt-tolerant PGPR isolated from halophytes (Gontia et al. 2011; Mapelli et al. 2013; Sgroy et al. 2009; Sharma et al. 2016; Shukla et al. 2012; Tiwari et al. 2011). According to the above studies, it can be concluded that salinity-resistant rhizobacteria or rhizobial bacteria with the ability to produce IAA may help to increase the resistance of plants to salinity stress and increase the amount of N in the plants.

One of the responses of plants under stress conditions such as salinity stress is the increase in the level of intracellular ethylene inside the plant (stress ethylene). Stress ethylene reduces the vegetative stage of plant growth and eventually decreases the plant yield (Etesami and Maheshwari 2018; Glick 2005). Siddikee et al. (2011) also stated that the ethylene produced in response to salinity stress reduced root growth and thus reduced water and nutrient uptake and finally decreased plant growth. Previous researchers found that formation and function of nitrogen-fixing nodules on legume roots were severely inhibited by addition of exogenous ethylene (Peters and Crist-Estes 1989). Ethylene inhibits the elongation of infection threads and, consequently, the formation of nodules in most legumes (Etesami et al. 2015b; Sugawara et al. 2006). In addition, production of ethylene, a signal molecule for induced systemic resistance in plants, decreases endophytic colonization in plants (Iniguez et al. 2005). The overproduction of ethylene can also cause the inhibition of root elongation, lateral root growth, and root hair formation (Belimov et al. 2009; Mayak et al. 2004b; Saleem et al. 2007), which subsequently results in decreasing the nodule number of on root. Some soil bacteria produce ACC deaminase enzyme. This enzyme converts ACC (a substrate for ethylene production) to α-ketobutyrate and ammonium. By modifying the ACC production in plants, ACC deaminase-producing bacteria can prevent the excess ethylene production in the plant (Glick 2014). The ACC deaminase trait has been extensively studied in numerous soil microbial species, that is, bacteria, fungi, and endophytes; however, this trait is most common among PGPR (Etesami and Maheshwari 2018; Glick 2005) such as the genera *Achromobacter*, *Acidovorax*, *Alcaligenes*, *Enterobacter*, *Klebsiella*, *Methylobacterium*, *Pseudomonas*, *Rhizobium*, and *Variovorax* (Esquivel-Cote et al. 2010). Ahmed et al. (2004) found that ACC deaminase-producing bacteria increased grain yield, root weight, root length, and more nitrogen uptake in straw and wheat grain by decreasing stress ethylene level compared to control.

Salinity has been shown to cause PGPR to lose their ACC deaminase enzyme production (Upadhyay et al. 2009). However, it has been found that ACC deaminase-producing salt-tolerant PGPR can survive well in a saline environment and that their beneficial properties help plants to overcome stress effects by reducing ethylene levels (Mayak et al. 2004a). The production of enzyme ACC deaminase has also been well reported in halotolerant bacteria isolated from halophytic plants and from saline environments (Jha et al. 2012; Siddikee et al. 2010; Zhou et al. 2017). The roots of halophytes may be associated with promising ACC deaminase-producing bacterial candidates for promoting growth and salt tolerance in crops (Etesami and Beattie 2018). For example, novel diazotrophic halotolerant bacteria from roots of *Salicornia brachiata* featured ACC deaminase activity and these isolates included *Brachybacterium saurashtrense*, *Brevibacterium casei*, *Cronobacter sakazakii*, *Haererehalobacter*, *Halomonas*, *Mesorhizobium*, *Pseudomonas*, *Rhizobium radiobacter*, *Vibrio*, and *Zhihengliuella* (Jha et al. 2012). Examples of halotolerant ACC deaminase-producing PGPR include *Micrococcus yunnanensis*, *Planococcus rifietoensis*, *Variovorax paradoxus*, *Brachybacterium saurashtrense*, *Klebsiella* sp., *Pseudomonas* sp., *P. stutzeri*, *P. putida*, *Agrobacterium tumefaciens*, *Ochrobactrum anthropi*, *Serratia marcescens*, *S. plymuthica*, *Thalassospira permensis*, *Microbacterium* sp., *M. oxydans*, *Bacillus* sp., *B. cereus*, *B. licheniformis*, *B. pumilus*, *B. subtilis*, *B. mycoides*, *S. plymuthica*, *M. oxydans*, *Microbacterium* sp., *Streptomyces* sp., *Rhodococcus* sp., *Rh. erythropolis*, *Nocardia* sp. *Halomonas taeheungii*, *H. xinjiangensis*, *Brevibacterium halotolerans*, *Achromobacter xylosoxidans*, *Zhihengliuella* sp., *Mesorhizobium* sp., *Brachybacterium* sp., *Vibrio alginolyticus*, *Brevibacterium casei*, *Cronobacter sakazakii*, *P. pseudoalcaligenes*, *Haererehalobacter* sp., *Pantoea agglomerans*, *P. oryzihabitans*, and *Halomonas* sp. (Gontia et al. 2011; Jha et al. 2012; Mapelli et al. 2013; Sgroy et al. 2009; Sharma et al. 2016; Szymańska et al. 2016; Teng et al. 2010; Zhou et al. 2017), which were isolated from species of halophytes. Generally, it seems that the production of IAA and ACC deaminase by salinity-tolerant PGPR might be an important tool in salt tolerance of plants/legumes in salt-affected soils (Etesami and Beattie 2018; Etesami and Maheshwari 2018).

5.4 PGPR and AM Fungi-Mediated Increase of P Availability Under Salinity Stress

Phosphorus (P) is an important and essential macronutrient for plants and microorganisms that its shortage restricts plant growth and development and, unlike N, does not have a large atmospheric source (Katznelson et al. 1962). Stem development, root and shoot firmness, fixation of nitrogen in legumes, macromolecular biosynthesis, improvement in crop quality, production and transfer of energy, photosynthesis, and resistance to plant diseases are related to P nutrition (Roychoudhury and Kaushik 1989). The deficiency of P supply and availability also remains a severe limitation on nitrogen fixation and symbiotic interactions (Pereira and Bliss 1989). In contrast to N, phosphate compounds are relatively insoluble and therefore are not easily leached from the soil profile. Usually, low amounts of P in soils and its tendency to react with soil constituents and the production of relatively insoluble compounds and therefore nonabsorbable compounds for plants make P very important for soil fertility. It is worth mentioning that the total amount of P in soil is much more than available P, but the importance of the available P in plant growth is higher. Phosphorus is predominantly found in mineral soil parts, either as adsorbed on mineral surfaces of soil or as relatively soluble sediments, or in organic forms that are either as adsorbed on mineral surfaces of soil, or as a part of microbial biomass or with organic matters. Phosphorus is extracted from phosphate rock (over 80% of phosphate rock reserves are utilized for fertilizer production) (Owen et al. 2015), which is a nonrenewable resource. This is very important for a number of countries without phosphate rock (Van Vuuren et al. 2010). The extraction of P minerals and spread of P fertilizers in agricultural land are not sustainable due to the limited resources of P, and this will make production of this fertilizer in the future difficult. In addition, in practice, P use efficiency (recovery of fertilizer P) is between 10% and 25%. That is, about 75% of this fertilizer in soil is converted into insoluble forms by reacting with metal cations (formation of complexes with Fe^{3+}, Al^{3+} and Ca^{2+}) (Gyaneshwar et al. 2002; Stevenson and Cole 1999).

Plants absorb P mainly as primary and secondary ions of orthophosphates ($H_2PO_4^-$ and HPO_4^{2-}), which are present in the soil solution. The concentration of these ions in the soil solution at any time is negligible (Marschner 1995) and often is less than 0.1 mg L^{-1}. The P in the soil solution should be renewed continuously; otherwise the plant will not have sufficient P to complete its growth. The concentration of various P ions in soil solution is closely related to the pH of environment. Acidic environments are suitable for $H_2PO_4^-$ ions and pH-containing environments higher than 7 for HPO_4^{2-} ions. According to a general experimental law, the maximum usable P for most agricultural plants is when the soil pH is in the range of 5.5–7. The P in calcareous soils is converted into insoluble calcium and magnesium compounds and in acid soils into iron and aluminum phosphates, thereby reducing its access to plants. In general, different forms of P in soil are controlled by the soil's natural properties, including pH, the soluble and exchangeable cations (Ca^{2+}, Mg^{2+}, Fe^{2+}, etc.), the type of soil particles, and their surface (Penfold 2001).

Salinity also reduces phosphorus (P) uptake/accumulation in crops grown in soils primarily by reducing P availability (Etesami and Maheshwari 2018; Paul and Lade 2014) because phosphate ions precipitate with Ca^{+2} ions (Navarro et al. 2001; Parida and Das 2005; Rogers et al. 2003). In addition, the reduction in P availability in saline soils was suggested to be a result of ionic strength effects that reduce the activity of phosphate and the tight control of P concentrations by sorption processes and by low solubility of Ca-P minerals. In saline soils, due to the presence of cations (e.g., Ca^{2+} and Mg^{2+}) in high concentrations, it is possible to form insoluble cation-phosphorus compounds in salt-affected soils (Etesami and Maheshwari 2018). Previous studies also showed that salinity could reduce the P uptake by plant (Barea et al. 2005; Patel et al. 2010). For example, Patel et al. (2010), in a study, showed that salinity reduces the P uptake by plant (*Acacia auriculiformis* A. Cunn. Ex Benth). In addition to forming insoluble cation-phosphorus compounds, decrease in the P uptake by plant under salinity stress can also be associated with the poor root development (Rawal and Kuligod 2014).

One of the ways to meet the P requirement of plants in saline or non-saline soils is to use chemical fertilizers; however, due to their limited resources, the increase in their cost, the environmental hazards resulting from their application, and the fixation of a large part of the P fertilizers consumed as insoluble forms for plant, attention to use of phosphate bio-fertilizers has been increasing in the past decades. It is well found that soil microorganisms (bacteria, fungi, actinomycetes, and even algae) interfere with some of the solubilizing processes of phosphate, which affect the transformation and maintenance of the status of P and ultimately its bioavailability for the plant (Richardson 2001; Sharma et al. 2013).

It is well known that various PGPR can increase the availability of nutrients to plant. For example, phosphate-solubilizing microorganisms (PSMs) provide P available to plants by different mechanisms (Sharma et al. 2013). There is evidence that soil phosphate-solubilizing bacteria (PSB) can solubilize and convert insoluble phosphorus into available forms to plant. Also, microbial biomass with absorption of soluble phosphorus prevents precipitation or fixation of P in the soil (Khan et al. 2007).

PSMs form a significant part of the population of soil microorganisms (Kucey et al. 1989). Bacteria and fungi account for the highest percentage of PSMs in soil; therefore they are divided into two groups of phosphate-solubilizing bacteria (PSB) and phosphate-solubilizing fungi (PSF) (Whitelaw et al. 1997). PSB can solubilize insoluble P compounds and release P from these compounds (Sharma et al. 2013). The most important types of PSB can be referred to *Enterobacter agglomerans*, *Bacillus circulans*, *B. subtilis*, *B. megaterium* var *phosphaticum*, *Pseudomonas putida*, *P. fluorescens*, *P. striata*, *Pantoea*, *Paenibacillus polymyxa*, *Rhizobium leguminosarum* bv. *trifolii*, *Xanthobacter agilis*, *Klebsiella*, and *Agrobacterium radiobacter*, and fungi *Aspergillus* and *Penicillium* are of the most important types of PSF (Sharma et al. 2013). It has been also reported that PSF produce more acids compared to PSB (Venkateswarlu et al. 1984) and are able to traverse long distances more easily than PSB and hence may be more important to P-solubilization in soils (Kucey 1983). Overall, due to having a higher population density, bacteria may be very effective at solubilizing phosphates compared to fungi.

The solubilization and mineralization of P in the rhizosphere is the most important mechanism of PGPR including PSB that provide plants with the nutrient. PSB play an important role in three main components of soil P cycles such as dissolution-precipitation, absorption and desorption (interactions between mineral surfaces and P), and mineralization-immobilization (transformation of organic P to inorganic P by biological transformations) (Gyaneshwar et al. 2002; Jones and Oburger 2011; Owen et al. 2015). PSB and their interactions with soil play a critical role in mediating the distribution of P between the soluble P reservoir in the soil solution and the total P of the soil through solubilization and mineralization reactions, the immobilization of P into the microbial biomass, or the formation of relatively available organic P and mineral forms. In general, PSB convert insoluble phosphate compounds into soluble forms by producing and releasing metabolites such as organic acids that, through their hydroxyl and carboxyl groups, chelate the cations (mainly calcium) bound to phosphate (Miller et al. 2010; Sagoe et al. 1998).

Some of the potential known mechanisms through which PSB could mobilize the soil P unavailable for plants include:

1. Acidification of rhizosphere medium by production of mineral acids and H^+ extrusion or $NH_4{}^+$ absorption (release of H^+ ions from the roots during the assimilation of $NH^+{}_4$) (Illmer and Schinner 1992).
2. Exchange reactions. In this process, the low molecular weight organic anions (e.g., succinic, citric, gluconic, α-ketogluconic, and oxalic acids) produced by PSB are exchanged for P on adsorption sites of soil (Chen et al. 2006; Jones and Oburger 2011; Zhang et al. 2014). The organic acids/organic anions (e.g., lactic acid, tartaric acid, aspartic acid, gluconic acid, oxalic acid, and citric acid) produced by PSB can also lower pH, compete with P for adsorption sites on the soil, and form soluble complexes with metal ions associated with insoluble P (Ca, Al, and Fe) (Sharma et al. 2013).
3. Releasing CO_2 during respiration. The formed carbonic acid caused by being dissolved CO_2 in water (present in the soil pores) can solubilize P by diminishing the mycorrhizosphere pH (Marschner 1995).
4. Stimulating the indirect dissolution of calcium phosphate (Ca-P) by removing and assimilating phosphate from the soil (sink theory) to re-establish the P equilibrium (Halvorson et al. 1990).
5. The release of chelating compounds (e.g., siderophores, enzymes, phenols, amino acids, sugars, and organic acid anions). These compounds reduce metals with variable oxidation states, which bound to phosphate, to a lower oxidation state, resulting in more soluble phosphate (Kim et al. 1997).
6. The secretion of extracellular alkaline and acid phosphatases or phytases (enzymatic hydrolysis), which convert organic phosphate into an inorganic form (Etesami and Maheshwari 2018; Franco-Correa et al. 2010; Gyaneshwar et al. 2002; Owen et al. 2015; Sharma et al. 2013).

It has been reported that the main mechanism for mineralization of organic P is the production of acid phosphatases (Khan et al. 2009; Sharma et al. 2013). Species of genera *Bacillus*, *Burkholderia*, *Enterobacter*, *Pseudomonas*, *Rhizobium*, *Serratia*, and *Staphylococcus* are the most important bacteria involved in P mineralization (Shedova et al. 2008).

Previous studies have shown that PSB have also other PGP traits (Etesami and Maheshwari 2018; Sharma et al. 2013). Based on scientific findings, plant growth-promoting bacteria (PGPB) probably have more than one mechanism for plant growth (Etesami and Maheshwari 2018). It has been shown that the increase in plant growth and yield was not associated with the application of PSB by increasing P uptake in plants (Poonguzhali et al. 2008). It is believed that PSMs, in fact, have potential for controlling plant pathogens as well as plant growth promotion through the production of siderophores, ACC deaminase, and IAA (Han and Lee 2006; Sharma et al. 2013). As an example, Gulati et al. (2010) reported that the phosphate-solubilizing strain BIHB723 belonging to *Acinetobacter* was capable of producing other PGP metabolites such as IAA, ACC deaminase, siderophore, and ammonia. In addition to solubilization of P in growth medium (rhizosphere), P absorption by the plant depends on root growth. The root system is especially important because it helps the plant to absorb water and nutrients (Etesami and Alikhani 2016a; Fageria and Moreira 2011). The absorption of P by plant is proportional to root density, so increasing rooting system increases the plant's ability to absorb P from soil (Grant et al. 2001). For example, Iqbal Hussain et al. (2013) stated that increased P uptake by plant could be due to increased root growth or root hair lengthening by specific microorganisms. It has been reported that low amounts of IAA result in prolonged initial roots, while high levels of IAA lead to an increase in the formation of lateral roots (Xie et al. 1996). Also, keeping the concentration of ethylene in low amounts can increase root growth. PSB with the ability to produce ACC deaminase can maintain the concentration of ethylene in low amounts through the hydrolysis of ethylene to ammonia and α-ketobutyrate (Pereira and Castro 2014). It seems that the combination of these mechanisms in PSB has the most effect on increasing plant growth indices under nutritional deficiency stress such as P (Etesami and Maheshwari 2018).

It has been found that the establishment and performance of PSMs are affected severely under environmental stresses including salinity (Johri et al. 1999). In other words, the solubilization ability of phosphate by PSMs decreases in the presence of salinity (decrease in the amount of inorganic P released by PSMs with increase in NaCl concentration) (Cherif-Silini et al. 2013; Sánchez-Porro et al. 2009; Srinivasan et al. 2012). However, it has been reported that the microorganisms (halotolerant microorganisms) isolated from saline environments or halophytes were able to maintain their ability in solubilizing phosphate in the presence of salinity (better performance of PSMs of salt-affected soils in releasing inorganic P from TCP (tricalcium phosphate) under salt stress conditions) (Etesami and Beattie 2018). It has been reported that PSB were also isolated from the stressed environments (Chen 2006; Etesami and Beattie 2018; Etesami and Maheshwari 2018). For example, the *Kushneria sinocarni*, a salt-loving strain, were isolated from salt lake sediments on the east coast of China and showed phosphate solubilization (Zhu et al. 2011). In another study, Srinivasan et al. (2012) showed that the amount of inorganic P released from TCP by the PSB strains (e.g., *Aerococcus* sp. strain PSBCRG1–1, *Pseudomonas aeruginosa* strain PSBI3–1, *Alteromonas* sp. PSBCRG, etc.) was found to increase with an increase in salt concentration up to 800 mM NaCl, whereas the reference strain (*P. striata*) showed a significant decrease in P-solubilization at NaCl concentrations. Phosphate-solubilizing salinity-resistant bacteria (e.g., *Klebsiella* sp., *Pseudomonas* sp., *P. stutzeri*, *Agrobacterium tumefaciens*, *Ochrobactrum anthropi*, *Halomonas elongata*, *H. eurihalina*, *H. sinaiensis*, *H. halmophila*, *H. ilicicola*, *H. indalina*, *H. variabilis*, *H. xinjiangensis*, *H. taeheungii*, *Zhihengliuella* sp., *Vibrio alginolyticus*, *Brevibacterium casei*, *Marinococcus halophilus*, *Halobacillus trueperi*, *Nesterenkonia halobia*, *Oceanobacillus picturae*, *Virgibacillus olivae*, *Chromohalobacter marismortui*, *Haererehalobacter* sp., *Cronobacter sakazakii*, and *Chromohalobacter salexigens* (Jha et al. 2012; Mapelli et al. 2013; Sharma et al. 2016) were also isolated from halophytes such as *Arthrocnemum indicum*, *Salicornia strobilacea*, and *Salicornia brachiate*. These bacteria were also able to supply the plant P and thus the growth of the plant under saline conditions. For example, following the inoculation of *Solanum lycopersicum* plant with phosphate-solubilizing *Achromobacter piechaudii*, the amount of plant P increased under salinity stress (Mayak et al. 2004a). In another study, the content of P in leaves of salinity-stressed wheat inoculated with phosphate-solubilizing *B. aquimaris* was significantly enhanced in a field experiment (Upadhyay and Singh 2015). A screen of the mangrove *Avicennia marina* rhizosphere identified 129 bacterial strains with the ability to solubilize rock phosphate up to 97% (El-Tarabily and Youssef 2010). Phosphate-solubilizing bacteria isolated from halophytes (e.g., *Arthrobacter*, *Bacillus*, *Azospirillum*, *Vibrio*, *Phyllobacterium*, and *Oceanobacillus picturae*) were able to solubilize $Ca_3(PO4)_2$, $AlPO_4$, and $FePO_4$ (Banerjee et al. 2010; Bashan et al. 2000; El-Tarabily and Youssef 2010; Yasmin and Bano 2011) and increase P content in both host plants and nonhost plants (halophytes and glycophytes) under salinity stress. In addition, when the oilseed halophyte *Salicornia bigelovii* was inoculated with phosphate-solubilizing halotolerant bacteria (i.e., *Vibrio*, *Azospirillum*, *Bacillus*, and *Phyllobacterium*), content of P in foliage also increased over non-inoculated plants (Bashan et al. 2000).

As mentioned above, in addition to bacteria, ability of solubilizing insoluble phosphate compounds in fungi has been well proven, constituting about 0.1–0.5% of total fungal populations (Kucey 1983). It is well known that the mycorrhizal association increases host nutrient acquisition, particularly P (Smith and Read 2010). In one study, the effects of AM fungi *Glomus intraradices* and *Gigaspora margarita* on *Plantago lanceolata* in poor soils showed that these fungi could increase the absorption of nutrients such as N, P, and K, but the effectiveness of each fungus was different (Veresoglou et al. 2011). The effect of mycorrhizal fungi on increasing P absorption by plant is one of the most important effects of these fungi. Mycorrhizal fungi extend beyond the root-depleted zone and play an important role in exploring and extracting P outside the root-depleted zone. In addition, other auxiliary mechanisms have been proposed for P uptake by mycorrhizal fungi. Mycorrhizal hyphae have a lower Km than the roots of plants. As a result, the minimum concentration or threshold concentration for fungal hyphae fungi is lower than roots of plant. Mycorrhizal hyphae have the ability to access microsites. These fungi can access the inorganic and organic P forms that are inaccessible to non-mycorrhizal plants (Zwetsloot et al. 2016). By secreting organic acids, which mobilize inorganic P from rock phosphate and phosphatase enzymes, fungi such as *Aspergillus* and *Penicillium* spp. and AM fungi belonging to the phylum Glomeromycota *Rhizophagus intraradices* and *Funneliformis mosseae* (Krüger et al. 2012) were able to increase the P availability and consequently enhance P uptake in diverse crop plants (Barrow and Osuna 2002; Cozzolino et al. 2013; Koide and Kabir 2000; Owen et al. 2015; Richardson and Simpson 2011; Smith et al. 2011; Vassilev et al. 1996; Williams et al. 2013). These fungi can also increase immobile nutrient absorption potential such as PO_4^{3-}, as a slowly diffusing ion, by forming an extensive mycelial network within the soil (Owen et al. 2015) that extends the plant's effective absorption surfaces beyond the nutrient-depleted zone that develops around the root caused by direct root uptake processes (Porcel et al. 2012). Fungal hyphae, which are functionally analogous to fine root hairs, also support a microbial community (Andrade et al. 1998) with a high potential in mineralizing and solubilizing recalcitrant nutrient complexes which are then transported to the plant via the mycelial network (Owen et al. 2015; Toljander et al. 2007).

Phosphate-solubilizing fungi have also been isolated from saline environments. For an example, PSF such as *Aspergillus* sp. PSFNRH-2, isolated from salt-affected soils, showed significant increase in P-solubilization with increase in NaCl concentration up to 1.0 M than the reference strain irrespective of NaCl concentrations (Srinivasan et al. 2012). According to previous studies (Al-Karaki 2000; Giri et al. 2003; Hirrel and Gerdemann 1980; Ojala et al. 1983; Pond et al. 1984; Poss et al. 1985), the improved growth of mycorrhizal plants was primarily related to mycorrhiza-mediated enhancement of host plant P nutrition (as the most important strategy of salinity stress tolerance in AM-colonized plants) in saline conditions. It has been reported that AM fungus moderates the negative effects of Na^+ and Cl^- on salinity-stressed plant by increasing the availability of P of the plant (Pfetffer and Bloss 1988; Poss et al. 1985) and maintaining vacuolar membrane integrity, which prevents these ions from interfering in growth metabolic pathways (Cantrell and Linderman 2001) and facilitates compartmentalization within vacuoles and selective ion intake (Rinaldelli and Mancuso 1996). The above findings can be interpreted as evidence of the salinity-resistant phosphate-solubilizing fungi and PGPR's ability to solubilize insoluble P compounds under salinity stress.

5.5 PGPR and AM Fungi-Mediated Increase of K Availability Under Salinity Stress

Potassium (K) is one of the essential macronutrients of plants and is the most abundant nutrient in plants after N. This nutrient is considered as a key parameter of soil fertility and plant growth as without adequate K, the plants will have poorly developed roots, grow slowly, produce small seeds, and have lower yields (Parmar and Sindhu 2013). Potassium plays a very important role in photosynthesis, cell division and growth, protein production, quantity and quality of products, and increased plant resistance to diseases and pests (Saber and Zanati 1984). Potassium is also required to activate over 80 different enzymes responsible for such plant and animal processes as energy metabolism, starch synthesis, nitrate reduction, photosynthesis, and sugar degradation. Without adequate amounts of K, water is lost from the cells, and the plant cells weaken and start to wilt. Potassium-deficient plants will have poorly developed cell walls with lower levels of stored protein and starch, and they become an easy meal for sucking insects and an easy target for invasion by fungal spores (Meena et al. 2014, 2015).

More than 90% of K in the soil exists in the form of insoluble rocks and silicate minerals (Etesami et al. 2017; Parmar and Sindhu 2013). Most of the K is unavailable for plant uptake. Muscovite, orthoclase, biotite, feldspar, illite, and mica are of the most important minerals including K, and the major amounts of K are present in the soil as a fixed form which is not directly taken up by plant. Plants obtain K from K originated from the addition of chemical fertilizers and/or K present in soil (Sparks and Huang 1985). Most soils have relatively high amounts of total K, but their available K is relatively low. Potassium is present in several forms in the soil, including mineral K (more than 90–98% of soil K), non-exchangeable K (approximately 1–10% of soil K), exchangeable K, and dissolved or solution K. Among different forms of K, its available (water-soluble K) and exchangeable forms are available for the use of plants, and the rest of the forms are almost non-available for the use of plants (Sparks and Huang 1985). Plants absorb K mainly in the form of K^+ and from the soil solution. The amount of K of soil solution is very variable and is typically between 1 and 10 mg/kg. The clay minerals contain a large amount of K, and often there is a positive correlation between available K (water-soluble K) and the amount of clay minerals in the soil (Sparks and Huang 1985). Non-exchangeable K plays a significant role in plant nutrition under intensive cultivation conditions, especially in cases where exchangeable K is less than the plant's adequacy limit (Mc Lean and Watson 1985). When soluble and exchangeable K of the soil decreases to less than the plant's adequacy limit, non-exchangeable K can be released from the layers of clay minerals (Tributh et al. 1987). The amount of K released in these conditions is influenced by the amount and type of K-bearing minerals such as muscovite, potassium feldspar, and vermiculite (Steffens and Sparks 1997). Many studies have shown that the release of non-exchangeable form and structural unavailable forms of K compounds in to soil solution can significantly contribute to the amount of K absorbed by the plant (Snapp et al. 1995).

In addition to imbalanced fertilizer application, the introduction of high-yielding crop varieties/hybrids, and the progressive intensification of agriculture, which deplete K reserve of soils at a faster rate (Parmar and Sindhu 2013), soil salinity has also caused K deficiency to be one of the major constraints in crop production. Sodium-induced K^+ deficiency has been implicated in various crops (Botella et al. 1997). It has been reported that salinity (Na^+ ions) affects K uptake by plant and prevents its absorption (Barea et al. 2005; Colla et al. 2008; Rawal and Kuligod 2014) by interfering with various transporters (competing with K for binding sites essential for various cellular functions) in the root plasma membrane, K^+-selective ion channels (Wild 1988). Na^+-induced decrease of K uptake has been attributed to the competitive intracellular influx of both ions (Cerda et al. 1995). The greater absorption of sodium by plant in saline soils leads to increasing Na^+/K^+ ratio in plant, disrupting various metabolic processes such as protein synthesis in the cytoplasm (Tester and Davenport 2003). As mentioned above, K has many roles in plant metabolism such as activating a range of enzymes. The high levels of Na or high Na/K ratios can not only substitute for K role in activating enzymes but can also disrupt various enzymatic processes in the cytoplasm (Bhandal and Malik 1988). Therefore, maintenance of adequate levels of K^+ is essential for plant survival in saline habitats (Botella et al. 1997). Providing K for salinity-stressed plants can partly reduce the negative effects of sodium on the plants. The K absorbed by the plants can influence the ionic balance of the cytoplasm or Na efflux from plants (Bach Allen and Cunningham 1983).

There are several ways to increase K solubilization and release K from K-bearing minerals, but the use of soil microorganisms can be a good complement to plant nutrition due to its ease of use and low cost (Etesami et al. 2017). The ability of roots of plants and microorganisms in the weathering of minerals and significant alterations in mineralogy has been reported in the past (Etesami et al. 2017; Nabiollahy et al. 2006). Biochemical processes that are involved in weathering of minerals mainly occur in the presence of soil microorganisms and are affected by soil microorganisms. Several studies have reported the effects of biological processes and secreted materials of the roots of plants and microorganisms on weathering of minerals in rhizosphere environment (Arocena et al. 2012; Wu et al. 2008).

Among the microorganisms of soil, AM fungi and bacteria are of great importance in terms of K availability to plants. It has been reported that some beneficial soil bacteria including potassium-solubilizing bacteria (KSB) (i.e., *Acidithiobacillus ferrooxidans*, *Enterobacter hormaechei*, *Arthrobacter* sp., *Pseudomonas* spp., *P. aeruginosa Paenibacillus* sp., *P. mucilaginosus*, *P. frequentans*, *P. glucanolyticus*, *Aminobacter*, *Sphingomonas*, *Burkholderia*, *Bacillus mucilaginosus*, *Bacillus edaphicus*, and *B. circulans*) could solubilize the insoluble K to soluble forms of K by various mechanisms including production of organic acids (i.e., succinic, citric, gluconic, α-ketogluconic and oxalic, lactic, propionic, glycolic, malonic, succinic, fumaric, and tartaric acids), which enhance the chelation of cations (e.g., Si and Al) bound to K and help for acidolysis of surrounding environment of microbes or directly dissolve rock K, and inorganic acids, acidolysis, polysaccharides, complexolysis, chelation, polysaccharides, and exchange reactions (Etesami et al. 2017; Meena et al. 2014; Parmar and Sindhu 2013; Uroz et al. 2009). Potasium-solubilizing bacteria also contribute directly to the growth of plants directly by increasing the solubilization of insoluble K compounds through increasing rooting system and improving symbiosis-mutually beneficial relationships with host plant at different stages of growth (Etesami et al. 2017). Johnston and Krauss (1998) showed that if the root density in the soil is low, the concentration of K in the soil solution should be so high that the root can easily absorb K. The PSB with production ability of IAA and ACC deaminase can also increase plant access to K by increasing the root system of plant under nutrient deficiency conditions (Etesami et al. 2017). Previous studies have also shown that these bacteria could exert beneficial effects on growth of crop plants (Badr et al. 2006; Basak and Biswas 2009, 2010; Han and Lee 2006; Nadeem et al. 2007; Sheng 2005; Sheng and He 2006). Nadeem et al. (2007) reported that plants inoculated with ACC deaminase-producing bacteria had higher growth and higher K^+/Na^+ and chlorophyll than non-inoculated plants in saline conditions. Mayak et al. (2004a) also reported that the absorption of P and K in tomato shoot increased under salinity stress by inoculation of *Achromobacter piechaudii* RV8.

In addition to KSB, mycorrhizal fungi can increase the plant growth and yield and absorption of nutrients (i.e., K) by increasing the absorption surfaces of plant roots (Meena et al. 2014). Fungi such as *Funneliformis mosseae*, *Aspergillus terreus*, *A. niger*, *Penicillium* sp., and *Rhizoglomus intraradices* can increase the K level in soil solution through the release of H^+, CO_2 and the production of organic acids/organic acid anions (e.g., citrate, malate, and oxalate), which solubilize K from insoluble K-bearing minerals such as feldspar and potassium aluminum silicates (Meena et al. 2014; Prajapati et al. 2012; Sangeeth et al. 2012; Sieverding et al. 2015; Wu et al. 2005). These fungi also influence K mobilization through mycelial transport (Rashid et al. 2016). It has been reported that AM fungi can also provide plants with K under salinity stress (Giri et al. 2007; Mohammad et al. 2003; Ojala et al. 1983; Porcel et al. 2012). For example, Giri et al. (2007), Colla et al. (2008), and Zuccarini and Okurowska (2008) showed that AM plants had a higher concentration of K in root and shoot tissues than non-AM plants under salinity stress. Higher K^+ accumulation by AM plants may help in maintaining a high K^+/Na^+ ratio, thus preventing the disruption of various enzymatic processes and inhibition of protein synthesis under salt stress conditions (Porcel et al. 2012). The above studies show that application of K-solubilizing microorganisms as bio-fertilizer for agriculture improvement can reduce the use of agrochemicals and support eco-friendly crop production in salt-affected soils. At the present time, there is a little information on salinity-resistant K-solubilizing microorganisms (KSMs). So, it is required to evaluate salinity-resistant microorganisms isolated from halophyte plants and saline environments in terms of solubilizing K-bearing minerals in the future.

5.6 PGPR and AM Fungi-Mediated Increase of Micronutrients Availability Under Salinity Stress

Most of the micronutrients are poorly available in saline-sodic soils, and plant growth is highly depressed on such types of soils (Yousfi et al. 2007). Plant growth on these soils is affected concurrently by salinity and micronutrients deficiency.

Iron (Fe), as a micronutrient, is needed by all kind of living organisms. This nutrient is a component of many enzymes which are involved in diverse biochemical processes including respiration, photosynthesis, and N_2 fixation (Kobayashi and Nishizawa 2012). Fe deficiency depressed nodule mass and particularly leghemoglobin content, number of bacteroids, and nitrogenase activity (Garcia et al. 2015; Tang et al. 1990). Fe predominantly exists in nature in ferric (Fe^{3+})

form. Plants absorb iron as ferrous (Fe^{2+}) form, which is oxidized to Fe^{3+}, thereby forming insoluble compounds and leaving a very low amount of iron for microbial or plant assimilation (Ma 2005). In general, the availability of Fe is very low in calcareous and saline-sodic soils throughout the world (Rabhi et al. 2007).

Microorganisms including PGPR can increase the availability of micronutrients for plants through various mechanisms, such as soil pH reduction and the production of chelating agents (Miransari 2013). PGPR can produce chelating agents in iron deficiency conditions, which are called siderophores. Siderophores are small, high-affinity Fe(III)-chelating compounds that can scavenge iron. These iron-siderophore complexes can be easily accessible to plants (Kloepper et al. 1980). The other theory about the supply of iron via siderophores is the ligand exchange. The iron provided through bacterial siderophores can interact with plant siderophores (phytosiderophores) in the reaction of ligand exchange, and eventually iron is absorbed through phytosiderophores (Latour et al. 2009). Siderophore-producing PGPR were able to fulfill iron requirement for normal growth of iron-starved tomato plants (Grobelak and Hiller 2017; Pii et al. 2015; Radzki et al. 2013; Scavino and Pedraza 2013). It is noteworthy that halotolerant PGPR (e.g., *Brachybacterium saurashtrense* sp., *Serratia marcescens, Acinetobacter, Pseudomonas* sp., *P. stutzeri, P. putida, Microbacterium, Streptomyces* sp., *S. umbrinus, S. griseoplanus, S. exfoliatus, S. avidinii, S. europaeiscabiei, Bacillus* sp., *B. cereus, B. pumilus, B. simplex, B. mycoides, Serratia plymuthica, Microbacterium* sp., *M. oxydans, Rhodococcus* sp., *R. equi, R. erythropolis, Micrococcus luteus, Nocardia* sp., *Agrobacterium tumefaciens, Zhihengliuella* sp., *Brachybacterium* sp., *Vibrio alginolyticus, Brevibacterium casei, Cronobacter sakazakii, P. pseudoalcaligenes, Haererehalobacter* sp., and *Halomonas* sp.) isolated from halophytes such as *Salicornia brachiate, Aster tripolium* L., and *Prosopis strombulifera* had a significant ability to produce siderophore that may also be used in saline soils (Etesami and Beattie 2018; Gontia et al. 2011; Jha et al. 2012; Sgroy et al. 2009; Szymańska et al. 2016). The potential evaluation of siderophore-producing salinity-resistant PGPR isolated from halophytes in terms of increasing the availability of other micronutrients (e.g., Zn, Mn, Cu, etc.) is also suggested for future research (Etesami and Beattie 2018).

It has been shown that some PGPR can increase the manganese (Mn) availability for plant. For example, PGPR such as *Bacillus, Pseudomonas*, and *Geobacter* can reduce Mn^{4+} to Mn^{2+}, which is a metabolically useful form of Mn for plants (Osorio Vega 2007). These bacteria affect the availability of Mn in soil mainly due to their effect on plant growth and thus root secretion (Etesami and Maheshwari 2018). Increasing root secretions caused by bacterial activity leads, in turn, to supply of electron (through decomposition of carbohydrate compounds in root exudates) and protons (through the proton-secretion system of root cells) needed to reduce MnO_2 to Mn^{2+}. In addition, PGPR can produce different chelating agents such as phenolic compounds and organic acids that form soluble complexes with Mn, Fe, and other micronutrients, preventing the precipitation of these micronutrients. The effects of soil microorganisms on the availability of Cu and Zn in soil are directly affected by the production of various compounds such as phenolic compounds and carboxylic acids or indirectly affected by plant growth and as a result of root exudates (Iqbal et al. 2010).

The role of mycorrhizal fungi in increasing the absorption of nutrients is important only in the ecosystems that have nutrient deficiency (Cardoso and Kuyper 2006). Enhanced acquisition of Zn, Cu, and Fe by mycorrhizal plants has been reported (Al-Karaki 2000; Al-Karaki and Clark 1998; Marschner and Dell 1994). The extensive mycelial system of the fungi can absorb nutrients such as N, P, K, Zn, Cu, and Fe. The nutrients absorbed by the fungi are not exclusive to the fungi, and some of them are transmitted to the plant (Cardoso and Kuyper 2006). Clark and Zeto (1996) showed that the use of AM fungi increased the Fe concentration in corn shoots. In addition, symbiosis of mycorrhiza increased Fe concentration in the sorghum plant (Caris et al. 1998). Zn availability is limited due to the low mobility and partial availability of this element in soil. Also, the absorption surface is an important parameter that affects the bioavailability of Zn. Therefore, increasing the absorption surface through colonization of arbuscular mycorrhizal fungi may have a special importance on Zn uptake in comparison with other plant nutrient elements (Hajiboland et al. 2009). Root colonization by AM fungi increases Zn absorption in root length unit due to increased root absorption surface caused by growth of fungal hyphae (Gao et al. 2007). In general, the above studies show that salinity-resistant microorganisms can increase the availability of micronutrients to salinity-stressed plants by the production of organic acids and chelating agents, increasing the rooting system of the plant (i.e., by production of IAA and ACC deaminase) and thus increasing the root exudates (Etesami et al. 2015a, b; Etesami and Maheshwari 2018).

5.7 Conclusions

Water and soil salinity is one of the most important agricultural problems in arid and semiarid regions. Utilization of soil halotolerant microorganisms is one of the strategies to deal with this stress. Since use of chemical fertilizers in salt-affected soils increases the salinity of the soil twice, and the application of chemical fertilizers is not common, or is carried out at the minimum, the use of halotolerant PGP microorganisms can reduce some limitations in the production of crop plants in saline soils. However, their application as bio-fertilizer in field conditions requires more research. A better understanding on the interaction of halotolerant bacteria and AM fungi when applied under field conditions is required. The clarification of the mechanisms used by halotolerant PGP microorganisms in improving the availability of micro- and macronutrients under salinity stress may help in the development of innovative and cost-effective management practices for improving the fertility and crop production capacity of salt-affected soils.

Acknowledgment We wish to thank the University of Tehran for providing the necessary facilities and funds for this study.

References

Abdelmoumen H, El Idrissi MM (2009) Germination, growth and nodulation of Trigonella foenum graecum (Fenu Greek) under salt stress. Afr J Biotechnol 8:2489–2496

Adesemoye AO, Kloepper JW (2009) Plant–microbes interactions in enhanced fertilizer-use efficiency. Appl Microbiol Biotechnol 85:1–12

Ahmad P (2013) Oxidative damage to plants, antioxidant networks and signaling. Academic, Elsevier, San Diego

Ahmed W, Shahroona B, Zahir ZA, Arshad M (2004) Inoculation with ACC-deaminase containing rhizobacteria for improving growth and yield of wheat. Pak J Agric Sci 41:119

Albacete A et al (2008) Hormonal changes in relation to biomass partitioning and shoot growth impairment in salinized tomato (Solanum lycopersicum L.) plants. J Exp Bot 59:4119–4131

Aliasgharzadeh N, Rastin SN, Towfighi H, Alizadeh A (2001) Occurrence of arbuscular mycorrhizal fungi in saline soils of the Tabriz Plain of Iran in relation to some physical and chemical properties of soil. Mycorrhiza 11:119–122

Al-Karaki GN (2000) Growth of mycorrhizal tomato and mineral acquisition under salt stress. Mycorrhiza 10:51–54

Al-Karaki GN (2006) Nursery inoculation of tomato with arbuscular mycorrhizal fungi and subsequent performance under irrigation with saline water. Sci Horticult 109:1–7

Al-Karaki GN, Clark RB (1998) Growth, mineral acquisition, and water use by mycorrhizal wheat grown under water stress. J Plant Nutr 21:263–276

America SSSo (2001) Glossary of soil science terms. Soil Science Society of America, Madison

Andrade G, Linderman RG, Bethlenfalvay GJ (1998) Bacterial associations with the mycorrhizosphere and hyphosphere of the arbuscular mycorrhizal fungus Glomus mosseae. Plant Soil 202:79–87

Arocena JM, Velde B, Robertson SJ (2012) Weathering of biotite in the presence of arbuscular mycorrhizae in selected agricultural crops. Appl Clay Sci 64:12–17

Asghari HR (2012) Vesicular-arbuscular (VA) mycorrhizae improve salinity tolerance in pre-inoculation subterranean clover (Trifolium subterraneum) seedlings. Int J Plant Prod 2:243–256

Ashraf M, Hasnain S, Berge O, Mahmood T (2004) Inoculating wheat seedlings with exopolysaccharide-producing bacteria restricts sodium uptake and stimulates plant growth under salt stress. Biol Fertil Soils 40:157–162

Aydi S, Sassi S, Abdelly C (2008) Growth, nitrogen fixation and ion distribution in Medicago truncatula subjected to salt stress. Plant Soil 312:59

Bach Allen E, Cunningham GL (1983) Effects of vesicular–arbuscular mycorrhizae on Distichlis spicata under three salinity levels. New Phytol 93:227–236

Badr MA, Shafei AM, Sharaf El-Deen SH (2006) The dissolution of K and P-bearing minerals by silicate dissolving bacteria and their effect on sorghum growth. Res J Agric Biol Sci 2:5–11

Banerjee S, Palit R, Sengupta C, Standing D (2010) Stress induced phosphate solubilization by Arthrobacter sp. and Bacillus sp. isolated from tomato rhizosphere. Aust J Crop Sci 4:378

Banik P, Midya A, Sarkar BK, Ghose SS (2006) Wheat and chickpea intercropping systems in an additive series experiment: advantages and weed smothering. Eur J Agron 24:325–332

Barassi CA, Ayrault G, Creus CM, Sueldo RJ, Sobrero MT (2006) Seed inoculation with Azospirillum mitigates NaCl effects on lettuce. Sci Horticult 109:8–14. https://doi.org/10.1016/j.scienta.2006.02.025

Barea JM, Azcón R, Azcón-Aguilar C (2005) Interactions between mycorrhizal fungi and bacteria to improve plant nutrient cycling and soil structure. In: Microorganisms in soils: roles in genesis and functions. Springer, pp 195–212

Barrow JR, Osuna P (2002) Phosphorus solubilization and uptake by dark septate fungi in fourwing saltbush, Atriplex canescens (Pursh) Nutt. J Arid Environ 51:449–459

Basak BB, Biswas DR (2009) Influence of potassium solubilizing microorganism (Bacillus mucilaginosus) and waste mica on potassium uptake dynamics by sudan grass (Sorghum vulgare Pers.) grown under two Alfisols. Plant Soil 317:235–255

Basak BB, Biswas DR (2010) Co-inoculation of potassium solubilizing and nitrogen fixing bacteria on solubilization of waste mica and their effect on growth promotion and nutrient acquisition by a forage crop. Biol Fertil Soils 46:641–648

Bashan Y, Moreno M, Troyo E (2000) Growth promotion of the seawater-irrigated oilseed halophyte Salicornia bigelovii inoculated with mangrove rhizosphere bacteria and halotolerant Azospirillum spp. Biol Fertil Soils 32:265–272

Belimov AA, Dodd IC, Hontzeas N, Theobald JC, Safronova VI, Davies WJ (2009) Rhizosphere bacteria containing 1-aminocyclopropane-1-carboxylate deaminase increase yield of plants grown in drying soil via both local and systemic hormone signalling. New Phytol 181:413–423. https://doi.org/10.1111/j.1469-8137.2008.02657.x

Bhandal IS, Malik CP (1988) Potassium estimation, uptake, and its role in the physiology and metabolism of flowering plants. In: International review of cytology, vol 110. Elsevier, pp 205–254

Bian G, Zhang Y, Qin S, Xing K, Xie H, Jiang J (2011) Isolation and biodiversity of heavy metal tolerant endophytic bacteria from halotolerant plant species located in coastal shoal of Nantong. Wei Sheng Wu Xue Bao = Acta Microbiol Sin 51:1538–1547

Botella MA, Martinez V, Pardines J, Cerdá A (1997) Salinity induced potassium deficiency in maize plants. J Plant Physiol 150:200–205. https://doi.org/10.1016/S0176-1617(97)80203-9

Cantrell IC, Linderman RG (2001) Preinoculation of lettuce and onion with VA mycorrhizal fungi reduces deleterious effects of soil salinity. Plant Soil 233:269–281

Cardoso IM, Kuyper TW (2006) Mycorrhizas and tropical soil fertility. Agric Ecosyst Environ 116:72–84

Caris C, Hördt W, Hawkins H-J, Römheld V, George E (1998) Studies of iron transport by arbuscular mycorrhizal hyphae from soil to peanut and sorghum plants. Mycorrhiza 8:35–39

Carmen B, Roberto D (2011) Soil bacteria support and protect plants against abiotic stresses. In: Shanker A (ed) Abiotic stress in plants mechanisms and adaptations. Pub InTech, Rijeka, pp 143–170

Cerda A, Pardines J, Botella MA, Martinez V (1995) Effect of potassium on growth, water relations, and the inorganic and organic solute contents for two maize cultivars grown under saline conditions. J Plant Nutr 18:839–851

Chakraborty U, Chakraborty B, Basnet M (2006) Plant growth promotion and induction of resistance in Camellia sinensis by Bacillus megaterium. J Basic Microbiol 46:186–195. https://doi.org/10.1002/jobm.200510050

Chakraborty D et al (2008) Effect of mulching on soil and plant water status, and the growth and yield of wheat (Triticum aestivum L.) in a semi-arid environment. Agric Water Manag 95:1323–1334

Chakraborty U, Chakraborty B, Dey P, Chakraborty AP (2015). 15) Role of microorganisms in alleviation of abiotic stresses for sustainable agriculture abiotic stresses in crop plants. CABI, Wallingford, p 232

Chalk PM, Souza RF, Urquiaga S, Alves BJR, Boddey RM (2006) The role of arbuscular mycorrhiza in legume symbiotic performance. Soil Biol Biochem 38:2944–2951

Chalk PM, Alves BJR, Boddey RM, Urquiaga S (2010) Integrated effects of abiotic stresses on inoculant performance, legume growth and symbiotic dependence estimated by 15 N dilution. Plant Soil 328:1–16

Chen J-H (2006) The combined use of chemical and organic fertilizers and/or biofertilizer for crop growth and soil fertility. Citeseer, p 20

Chen YP, Rekha PD, Arun AB, Shen FT, Lai WA, Young CC (2006) Phosphate solubilizing bacteria from subtropical soil and their tricalcium phosphate solubilizing abilities. Appl Soil Ecol 34:33–41

Cheng-Song XIN, He-Zhong D, Zhen L, Wei T, Zhang D-M, Wei-Jiang LI, Xiang-Qiang K (2010) Effects of N, P, and K fertilizer application on cotton growing in saline soil in yellow river delta. Acta Agron Sin 36:1698–1706

Cherif-Silini H, Silini A, Ghoul M, Yahiaoui B, Arif F (2013) Solubilization of phosphate by the Bacillus under salt stress and in the presence of osmoprotectant compounds. Afr J Microbiol Res 7:4562–4571

Choudhary D (2012) Microbial rescue to plant under habitat-imposed abiotic and biotic stresses. Appl Microbiol Biotechnol 96:1137–1155

Choudhary DK, Kasotia A, Jain S, Vaishnav A, Kumari S, Sharma KP, Varma A (2015) Bacterial-mediated tolerance and resistance to plants under abiotic and biotic stresses. J Plant Growth Regul 1–25

Clark RB, Zeto SK (1996) Mineral acquisition by mycorrhizal maize grown on acid and alkaline soil. Soil Biol Biochem 28:1495–1503

Colla G, Rouphael Y, Cardarelli M, Tullio M, Rivera CM, Rea E (2008) Alleviation of salt stress by arbuscular mycorrhizal in zucchini plants grown at low and high phosphorus concentration. Biol Fertil Soils 44:501–509

Contreras-Cornejo HA, Macías-Rodríguez L, Cortés-Penagos C, López-Bucio J (2009) Trichoderma virens, a plant beneficial fungus, enhances biomass production and promotes lateral root growth through an auxin-dependent mechanism in Arabidopsis. Plant Physiol 149:1579–1592

Cortina C, Culiáñez-Macià FA (2005) Tomato abiotic stress enhanced tolerance by trehalose biosynthesis. Plant Sci 169:75–82

Cozzolino V, Di Meo V, Piccolo A (2013) Impact of arbuscular mycorrhizal fungi applications on maize production and soil phosphorus availability. J Geochem Explor 129:40–44

Dahmardeh M, Ghanbari A, Syasar B, Ramroudi M (2009) Effect of intercropping maize (Zea mays L.) with cow pea (Vigna unguiculata L.) on green forage yield and quality evaluation. Asian J Plant Sci 8:235

Damodaran T et al (2014) Rhizosphere and endophytic bacteria for induction of salt tolerance in gladiolus grown in sodic soils. J Plant Interact 9:577–584

Davies WJ, Zhang J, Yang J, Dodd IC (2011) Novel crop science to improve yield and resource use efficiency in water-limited agriculture. J Agric Sci 149:123–131

del Amor Francisco M, Cuadra-Crespo P (2012) Plant growth-promoting bacteria as a tool to improve salinity tolerance in sweet pepper. Funct Plant Biol 39:82–90

Delavechia C, Hampp E, Fabra A, Castro S (2003) Influence of pH and calcium on the growth, polysaccharide production and symbiotic association of Sinorhizobium meliloti SEMIA 116 with alfalfa roots. Biol Fertil Soils 38:110–114

Delgado-García M, De la Garza-Rodríguez I, Cruz-Hernández MA, Balagurusamy N, Aguilar C, Rodríguez-Herrera R (2013) Characterization and selection of halophilic microorganisms isolated from Mexican soils. J Agric Biol Sci 8:457–464

Diby P, Bharathkumar S, Sudha N (2005a) Osmotolerance in biocontrol strain of pseudomonas pseudoalcaligenes MSP-538: a study using osmolyte, protein and gene expression profiling. Ann Microbiol 55:243–247

Diby P, Sarma YR, Srinivasan V, Anandaraj M (2005b) Pseudomonas fluorescens mediated vigour in black pepper (piper nigrum L.) under green house cultivation. Ann Microbiol 55:171–174

Dimkpa C, Weinand T, Asch F (2009) Plant-rhizobacteria interactions alleviate abiotic stress conditions. Plant Cell Environ 32:1682–1694. https://doi.org/10.1111/j.1365-3040.2009.02028.x

Dolkar D, Dolkar P, Angmo S, Chaurasia OP, Stobdan T (2018) Stress tolerance and plant growth promotion potential of Enterobacter ludwigii PS1 isolated from Seabuckthorn rhizosphere. Biocatal Agric Biotechnol. https://doi.org/10.1016/j.bcab.2018.04.012

Dulormne M, Musseau O, Muller F, Toribio A, Bâ A (2010) Effects of NaCl on growth, water status, N 2 fixation, and ion distribution in Pterocarpus officinalis seedlings. Plant Soil 327:23–34

Dunlap JR, Binzel ML (1996) NaCI reduces indole-3-acetic acid levels in the roots of tomato plants independent of stress-induced abscisic acid. Plant Physiol 112:379–384

Duzan HM, Zhou X, Souleimanov A, Smith DL (2004) Perception of Bradyrhizobium japonicum Nod factor by soybean [Glycine max (L.) Merr.] root hairs under abiotic stress conditions. J Exp Bot 55:2641–2646

Egamberdieva D, Kucharova Z (2009) Selection for root colonising bacteria stimulating wheat growth in saline soils. Biol Fertil Soils 45:563–571

Egamberdieva D, Kamilova F, Validov S, Gafurova L, Kucharova Z, Lugtenberg B (2008) High incidence of plant growth-stimulating bacteria associated with the rhizosphere of wheat grown on salinated soil in Uzbekistan. Environ Microbiol 10:1–9. https://doi.org/10.1111/j.1462-2920.2007.01424.x

Egamberdieva D, Berg G, Lindström K, Räsänen L (2013) Alleviation of salt stress of symbiotic Galega officinalis L. (goat's rue) by co-inoculation of Rhizobium with root-colonizing Pseudomonas. Plant Soil 369:453–465. https://doi.org/10.1007/s11104-013-1586-3

Egamberdiyeva D (2005) Plant-growth-promoting rhizobacteria isolated from a Calcisol in a semi-arid region of Uzbekistan: biochemical characterization and effectiveness. J Plant Nutr Soil Sci 168:94–99

Elbein AD, Pan YT, Pastuszak I, Carroll D (2003) New insights on trehalose: a multifunctional molecule. Glycobiology 13:17R–27R

El-Tarabily KA, Youssef T (2010) Enhancement of morphological, anatomical and physiological characteristics of seedlings of the mangrove Avicennia marina inoculated with a native phosphate-solubilizing isolate of Oceanobacillus picturae under greenhouse conditions. Plant Soil 332:147–162

Esquivel-Cote R, Ramírez-Gama RM, Tsuzuki-Reyes G, Orozco-Segovia A, Huante P (2010) Azospirillum lipoferum strain AZm5 containing 1-aminocyclopropane-1-carboxylic acid deaminase improves early growth of tomato seedlings under nitrogen deficiency. Plant Soil 337:65–75. https://doi.org/10.1007/s11104-010-0499-7

Etesami H (2018) Can interaction between silicon and plant growth promoting rhizobacteria benefit in alleviating abiotic and biotic stresses in crop plants? Agric Ecosyst Environ 253:98–112. https://doi.org/10.1016/j.agee.2017.11.007

Etesami H, Alikhani HA (2016a) Co-inoculation with endophytic and rhizosphere bacteria allows reduced application rates of N-fertilizer for rice plant. Rhizosphere 2:5–12

Etesami H, Alikhani HA (2016b) Rhizosphere and endorhiza of oilseed rape (Brassica napus L.) plant harbor bacteria with multifaceted beneficial effects. Biol Control 94:11–24. https://doi.org/10.1016/j.biocontrol.2015.12.003

Etesami H, Beattie GA (2017) Plant-microbe interactions in adaptation of agricultural crops to abiotic stress conditions. In: Probiotics and plant health. Springer, pp 163–200

Etesami H, Beattie G (2018) Mining halophytes for plant growth-promoting halotolerant bacteria to enhance the salinity tolerance of non-halophytic crops. Front Microbiol 9:148. https://doi.org/10.3389/fmicb.2018.00148

Etesami H, Maheshwari DK (2018) Use of plant growth promoting rhizobacteria (PGPRs) with multiple plant growth promoting traits in stress agriculture: action mechanisms and future prospects. Ecotoxicol Environ Saf 156:225–246. https://doi.org/10.1016/j.ecoenv.2018.03.013

Etesami H, Alikhani HA, Hosseini HM (2015a) Indole-3-acetic acid (IAA) production trait, a useful screening to select endophytic and rhizosphere competent bacteria for rice growth promoting agents. MethodsX 2:72–78

Etesami H, Alikhani HA, Mirseyed Hosseini H (2015b) Indole-3-acetic acid and 1-aminocyclopropane-1-carboxylate deaminase: bacterial traits required in rhizosphere, rhizoplane and/or endophytic competence by beneficial bacteria. In: Maheshwari DK (ed) Bacterial metabolites in sustainable agroecosystem. Springer, Cham, pp 183–258. https://doi.org/10.1007/978-3-319-24654-3_8

Etesami H, Emami S, Alikhani HA (2017) Potassium solubilizing bacteria (KSB): mechanisms, promotion of plant growth, and future prospects – a review. J Soil Sci Plant Nutr 17:897–911

Evelin H, Kapoor R, Giri B (2009) Arbuscular mycorrhizal fungi in alleviation of salt stress: a review. Ann Bot 104:1263–1280

Fageria NK, Moreira A (2011). 4) The role of mineral nutrition on root growth of crop plants. Adv Agron 110:251–331

FAO (2005) Available on URL: http://www.fao.org

FAO (2008) Land and plant nutrition management service. http://www.fao.org/ag/agl/agll/spush

Farías-Rodríguez R, Mellor RB, Arias C, Peña-Cabriales JJ (1998) The accumulation of trehalose in nodules of several cultivars of common bean (Phaseolus vulgaris) and its correlation with resistance to drought stress. Physiol Plant 102:353–359

Feigin A (1985) Fertilization management of crops irrigated with saline water. Plant Soil 89:285–299. https://doi.org/10.1007/BF02182248

Feng G, Zhang F, Li X, Tian C, Tang C, Rengel Z (2002) Improved tolerance of maize plants to salt stress by arbuscular mycorrhiza is related to higher accumulation of soluble sugars in roots. Mycorrhiza 12:185–190

Fernandez O, Béthencourt L, Quero A, Sangwan RS, Clément C (2010) Trehalose and plant stress responses: friend or foe? Trends Plant Sci 15:409–417

Flowers TJ (2004) Improving crop salt tolerance. J Exp Bot 55:307–319

Founoune H, Duponnois R, Bâ AM, El Bouami F (2002) Influence of the dual arbuscular endomycorrhizal/ectomycorrhizal symbiosis on the growth of Acacia holosericea (A. Cunn. ex G. Don) in glasshouse conditions. Ann For Sci 59:93–98

Franco-Correa M, Quintana A, Duque C, Suarez C, Rodríguez MX, Barea J-M (2010) Evaluation of actinomycete strains for key traits related with plant growth promotion and mycorrhiza helping activities. Appl Soil Ecol 45:209–217

Franzini VI, Azcon R, Mendes FL, Aroca R (2010) Interactions between Glomus species and Rhizobium strains affect the nutritional physiology of drought-stressed legume hosts. J Plant Physiol 167:614–619

Fu Q, Liu C, Ding N, Lin Y, Guo B (2010) Ameliorative effects of inoculation with the plant growth-promoting rhizobacterium Pseudomonas sp. DW1 on growth of eggplant (Solanum melongena L.) seedlings under salt stress. Agric Water Manag 97:1994–2000. https://doi.org/10.1016/j.agwat.2010.02.003

Gamalero E, Glick BR (2011) Mechanisms used by plant growth-promoting bacteria. In: Bacteria in agrobiology: plant nutrient management. Springer, pp 17–46

Gao X, Kuyper TW, Zou C, Zhang F, Hoffland E (2007) Mycorrhizal responsiveness of aerobic rice genotypes is negatively correlated with their zinc uptake when nonmycorrhizal. Plant Soil 290:283–291

Garcia NS, Fu F, Sedwick PN, Hutchins DA (2015) Iron deficiency increases growth and nitrogen-fixation rates of phosphorus-deficient marine cyanobacteria. ISME J 9:238–245. https://doi.org/10.1038/ismej.2014.104

Garg N, Chandel S (2011) The effects of salinity on nitrogen fixation and trehalose metabolism in mycorrhizal Cajanus cajan (L.) millsp. plants. J Plant Growth Regul 30:490–503

Garg N, Manchanda G (2008) Effect of arbuscular mycorrhizal inoculation on salt-induced nodule senescence in Cajanus cajan (pigeonpea). J Plant Growth Regul 27:115

Garg N, Manchanda G (2009) Role of arbuscular mycorrhizae in the alleviation of ionic, osmotic and oxidative stresses induced by salinity in Cajanus cajan (L.) Millsp.(pigeonpea). J Agron Crop Sci 195:110–123

Giri B, Mukerji KG (2004) Mycorrhizal inoculant alleviates salt stress in Sesbania aegyptiaca and Sesbania grandiflora under field conditions: evidence for reduced sodium and improved magnesium uptake. Mycorrhiza 14:307–312

Giri B, Kapoor R, Mukerji KG (2003) Influence of arbuscular mycorrhizal fungi and salinity on growth, biomass, and mineral nutrition of Acacia auriculiformis. Biol Fertil Soils 38:170–175

Giri B, Kapoor R, Mukerji KG (2007) Improved tolerance of Acacia nilotica to salt stress by arbuscular mycorrhiza, Glomus fasciculatum may be partly related to elevated K/Na ratios in root and shoot tissues. Microb Ecol 54:753–760

Glick BR (2005) Modulation of plant ethylene levels by the bacterial enzyme ACC deaminase. FEMS Microbiol Lett 251:1–7

Glick BR (2012) Plant growth-promoting bacteria: mechanisms and applications. Scientifica 2012:963401

Glick BR (2014) Bacteria with ACC deaminase can promote plant growth and help to feed the world. Microbiol Res 169:30–39

Gontia I, Kavita K, Schmid M, Hartmann A, Jha B (2011) Brachybacterium saurashtrense sp. nov., a halotolerant root-associated bacterium with plant growth-promoting potential. Int J Syst Evol Microbiol 61:2799–2804

Gosling P, Hodge A, Goodlass G, Bending GD (2006) Arbuscular mycorrhizal fungi and organic farming. Agric Ecosyst Environ 113:17–35

Goss MJ, De Varennes A (2002) Soil disturbance reduces the efficacy of mycorrhizal associations for early soybean growth and N2 fixation. Soil Biol Biochem 34:1167–1173

Grant CA, Flaten DN, Tomasiewicz DJ, Sheppard SC (2001) The importance of early season phosphorus nutrition. Can J Plant Sci 81:211–224

Gray EJ, Smith DL (2005) Intracellular and extracellular PGPR: commonalities and distinctions in the plant-bacterium signaling processes. Soil Biol Biochem 37:395–412. https://doi.org/10.1016/j.soilbio.2004.08.030

Greaves JE (1922) Influence of salts on bacterial activities of soil. Bot Gaz 73:161–180

Grobelak A, Hiller J (2017) Bacterial siderophores promote plant growth: screening of catechol and hydroxamate siderophores. Int J Phytoremediation 19:825–833

Grover M, Ali SZ, Sandhya V, Rasul A, Venkateswarlu B (2011) Role of microorganisms in adaptation of agriculture crops to abiotic stresses. World J Microbiol Biotechnol 27:1231–1240

Gulati A, Sharma N, Vyas P, Sood S, Rahi P, Pathania V, Prasad R (2010) Organic acid production and plant growth promotion as a function of phosphate solubilization by Acinetobacter rhizosphaerae strain BIHB 723 isolated from the cold deserts of the trans-Himalayas. Arch Microbiol 192:975–983

Gyaneshwar P, Kumar GN, Parekh LJ, Poole PS (2002) Role of soil microorganisms in improving P nutrition of plants. In: Food security in nutrient-stressed environments: exploiting plants' genetic capabilities. Springer, pp 133–143

Hajiboland R, Aliasgharzad N, Barzeghar R (2009) Influence of arbuscular mycorrhizal fungi on uptake of Zn and P by two contrasting rice genotypes. Plant Soil Environ 55:93–100

Hajiboland R, Aliasgharzadeh N, Laiegh SF, Poschenrieder C (2010) Colonization with arbuscular mycorrhizal fungi improves salinity tolerance of tomato (Solanum lycopersicum L.) plants. Plant Soil 331:313–327

Halvorson HO, Keynan A, Kornberg HL (1990) Utilization of calcium phosphates for microbial growth at alkaline pH. Soil Biol Biochem 22:887–890

Hamdia MAE-S, Shaddad MAK, Doaa MM (2004) Mechanisms of salt tolerance and interactive effects of Azospirillum brasilense inoculation on maize cultivars grown under salt stress conditions. Plant Growth Regul 44:165–174. https://doi.org/10.1023/B:GROW.0000049414.03099.9b

Hamilton CE, Bever JD, Labbé J, Yang X, Yin H (2016) Mitigating climate change through managing constructed-microbial communities in agriculture. Agric Ecosyst Environ 216:304–308

Han H-S, Lee KD (2006) Effect of co-inoculation with phosphate and potassium solubilizing bacteria on mineral uptake and growth of pepper and cucumber. Plant Soil Environ 52:130

Hingole SS, Pathak AP (2016) Saline soil microbiome: a rich source of halotolerant PGPR. J Crop Sci Biotechnol 19:231–239

Hirrel MC, Gerdemann JW (1980) Improved growth of onion and bell pepper in saline soils by two vesicular-arbuscular mycorrhizal fungi. 1. Soil Sci Soc Am J 44:654–655

Illmer P, Schinner F (1992) Solubilization of inorganic phosphates by microorganisms isolated from forest soils. Soil Biol Biochem 24:389–395

Iniguez AL, Dong Y, Carter HD, Ahmer BM, Stone JM, Triplett EW (2005) Regulation of enteric endophytic bacterial colonization by plant defenses. Mol Plant-Microbe Interact MPMI 18:169–178. https://doi.org/10.1094/mpmi-18-0169

Iqbal Hussain M, Naeem Asghar H, Javed Akhtar M, Arshad M (2013) Impact of phosphate solubilizing bacteria on growth and yield of maize. Soil Environ 32:71–78

Iqbal U, Jamil N, Ali I, Hasnain S (2010) Effect of zinc-phosphate-solubilizing bacterial isolates on growth of Vigna radiata. Ann Microbiol 60:243–248

Jamil A, Riaz S, Ashraf M, Foolad MR (2011) Gene expression profiling of plants under salt stress. Crit Rev Plant Sci 30:435–458. https://doi.org/10.1080/07352689.2011.605739

Jebara S, Drevon JJ, Jebara M (2010) Modulation of symbiotic efficiency and nodular antioxidant enzyme activities in two Phaseolus vulgaris genotypes under salinity. Acta Physiol Plant 32:925–932

Jha B, Gontia I, Hartmann A (2012) The roots of the halophyte Salicornia brachiata are a source of new halotolerant diazotrophic bacteria with plant growth-promoting potential. Plant Soil 356:265–277

Joe MM, Devaraj S, Benson A, Sa T (2016) Isolation of phosphate solubilizing endophytic bacteria from Phyllanthus amarus Schum & Thonn: evaluation of plant growth promotion and antioxidant activity under salt stress. J Appl Res Med Aromat Plants 3:71–77

Johnston AE, Krauss A (1998) The essential role of potassium in diverse cropping systems: future research needs and benefits. In: Essential role of potassium in diverse cropping systems. International Potash Institute, Basel, pp 101–120

Johri JK, Surange S, Nautiyal CS (1999) Occurrence of salt, pH, and temperature-tolerant, phosphate-solubilizing bacteria in alkaline soils. Curr Microbiol 39:89–93

Jones DL, Oburger E (2011) Solubilization of phosphorus by soil microorganisms. In: Phosphorus in action. Springer, pp 169–198

Karlidag H, Yildirim E, Turan M, Pehluvan M, Donmez F (2013) Plant growth-promoting rhizobacteria mitigate deleterious effects of salt stress on strawberry plants (Fragaria× ananassa). Hortscience 48:563–567

Katznelson H, Peterson EA, Rouatt JW (1962) Phosphate-dissolving microorganisms on seed and in the root zone of plants. Can J Bot 40:1181–1186

Kaya C, Ashraf M, Sonmez O, Aydemir S, Tuna AL, Cullu MA (2009a) The influence of arbuscular mycorrhizal colonisation on key growth parameters and fruit yield of pepper plants grown at high salinity. Sci Horticult 121:1–6

Kaya C, Tuna AL, Yokaş I (2009b) The role of plant hormones in plants under salinity stress. In: Salinity and water stress. Springer, pp 45–50

Kaymak HC, Guvenc I, Yarali F, Donmez MF (2009) The effects of bio-priming with PGPR on germination of radish (Raphanus sativus L.) seeds under saline conditions. Turk J Agric For 33:173–179

Khadri M, Tejera NA, Lluch C (2006) Alleviation of salt stress in common bean (Phaseolus vulgaris) by exogenous abscisic acid supply. J Plant Growth Regul 25:110–119

Khan MS, Zaidi A, Wani PA (2007) Role of phosphate-solubilizing microorganisms in sustainable agriculture—a review. Agron Sustain Dev 27:29–43

Khan MS, Zaidi A, Wani PA, Ahemad M, Oves M (2009) Functional diversity among plant growth-promoting rhizobacteria: current status. In: Microbial strategies for crop improvement. Springer, pp 105–132

Khan AL, Hamayun M, Kang S-M, Kim Y-H, Jung H-Y, Lee J-H, Lee I-J (2012) Endophytic fungal association via gibberellins and indole acetic acid can improve plant growth under abiotic stress: an example of Paecilomyces formosus LHL10. BMC Microbiol 12:1

Kim J, Rees DC (1994) Nitrogenase and biological nitrogen fixation. Biochemistry 33:389–397. https://doi.org/10.1021/bi00168a001

Kim KY, Jordan D, McDonald GA (1997) Effect of phosphate-solubilizing bacteria and vesicular-arbuscular mycorrhizae on tomato growth and soil microbial activity. Biol Fertil Soils 26:79–87

Kloepper JW, Leong J, Teintze M, Schroth MN (1980) Enhanced plant growth by siderophores produced by plant growth-promoting rhizobacteria. Nature 286:885–886
Kobayashi T, Nishizawa NK (2012) Iron uptake, translocation, and regulation in higher plants. Annu Rev Plant Biol 63:131–152
Koide RT, Kabir Z (2000) Extraradical hyphae of the mycorrhizal fungus Glomus intraradices can hydrolyse organic phosphate. New Phytol 148:511–517
Krüger M, Krüger C, Walker C, Stockinger H, Schüßler A (2012) Phylogenetic reference data for systematics and phylotaxonomy of arbuscular mycorrhizal fungi from phylum to species level. New Phytol 193:970–984
Kucey RMN (1983) Phosphate-solubilizing bacteria and fungi in various cultivated and virgin Alberta soils. Can J Soil Sci 63:671–678
Kucey RMN, Janzen HH, Leggett ME (1989) Microbially mediated increases in plant-available phosphorus. In: Advances in agronomy, vol 42. Elsevier, pp 199–228
Kumar A, Sharma S, Mishra S (2010) Influence of arbuscular mycorrhizal (AM) fungi and salinity on seedling growth, solute accumulation, and mycorrhizal dependency of Jatropha curcas L. J Plant Growth Regul 29:297–306
Latour X, Delorme S, Mirleau P, Lemanceau P (2009) Identification of traits implicated in the rhizosphere competence of fluorescent pseudomonads: description of a strategy based on population and model strain studies. In: Sustainable agriculture. Springer, pp 285–296
Long RL et al (2008) Seed persistence in the field may be predicted by laboratory-controlled aging. Weed Sci 56:523–528
López M, Lluch C (2008) Nitrogen fixation is synchronized with carbon metabolism in Lotus japonicus and Medicago truncatula nodules under salt stress. J Plant Interact 3:137–144
López M, Tejera NA, Iribarne C, Lluch C, Herrera-Cervera JA (2008) Trehalose and trehalase in root nodules of Medicago truncatula and Phaseolus vulgaris in response to salt stress. Physiol Plant 134:575–582
Ma JF (2005) Plant root responses to three abundant soil minerals: silicon, aluminum and iron. Crit Rev Plant Sci 24:267–281
Ma Y, Galinski EA, Grant WD, Oren A, Ventosa A (2010) Halophiles 2010: life in saline environments. Appl Environ Microbiol 76:6971–6981
Mapelli F et al (2013) Potential for plant growth promotion of rhizobacteria associated with Salicornia growing in Tunisian hypersaline soils. BioMed Res Int 2013:248078
Marasco R et al (2012) A drought resistance-promoting microbiome is selected by root system under desert farming. PLoS One 7:e48479
Margesin R, Schinner F (2001) Potential of halotolerant and halophilic microorganisms for biotechnology. Extremophiles 5:73–83
Marschner H (1995) Mineral nutrition of higher plants, 2nd edn. Academic Press, London
Marschner H, Dell B (1994) Nutrient uptake in mycorrhizal symbiosis. Plant Soil 159:89–102
Mayak S, Tirosh T, Glick BR (2004a) Plant growth-promoting bacteria confer resistance in tomato plants to salt stress. Plant Physiol Biochem 42:565–572
Mayak S, Tirosh T, Glick BR (2004b) Plant growth-promoting bacteria that confer resistance to water stress in tomatoes and peppers. Plant Sci 166:525–530
McLean EO, Watson ME (1985) Soil measurements of plant-available potassium. In: Potassium in agriculture, pp 277–308
McMillen BG, Juniper S, Abbott LK (1998) Inhibition of hyphal growth of a vesicular-arbuscular mycorrhizal fungus in soil containing sodium chloride limits the spread of infection from spores. Soil Biol Biochem 30:1639–1646
Meena VS, Maurya BR, Verma JP (2014) Does a rhizospheric microorganism enhance K+ availability in agricultural soils? Microbiol Res 169:337–347
Meena VS, Maurya BR, Verma JP, Aeron A, Kumar A, Kim K, Bajpai VK (2015) Potassium solubilizing rhizobacteria (KSR): isolation, identification, and K-release dynamics from waste mica. Ecol Eng 81:340–347

Mensah JK, Ihenyen J (2009) Effects of salinity on germination, seedling establishment and yield of three genotypes of mung bean (Vigna mungo L. Hepper) in Edo State, Nigeria. Niger Ann Nat Sci 8:17–24
Miller SH, Browne P, Prigent-Combaret C, Combes-Meynet E, Morrissey JP, O'Gara F (2010) Biochemical and genomic comparison of inorganic phosphate solubilization in Pseudomonas species. Environ Microbiol Rep 2:403–411
Mills HAJJ et al (1996) Plant analysis handbook II: a practical preparation, analysis, and interpretation guide. Potash and Phosphate Institute
Milošević NA, Marinković JB, Tintor BB (2012) Mitigating abiotic stress in crop plants by microorganisms. Zbornik Matice Srpske Za Prirodne Nauke 17–26
Miransari M (2013) Soil microbes and the availability of soil nutrients. Acta Physiol Plant 35:3075–3084
Miransari M, Smith D (2008) Using signal molecule genistein to alleviate the stress of suboptimal root zone temperature on soybean-Bradyrhizobium symbiosis under different soil textures. J Plant Interact 3:287–295
Mohammad MJ, Malkawi HI, Shibli R (2003) Effects of arbuscular mycorrhizal fungi and phosphorus fertilization on growth and nutrient uptake of barley grown on soils with different levels of salts. J Plant Nutr 26:125–137
Moradi A, Tahmourespour A, Hoodaji M, Khors F (2011) Effect of salinity on free living-diazotroph and total bacterial populations of two saline soils. Afr J Microbiol Res 5:144–148
Müller J, Boller T, Wiemken A (2001) Trehalose becomes the most abundant non-structural carbohydrate during senescence of soybean nodules. J Exp Bot 52:943–947
Munns R (2005) Genes and salt tolerance: bringing them together. New Phytol 167:645–663
Munns R, Tester M (2008) Mechanisms of salinity tolerance. Annu Rev Plant Biol 59:651–681. https://doi.org/10.1146/annurev.arplant.59.032607.092911
Nabiollahy K, Khormali F, Bazargan K, Ayoubi S (2006) Forms of K as a function of clay mineralogy and soil development. Clay Miner 41:739–749
Nadeem SM, Zahir ZA, Naveed M, Arshad M (2007) Preliminary investigations on inducing salt tolerance in maize through inoculation with rhizobacteria containing ACC deaminase activity. Can J Microbiol 53:1141–1149
Nadeem SM, Zahir ZA, Naveed M, Arshad M (2009) Rhizobacteria containing ACC-deaminase confer salt tolerance in maize grown on salt-affected fields. Can J Microbiol 55:1302–1309
Nadeem SM, Zahir ZA, Naveed M, Nawaz S (2013) Mitigation of salinity-induced negative impact on the growth and yield of wheat by plant growth-promoting rhizobacteria in naturally saline conditions. Ann Microbiol 63:225–232
Navarro JM, Botella MA, Cerdá A, Martinez V (2001) Phosphorus uptake and translocation in salt-stressed melon plants. J Plant Physiol 158:375–381
Naz I, Bano A, Ul-Hassan T (2009) Isolation of phytohormones producing plant growth promoting rhizobacteria from weeds growing in Khewra salt range, Pakistan and their implication in providing salt tolerance to Glycine max L. Afr J Biotechnol 8:5762–5768
Ocón A, Hampp R, Requena N (2007) Trehalose turnover during abiotic stress in arbuscular mycorrhizal fungi. New Phytol 174:879–891
Ojala JC, Jarrell WM, Menge JA, Johnson ELV (1983) Influence of mycorrhizal fungi on the mineral nutrition and yield of onion in saline soil. 1. Agron J 75:255–259
Orhan F, Gulluce M (2015) Isolation and characterization of salt-tolerant bacterial strains in salt-affected soils of Erzurum, Turkey. Geomicrobiol J 32:521–529
Osorio Vega NW (2007) A review on beneficial effects of rhizosphere bacteria on soil nutrient availability and plant nutrient uptake. Rev Fac Nac Agron Medellin 60:3621–3643
Owen D, Williams AP, Griffith GW, Withers PJA (2015) Use of commercial bio-inoculants to increase agricultural production through improved phosphorus acquisition. Appl Soil Ecol 86:41–54
Parida AK, Das AB (2005) Salt tolerance and salinity effects on plants: a review. Ecotoxicol Environ Saf 60:324–349. https://doi.org/10.1016/j.ecoenv.2004.06.010

Parmar P, Sindhu SS (2013) Potassium solubilization by rhizosphere bacteria: influence of nutritional and environmental conditions. J Microbiol Res 3:25–31

Patel AD, Jadeja H, Pandey AN (2010) Effect of salinization of soil on growth, water status and nutrient accumulation in seedlings of Acacia auriculiformis (Fabaceae). J Plant Nutr 33:914–932

Patreze CM, Cordeiro L (2004) Nitrogen-fixing and vesicular–arbuscular mycorrhizal symbioses in some tropical legume trees of tribe Mimoseae. For Ecol Manag 196:275–285

Paul D (2012) Osmotic stress adaptations in rhizobacteria. J Basic Microbiol 52:1–10

Paul D (2013) Osmotic stress adaptations in rhizobacteria. J Basic Microbiol 53:101–110. https://doi.org/10.1002/jobm.201100288

Paul D, Lade H (2014) Plant-growth-promoting rhizobacteria to improve crop growth in saline soils: a review. Agron Sustain Dev 34:737–752

Paul D, Nair S (2008) Stress adaptations in a Plant Growth Promoting Rhizobacterium (PGPR) with increasing salinity in the coastal agricultural soils. J Basic Microbiol 48:378–384. https://doi.org/10.1002/jobm.200700365

Paul D, Sarma YR (2006) Plant growth promoting rhizobacteria (PGPR)-mediated root proliferation in black pepper (Piper nigrum L.) as evidenced through GS Root software. Arch Phytopathol Plant Protect 39:311–314

Penfold C (2001) 32. In: May R (ed) Phosphorus management in broadacre organic farming systems, p 179

Pereira PAA, Bliss FA (1989) Selection of common bean (Phaseolus vulgaris L.) for N2 fixation at different levels of available phosphorus under field and environmentally-controlled conditions. Plant Soil 115:75–82

Pereira SIA, Castro PML (2014) Phosphate-solubilizing rhizobacteria enhance Zea mays growth in agricultural P-deficient soils. Ecol Eng 73:526–535

Peters NK, Crist-Estes DK (1989) Nodule formation is stimulated by the ethylene inhibitor aminoethoxyvinylglycine. Plant Physiol 91:690–693

Peterson RL, Massicotte HB, Melville LH (2004) Mycorrhizas: anatomy and cell biology. NRC Research Press, Ottawa

Pfetffer CM, Bloss HE (1988) Growth and nutrition of guayule (Parthenium argentatum) in a saline soil as influenced by vesicular–arbuscular mycorrhiza and phosphorus fertilization. New Phytol 108:315–321

Piccoli P, Travaglia C, Cohen A, Sosa L, Cornejo P, Masuelli R, Bottini R (2011) An endophytic bacterium isolated from roots of the halophyte Prosopis strombulifera produces ABA, IAA, gibberellins A1 and A3 and jasmonic acid in chemically-defined culture medium. Plant Growth Regul 64:207–210

Pii Y, Penn A, Terzano R, Crecchio C, Mimmo T, Cesco S (2015) Plant-microorganism-soil interactions influence the Fe availability in the rhizosphere of cucumber plants. Plant Physiol Biochem 87:45–52

Pond EC, Menge JA, Jarrell WM (1984) Improved growth of tomato in salinized soil by vesicular-arbuscular mycorrhizal fungi collected from saline soils. Mycologia 76:74–84

Poonguzhali S, Madhaiyan M, Sa T (2008) Isolation and identification of phosphate solubilizing bacteria from Chinese cabbage and their effect on growth and phosphorus utilization of plants. J Microbiol Biotechnol 18:773–777

Porcel R, Aroca R, Ruiz-Lozano JM (2012) Salinity stress alleviation using arbuscular mycorrhizal fungi. A review. Agron Sustain Dev 32:181–200

Porras-Soriano A, Soriano-Martín ML, Porras-Piedra A, Azcón R (2009) Arbuscular mycorrhizal fungi increased growth, nutrient uptake and tolerance to salinity in olive trees under nursery conditions. J Plant Physiol 166:1350–1359

Poss JA, Pond E, Menge JA, Jarrell WM (1985) Effect of salinity on mycorrhizal onion and tomato in soil with and without additional phosphate. Plant Soil 88:307–319

Prajapati K, Sharma MC, Modi HA (2012) Isolation of two potassium solubilizing fungi from ceramic industry soils. Life Sci Leafl 5:71–75

Rabhi M, Barhoumi Z, Ksouri R, Abdelly C, Gharsalli M (2007) Interactive effects of salinity and iron deficiency in Medicago ciliaris. C R Biol 330:779–788
Rabie G, Almadini AM (2005) Role of bioinoculants in development of salt-tolerance of Vicia faba plants under salinity stress. Afr J Biotechnol 4:210–222
Radzki W, Mañero FJG, Algar E, García JAL, García-Villaraco A, Solano BR (2013) Bacterial siderophores efficiently provide iron to iron-starved tomato plants in hydroponics culture. Antonie Van Leeuwenhoek 104:321–330
Ramadoss D, Lakkineni VK, Bose P, Ali S, Annapurna K (2013) Mitigation of salt stress in wheat seedlings by halotolerant bacteria isolated from saline habitats. SpringerPlus 2:1–6
Rashid MI, Mujawar LH, Shahzad T, Almeelbi T, Ismail IMI, Oves M (2016) Bacteria and fungi can contribute to nutrients bioavailability and aggregate formation in degraded soils. Microbiol Res 183:26–41
Rawal R, Kuligod VB (2014) Influence of graded doses of nitrogen on nutrient uptake and grain yield of maize (Zea mays) under varying levels of soil salinity. Karnataka J Agric Sci 27:22–24
Rawat R, Tewari L (2011) Effect of abiotic stress on phosphate solubilization by biocontrol fungus Trichoderma sp. Curr Microbiol 62:1521–1526
Rehman A, Nautiyal CS (2002) Effect of drought on the growth and survival of the stress-tolerant bacterium Rhizobium sp. NBRI2505 sesbania and its drought-sensitive transposon Tn5 mutant. Curr Microbiol 45:368–377
Richardson AE (2001) Prospects for using soil microorganisms to improve the acquisition of phosphorus by plants. Funct Plant Biol 28:897–906
Richardson AE, Simpson RJ (2011) Soil microorganisms mediating phosphorus availability update on microbial phosphorus. Plant Physiol 156:989–996
Rinaldelli E, Mancuso S (1996) Response of young mycorrhizal and non-mycorrhizal plants of olive tree (Olea europaea L.) to saline conditions. I. Short-term electrophysiological and long-term vegetative salt effects. Adv Hortic Sci 10:126–134
Rogers ME, Grieve CM, Shannon MC (2003) Plant growth and ion relations in lucerne (Medicago sativa L.) in response to the combined effects of NaCl and P. Plant Soil 253:187–194
Rojas-Tapias D, Moreno-Galván A, Pardo-Díaz S, Obando M, Rivera D, Bonilla R (2012) Effect of inoculation with plant growth-promoting bacteria (PGPB) on amelioration of saline stress in maize (Zea mays). Appl Soil Ecol 61:264–272
Roychoudhury P, Kaushik BD (1989) Solubilization of Mussoorie rock phosphate by cyanobacteria. Curr Sci 58:569–570
Ruiz-Lozano JM (2003) Arbuscular mycorrhizal symbiosis and alleviation of osmotic stress. New perspectives for molecular studies. Mycorrhiza 13:309–317
Ruppel S, Franken P, Witzel K (2013) Properties of the halophyte microbiome and their implications for plant salt tolerance. Funct Plant Biol 40:940–951
Saber MSM, Zanati MR (1984) Effectiveness of inoculation with silicate bacteria in relation to the potassium content of plants using the intensive cropping technique. Agric Res Rev
Sadeghi A, Karimi E, Dahaji PA, Javid MG, Dalvand Y, Askari H (2012) Plant growth promoting activity of an auxin and siderophore producing isolate of Streptomyces under saline soil conditions. World J Microbiol Biotechnol 28:1503–1509. https://doi.org/10.1007/s11274-011-0952-7
Sagoe CI, Ando T, Kouno K, Nagaoka T (1998) Relative importance of protons and solution calcium concentration in phosphate rock dissolution by organic acids. Soil Sci Plant Nutr 44:617–625
Saleem M, Arshad M, Hussain S, Bhatti AS (2007) Perspective of plant growth promoting rhizobacteria (PGPR) containing ACC deaminase in stress agriculture. J Ind Microbiol Biotechnol 34:635–648
Sánchez-Porro C, Rafael R, Soto-Ramírez N, Márquez MC, Montalvo-Rodríguez R, Ventosa A (2009) Description of Kushneria aurantia gen. nov., sp. nov., a novel member of the family Halomonadaceae, and a proposal for reclassification of Halomonas marisflavi as Kushneria marisflavi comb. nov., of Halomonas indalinina as Kushneria indalinina comb. nov. and of Halomonas avicenniae as Kushneria avicenniae comb. nov. Int J Syst Evol Microbiol 59:397–405

Sangeeth KP, Bhai RS, Srinivasan V (2012) Paenibacillus glucanolyticus, a promising potassium solubilizing bacterium isolated from black pepper (Piper nigrum L.) rhizosphere. J Spices Aromat Crop 21:118–124
Sannazzaro AI, Echeverría M, Albertó EO, Ruiz OA, Menéndez AB (2007) Modulation of polyamine balance in Lotus glaber by salinity and arbuscular mycorrhiza. Plant Physiol Biochem 45:39–46
Scavino AF, Pedraza RO (2013) The role of siderophores in plant growth-promoting bacteria. In: Bacteria in agrobiology: crop productivity. Springer, pp 265–285
Sengupta A, Chaudhuri S (2002) Arbuscular mycorrhizal relations of mangrove plant community at the Ganges river estuary in India. Mycorrhiza 12:169–174
Sgroy V, Cassán F, Masciarelli O, Del Papa MF, Lagares A, Luna V (2009) Isolation and characterization of endophytic plant growth-promoting (PGPB) or stress homeostasis-regulating (PSHB) bacteria associated to the halophyte Prosopis strombulifera. Appl Microbiol Biotechnol 85:371–381
Sha Valli Khan PS, Nagamallaiah GV, Dhanunjay Rao M, Sergeant K, Hausman JF (2014) Chapter 2: Abiotic stress tolerance in plants: insights from proteomics. In: Emerging technologies and management of crop stress tolerance. Academic, San Diego, pp 23–68. https://doi.org/10.1016/B978-0-12-800875-1.00002-8
Shamseldin A, Werner D (2004) Selection of competitive strains of Rhizobium nodulating Phaseolus vulgaris and adapted to environmental conditions in Egypt, using the gus-reporter gene technique. World J Microbiol Biotechnol 20:377–382
Shamseldin A, Werner D (2005) High salt and high pH tolerance of new isolated Rhizobium etli strains from Egyptian soils. Curr Microbiol 50:11–16
Shannon MC, Grieve CM (1998) Tolerance of vegetable crops to salinity. Sci Horticult 78:5–38
Sharifi M, Ghorbanli M, Ebrahimzadeh H (2007) Improved growth of salinity-stressed soybean after inoculation with salt pre-treated mycorrhizal fungi. J Plant Physiol 164:1144–1151
Sharma SB, Sayyed RZ, Trivedi MH, Gobi TA (2013) Phosphate solubilizing microbes: sustainable approach for managing phosphorus deficiency in agricultural soils. SpringerPlus 2:587
Sharma S, Kulkarni J, Jha B (2016) Halotolerant rhizobacteria promote growth and enhance salinity tolerance in peanut. Front Microbiol 7:1600
Shedova E, Lipasova V, Velikodvorskaya G, Ovadis M, Chernin L, Khmel I (2008) Arbuscular mycorrhizal relations of mangrove plant community at the Ganges river estuary in India. Mycorrhiza 53:110–114
Sheng XF (2005) Growth promotion and increased potassium uptake of cotton and rape by a potassium releasing strain of Bacillus edaphicus. Soil Biol Biochem 37:1918–1922
Sheng XF, He LY (2006) Solubilization of potassium-bearing minerals by a wild-type strain of Bacillus edaphicus and its mutants and increased potassium uptake by wheat. Can J Microbiol 52:66–72
Sheng M, Tang M, Chen H, Yang B, Zhang F, Huang Y (2008) Influence of arbuscular mycorrhizae on photosynthesis and water status of maize plants under salt stress. Mycorrhiza 18:287–296
Shrivastava P, Kumar R (2015) Soil salinity: a serious environmental issue and plant growth promoting bacteria as one of the tools for its alleviation. Saudi J Biol Sci 22:123–131
Shukla PS, Agarwal PK, Jha B (2012) Improved salinity tolerance of Arachis hypogaea (L.) by the interaction of halotolerant plant-growth-promoting rhizobacteria. J Plant Growth Regul 31:195–206
Siddikee MA, Chauhan PS, Anandham R, Han G-H, Sa T (2010) Isolation, characterization, and use for plant growth promotion under salt stress, of ACC deaminase-producing halotolerant bacteria derived from coastal soil. J Microbiol Biotechnol 20:1577–1584
Siddikee MA, Glick BR, Chauhan PS, Jong Yim W, Sa T (2011) Enhancement of growth and salt tolerance of red pepper seedlings (Capsicum annuum L.) by regulating stress ethylene synthesis with halotolerant bacteria containing 1-aminocyclopropane-1-carboxylic acid deaminase activity. Plant Physiol Biochem 49:427–434

Siddiqui MH, Mohammad F, Khan MN (2009) Morphological and physio-biochemical characterization of Brassica juncea L. Czern. & Coss. genotypes under salt stress. J Plant Interact 4:67–80

Sieverding E, da Silva GA, Berndt R, Oehl F (2015) Rhizoglomus, a new genus of the Glomeraceae. Mycotaxon 129:373–386

Singh RP, Jha PN (2016) A halotolerant bacterium Bacillus licheniformis HSW-16 augments induced systemic tolerance to salt stress in wheat plant (Triticum aestivum). Front Plant Sci 7:1890

Smith SE, Read DJ (2010) Mycorrhizal symbiosis. Academic

Smith SE, Jakobsen I, Grønlund M, Smith FA (2011) Roles of arbuscular mycorrhizas in plant phosphorus nutrition: interactions between pathways of phosphorus uptake in arbuscular mycorrhizal roots have important implications for understanding and manipulating plant phosphorus acquisition. Plant Physiol 156:1050–1057

Snapp S, Koide R, Lynch J (1995) Exploitation of localized phosphorus-patches by common bean roots. Plant Soil 177:211–218

Soleimani M, Akbar S, Hajabbasi MA (2011) Enhancing phytoremediation efficiency in response to environmental pollution stress. Plants Environ 23:10–14

Sparks DL, Huang PM (1985) Physical chemistry of soil potassium. In: Potassium in agriculture, pp 201–276

Srinivasan R, Yandigeri MS, Kashyap S, Alagawadi AR (2012) Effect of salt on survival and P-solubilization potential of phosphate solubilizing microorganisms from salt affected soils. Saudi J Biol Sci 19:427–434

Steffens D, Sparks DL (1997) Kinetics of nonexchangeable ammonium release from soils. Soil Sci Soc Am J 61:455–462

Stevenson FJ, Cole MA (1999) Cycles of soils: carbon, nitrogen, phosphorus, sulfur, micronutrients. Wiley, Hoboken

Sugawara M, Okazaki S, Nukui N, Ezura H, Mitsui H, Minamisawa K (2006) Rhizobitoxine modulates plant-microbe interactions by ethylene inhibition. Biotechnol Adv 24:382–388. https://doi.org/10.1016/j.biotechadv.2006.01.004

Szymańska S, Płociniczak T, Piotrowska-Seget Z, Złoch M, Ruppel S, Hrynkiewicz K (2016) Metabolic potential and community structure of endophytic and rhizosphere bacteria associated with the roots of the halophyte Aster tripolium L. Microbiol Res 182:68–79

Tang C, Robson AD, Dilworth MJ (1990) A split-root experiment shows that iron is required for nodule initiation in Lupinus angustifolius L. New Phytol 115:61–67

Teng S, Liu Y, Zhao L (2010) Isolation, identification and characterization of ACC deaminase-containing endophytic bacteria from halophyte Suaeda salsa. Wei Sheng Wu Xue Bao = Acta Microbiol Sin 50:1503–1509

Tester M, Davenport R (2003) Na+ tolerance and Na+ transport in higher plants. Ann Bot 91:503–527

Theunis M (2005) IAA biosynthesis in rhizobia and its potential role in symbiosis. PhD thesis, Universiteit Antwerpen

Tiwari S, Singh P, Tiwari R, Meena KK, Yandigeri M, Singh DP, Arora DK (2011) Salt-tolerant rhizobacteria-mediated induced tolerance in wheat (Triticum aestivum) and chemical diversity in rhizosphere enhance plant growth. Biol Fertil Soils 47:907

Toljander JF, Lindahl BD, Paul LR, Elfstrand M, Finlay RD (2007) Influence of arbuscular mycorrhizal mycelial exudates on soil bacterial growth and community structure. FEMS Microbiol Ecol 61:295–304

Tributh HV, Boguslawski EV, Lieres AV, Steffens D, Mengel K (1987) Effect of potassium removal by crops on transformation of illitic clay minerals. Soil Sci 143:404–409

Upadhyay SK, Singh DP (2015) Effect of salt-tolerant plant growth-promoting rhizobacteria on wheat plants and soil health in a saline environment. Plant Biol 17:288–293

Upadhyay S, Singh D, Saikia R (2009) Genetic diversity of plant growth promoting rhizobacteria isolated from rhizospheric soil of wheat under saline condition. Curr Microbiol 59:489–496. https://doi.org/10.1007/s00284-009-9464-1

Uroz S, Calvaruso C, Turpault M-P, Frey-Klett P (2009) Mineral weathering by bacteria: ecology, actors and mechanisms. Trends Microbiol 17:378–387

van Hoorn JW, Katerji N, Hamdy A, Mastrorilli M (2001) Effect of salinity on yield and nitrogen uptake of four grain legumes and on biological nitrogen contribution from the soil. Agric Water Manag 51:87–98. https://doi.org/10.1016/S0378-3774(01)00114-7

Van Vuuren DP, Bouwman AF, Beusen AHW (2010) Phosphorus demand for the 1970–2100 period: a scenario analysis of resource depletion. Glob Environ Chang 20:428–439

Vassilev N, Fenice M, Federici F (1996) Rock phosphate solubilization with gluconic acid produced by immobilized Penicillium variabile P16. Biotechnol Tech 10:585–588

Venkateswarlu B, Rao AV, Raina P (1984) Evaluation of phosphorus solubilisation by microorganisms isolated from Aridisols. J Indian Soc Soil Sci 32:273–277

Venkateswarlu B, Desai S, Prasad YG (2008) Agriculturally important microorganisms for stressed ecosystems: challenges in technology development and application. In: Khachatourians GG, Arora DK, Rajendran TP, Srivastava AK (eds) Agriculturally important, Microorganisms, vol 1. Academic World, Bhopal, pp 225–246

Veresoglou SD, Shaw LJ, Sen R (2011) Glomus intraradices and Gigaspora margarita arbuscular mycorrhizal associations differentially affect nitrogen and potassium nutrition of Plantago lanceolata in a low fertility dune soil. Plant Soil 340:481–490

Vreeland RH (1987) Mechanisms of halotolerance in microorganisms. CRC Crit Rev Microbiol 14:311–356

Waller F et al (2005) The endophytic fungus Piriformospora indica reprograms barley to salt-stress tolerance, disease resistance, and higher yield. Proc Natl Acad Sci U S A 102:13386–13391

Wang Q, Dodd IC, Belimov AA, Jiang F (2016) Rhizosphere bacteria containing 1-aminocyclopropane-1-carboxylate deaminase increase growth and photosynthesis of pea plants under salt stress by limiting Na+ accumulation. Funct Plant Biol 43:161–172

Werner D (1992) Symbiosis of plants and microbes. Chapman & Hall, London

Whitelaw MA, Harden TJ, Bender GL (1997) Plant growth promotion of wheat inoculated with Penicillium radicum sp. nov. Soil Res 35:291–300

Wild A (1988) Plant nutrients in soil: phosphate. In: Russell's soil conditions and plant growth

Williams A, Ridgway HJ, Norton DA (2013) Different arbuscular mycorrhizae and competition with an exotic grass affect the growth of Podocarpus cunninghamii Colenso cuttings. New For 44:183–195

Wu SC, Cao ZH, Li ZG, Cheung KC, Wong MH (2005) Effects of biofertilizer containing N-fixer, P and K solubilizers and AM fungi on maize growth: a greenhouse trial. Geoderma 125:155–166

Wu Q-S, Xia R-X, Zou Y-N (2008) Improved soil structure and citrus growth after inoculation with three arbuscular mycorrhizal fungi under drought stress. Eur J Soil Biol 44:122–128

Wu Q-S, Zou Y-N, He X-H (2010) Contributions of arbuscular mycorrhizal fungi to growth, photosynthesis, root morphology and ionic balance of citrus seedlings under salt stress. Acta Physiol Plant 32:297–304

Wu Z, Yue H, Lu J, Li C (2012) Characterization of rhizobacterial strain Rs-2 with ACC deaminase activity and its performance in promoting cotton growth under salinity stress. World J Microbiol Biotechnol 28:2383–2393

Xie H, Pasternak JJ, Glick BR (1996) Isolation and characterization of mutants of the plant growth-promoting rhizobacterium Pseudomonas putida GR12-2 that overproduce indoleacetic acid. Curr Microbiol 32:67–71

Yang J, Kloepper JW, Ryu C-M (2009) Rhizosphere bacteria help plants tolerate abiotic stress. Trends Plant Sci 14:1–4

Yao L, Wu Z, Zheng Y, Kaleem I, Li C (2010) Growth promotion and protection against salt stress by Pseudomonas putida Rs-198 on cotton. Eur J Soil Biol 46:49–54

Yasmin H, Bano A (2011) Isolation and characterization of phosphate solubilizing bacteria from rhizosphere soil of weeds of Khewra salt range and Attock. Pak J Bot 43:1663–1668

Yildirim E, Taylor AG (2005) Effect of biological treatments on growth of bean plans under salt stress. Ann Rep Bean Improv Coop 48:176–177

Yoon J-H, Choi SH, Lee K-C, Kho YH, Kang KH, Park Y-H (2001) Halomonas marisflavae sp. nov., a halophilic bacterium isolated from the Yellow Sea in Korea. Int J Syst Evol Microbiol 51:1171–1177

Yousfi S, Mahmoudi H, Abdelly C, Gharsalli M (2007) Effect of salt on physiological responses of barley to iron deficiency. Plant Physiol Biochem 45:309–314

Zacarías JJJ, Altamirano-Hernández J, Cabriales JJP (2004) Nitrogenase activity and trehalose content of nodules of drought-stressed common beans infected with effective (Fix+) and ineffective (Fix−) rhizobia. Soil Biol Biochem 36:1975–1981

Zhang L, Fan J, Ding X, He X, Zhang F, Feng G (2014) Hyphosphere interactions between an arbuscular mycorrhizal fungus and a phosphate solubilizing bacterium promote phytate mineralization in soil. Soil Biol Biochem 74:177–183

Zhou N, Zhao S, Tian CY (2017) Effect of halotolerant rhizobacteria isolated from halophytes on the growth of sugar beet (Beta vulgaris L.) under salt stress. FEMS Microbiol Lett 364. https://doi.org/10.1093/femsle/fnx091

Zhu F, Qu L, Hong X, Sun X (2011) Isolation and characterization of a phosphate-solubilizing halophilic bacterium Kushneria sp. YCWA18 from Daqiao Saltern on the coast of Yellow Sea of China. Evid Based Complement Alternat Med 2011

Zuccarini P, Okurowska P (2008) Effects of mycorrhizal colonization and fertilization on growth and photosynthesis of sweet basil under salt stress. J Plant Nutr 31:497–513

Zwetsloot MJ, Lehmann J, Bauerle T, Vanek S, Hestrin R, Nigussie A (2016) Phosphorus availability from bone char in a P-fixing soil influenced by root-mycorrhizae-biochar interactions. Plant Soil 408:95–105

Bacterial Endophytes from Halophytes: How Do They Help Plants to Alleviate Salt Stress?

6

Ignacio D. Rodríguez-Llorente, Eloisa Pajuelo, Salvadora Navarro-Torre, Jennifer Mesa-Marín, and Miguel A. Caviedes

Abstract

Soil salinization is one of the main factors influencing the reduction of crop yield all over the world. Solutions for saline agriculture based on biological systems involving plants or cultivars well-adapted to grow in soils affected by salinity and associated microorganisms are gaining interest. Halophytes can survive and reproduce in environments with high salt concentration and are excellent models to study plant adaptation to saline environments. Plant growth and adaptation, particularly in stress conditions, is highly influenced by microorganisms that colonize rhizosphere and endosphere. Halophyte microbiome has to be adapted to soil salinity and contribute to plant growth in the presence of high concentrations of NaCl. Endophytic bacteria isolated from halophytes growing in saline soils may help to alleviate plant by altering plant hormone status and uptake of nutrient elements and/or modulating the production of reactive oxygen species (ROS) through different mechanism, including 1-aminocyclopropane-1-carboxylic acid (ACC)-deaminase activity, phosphate solubilization, nitrogen fixation, and the production of indole-3-acetic acid (IAA), abscisic acid (ABA), siderophores, and volatiles, among others. In this chapter, the main genera of endophytes isolated from halophytes, the presence of plant growth-promoting properties in these endophytes, and how they could help to alleviate salt stress in plants are reviewed and discussed.

Keywords

Bacterial endophytes · Halophytes · Salt stress · PGPB

I. D. Rodríguez-Llorente (✉) · E. Pajuelo · S. Navarro-Torre · J. Mesa-Marín · M. A. Caviedes
Departamento de Microbiología y Parasitología, Facultad de Farmacia, Universidad de Sevilla, Sevilla, Spain
e-mail: irodri@us.es

M. Kumar et al. (eds.), *Saline Soil-based Agriculture by Halotolerant Microorganisms*, https://doi.org/10.1007/978-981-13-8335-9_6

6.1 Introduction

Soil salinization is a worldwide problem that could affect 1–10 billion hectares with a potential increase of around 15% per year (Yensen and Biel 2006). It is also estimated that up to 50% of the irrigated lands could be affected by salinity or sodicity (Pitman and Lauchli 2002). This growing problem is one of the main factors influencing the reduction in crop yields all over the world, since salinity severely affects plant growth (Gerhardt et al. 2017). Revegetation and remediation of ecosystems affected by salinity are mandatory to feed the world growing population.

Traditional systems to reduce soil salinity include lixiviation, which requires high amounts of water, and the addition of organic and chemical compounds, which are expensive and impoverish soils (Jesus et al. 2015). In that way, biological systems involving plants and associated microorganisms are gaining interest in salinity soil remediation (Qin et al. 2016).

Physiological and molecular mechanisms undergoing salinity tolerance in plants have been widely studied and reviewed (Deinlein et al. 2014; Gupta and Huang 2014; Mickelbart et al. 2015; Zhu 2016). This information has been used to develop strategies for alleviating crop salinity stress using plants (Ismail and Horie 2017). A group of these strategies is based on the construction of transgenic plants overexpressing genes related with plant salt adaptation (Roy et al. 2014). Although these approaches are still being developed, several aspects limit their success including the following: (i) they require a lot of time and effort (Coleman-Derr and Tringe 2014); (ii) the mutants sometimes are unstable (Jewell et al. 2010); (iii) there are important tetraploid and hexaploid crop species for which molecular techniques are not applicable (Kumar et al. 2015); and (iv) actually they do not have a public acceptance, and this acceptance in the future is still uncertain (Fedoroff et al. 2010). In addition, saline stress is frequently linked in nature to other abiotic stresses, such as alkaline stress (Bui et al. 2014) or stress by organic and metal contaminants.

Due to all these limitations, alternative ecological strategies involving plants or cultivars well-adapted to grow in soils affected by salinity and microorganisms associated with these plants are gaining interest to fight against soil salinization. In this context, halophytes, plants that can survive and reproduce in environments where the salt concentration exceeds 200 mM NaCl (Flowers and Colmer 2008), are excellent models to study plant adaptation to saline environments and develop strategies to improve salt stress tolerance in plants (Shabala 2013).

It is well-accepted that plant growth and adaptation, particularly in stress conditions, is influenced by microorganisms that colonize plant rhizosphere and endosphere, that is, the plant microbiome (Bhattacharyya and Jha 2012). Most of these microorganisms have plant growth-promoting (PGP) properties, a group of direct and indirect mechanisms that promote plant growth. Although it is not easy to differentiate rhizospheric and endophytic microorganisms (most rhizobacteria colonize the root surface and even penetrate into the root cortex), this chapter will be focused on the mechanisms by which endophytes, particularly those isolated from halophytes, could help plants to alleviate salt stress.

6.2 Halophytes and Their Associated Microorganisms

Comparing with human microbiome research, plant microbiome research is still being born (Bai et al. 2015). Concerning halophyte microbiomes, only few have been studied in the last decade, and most of them were works merely descriptive that tried to classify and count microbial species (Ruppel et al. 2013). More recently, microorganisms (particularly bacteria) have been isolated from the rhizosphere of halophytes and screened for PGP properties and tolerance to NaCl. Some examples are the isolation of plant growth-promoting bacteria (PGPB) from the rhizospheres of *Sesuvium verrucosum* in Saudi Arabia (El-Awady et al. 2015), *Spartina maritima* and *Arthrocnemum macrostachyum* in Spain (Mesa et al. 2015a; Navarro-Torre et al. 2016), or *Salsola grandis* in Turkey (Kataoka et al. 2017). There are also a lower number of examples of endophytes isolated from halophytes such as *Salicornia brachiata*, *Sphaeranthus indicus*, *Cressa cretica*, and *Suaeda nudiflora* growing in India (Arora et al. 2014); *Halimione portulacoides* in Portugal (Fidalgo et al. 2016); or *Spartina maritima* and *Arthrocnemum macrostachyum* in Spain (Mesa et al. 2015b; Navarro-Torre et al. 2016; Navarro-Torre et al. 2017a).

As expected, most of bacteria isolated from halophyte microbiomes were halotolerant or halophiles, that is, they were able to grow at high concentrations of NaCl or required NaCl to grow. How these microorganisms influence halophytes adaptation to saline environments is an interesting unexplored task that is gaining interest in the scientific community. The growing interest in these microbiomes results in the recent isolation of a good amount of novel species with PGP properties (Table 6.1).

Description of these new species most of the times requires sequencing their complete genomes, providing good genomic resources to study the mechanisms underlying plant growth promotion. In the same context, the study of the complete microbiome of the halophyte *Suaeda salsa* revealed the presence of a high number of genes contributing to salt stress acclimatization and nutrient solubilization (Yuan et al. 2016). Halophyte microbiome has to be adapted to soil salinity, and this advantage may be exploited to design bacterial bioinoculants for salt phytoremediation or to sustain crop production in salty soils. In this sense, an exhaustive revision on rhizobacteria isolated from halophytes and examples of how they could help plants deal with salinity have been recently published (Etesami and Beattie 2018).

6.3 Halophyte-Microbiome Interactions and Abiotic Stress

Halophytes have evolved to tolerate high salt concentrations employing several well-described mechanism (Kumari et al. 2015a, b; Shabala 2013). These mechanisms may be not exclusive to sodium and chloride and may confer tolerance to other toxic ions and compounds. Indeed, some halophytes have showed their ability to tolerate and/or accumulate high levels of toxic pollutants. For example, several species of the genus *Spartina* are able to tolerate and accumulate in their tissues organic and

Table 6.1 New bacteria species with PGP properties isolated from halophyte microbiomes in the last 2 years

Halophyte	New species	Location	Reference
Spartina maritima	*Marinomonas spartinae*	Endophyte	Lucena et al. (2016)
	Vibrio spartinae	Endophyte	Lucena et al. (2017)
Arthrocnemum macrostachyum	*Microbulbifer rhizosphaerae*	Rhizosphere	Camacho et al. (2016a)
	Labrenzia salina	Rhizosphere	Camacho et al. (2016b)
	Vibrio palustris	Endophyte	Lucena et al. (2017)
	Kocuria salina	Rhizosphere	Camacho et al. (2017)
Halimione portulacoides	*Zunongwangia endophytica*	Endophyte	Fidalgo et al. (2017a)
	Altererythrobacter halimionae	Endophytes	Fidalgo et al. (2017b)
	A. endophyticus		
	Saccharospirillum correiae	Endophyte	Fidalgo et al. (2017c)
Plantago winteri	*Ancylobacter pratisalsi*	Rhizosphere	Suarez et al. (2017)
Tamarix chinensis	*Salinicola tamaricis*	Endophyte	Zhao et al. (2017)
Limonium sinense	*Glutamicibacter halophytocola*	Endophyte	Feng et al. (2017)
Anabasis elatior	*Aurantimonas endophytica*	Endophyte	Liu et al. (2016)
Suaeda maritime	*Martelella suaedae*	Endophyte	Chung et al. (2016)
Limonium tetragonum	*Martelella limonii*	Endophyte	Chung et al. (2016)

inorganic contaminants (Mackova et al. 2006; Redondo-Gómez et al. 2011). Particularly, *S. densiflora* and *S. maritima* were able to bioaccumulate high levels of heavy metals in their tissues and mainly in roots (Redondo-Gómez 2013). In the same way, the ability of *A. macrostachyum* to accumulate high amounts of Cd has been reported (Redondo-Gómez et al. 2010).

Microorganisms associated with halophytes could promote plant phytoremediation capacity by reducing plant stress response (Mittler et al. 2004), producing detoxification of pollutants, or acting over pollutant bioavailability. Several recent works have reported the mitigation of toxic effects of pollutants on halophytes by microorganisms of their microbiomes. For example, bacteria from the rhizospheres of both *Spartina densiflora* and *S. maritima* were able to improve plant growth under salt and metal stress (Mateos-Naranjo et al. 2015; Mesa et al. 2015c; Paredes-Páliz et al. 2017), and endophytes isolated from *S. maritima* also improved plant growth in polluted marsh soils (Mesa et al. 2015b). Similarly, both endophytic and rhizospheric bacteria associated with *A. macrostachyum* promoted plant growth in heavy metal-contaminated marsh soils (Navarro-Torre et al. 2017b). Concerning organic pollutants, halophytes such as *Juncus acutus*, *Halocnemum strobilaceum*, or

Scirpus triqueter assisted by their associated bacteria (both rhizospheric and endophytic) have proven to be an interesting tool for hydrocarbon remediation in oil-contaminated soils (Al-Mailem et al. 2010; Liu et al. 2011; Syranidou et al. 2017; Zhao et al. 2013). Based on the results published in these works, the couple halophyte-microbiome has been proposed as an ecological and efficient tool for the restoration of contaminated and degraded marsh soils (Zhao et al. 2013).

6.4 How Do Endophytes Isolated from Halophytes Help Plants to Alleviate Salt Stress?

Plant-associated microbes are essential for plant adaptation to salinity (Munns and Gilliham 2015; Tkacz and Poole 2015). In general, the induction of plant tolerance to different abiotic stresses is referred to as induced systemic tolerance (IST) (Yang et al. 2009). PGPB elicit plant tolerance to salt stress (and other abiotic stresses) by altering plant hormone status (Vurukonda et al. 2016) and uptake of nutrient elements (Gerhardt et al. 2017) and/or modulate the production of reactive oxygen species (ROS) (Gururani et al. 2013). Several bacterial traits seem to be involved in these processes, including 1-aminocyclopropane-1-carboxylic acid (ACC)-deaminase activity, phosphate solubilization, nitrogen fixation, and the production of hormones, particularly indole-3-acetic acid (IAA) and abscisic acid (ABA), siderophores, and volatiles (Fig. 6.1) (Farag et al. 2013; Kumari et al. 2015; Nadeem et al. 2016).

6.4.1 Endophytes and Plant Hormone Status

Plants subjected to salinity stress produce high levels of ethylene, which retard root development (Mahajan and Tuteja 2005). Endophytes with the ability to produce ACC deaminase, enzyme that catalyzes the conversion of the ethylene precursor ACC to ammonia and α-ketobutyrate, could reduce plant ethylene levels helping to overcome salt-induced growth inhibition (Senthilkumar et al. 2009). In addition, these endophytes could provide a nitrogen source (ammonia) to plants (Hardoim et al. 2008). ACC deaminase-producing endophytes have been isolated from the halophytes *H. portulacoides*, *Limonium sinense*, *Prosopis strombulifera*, *S. maritima*, or *Salicornia europaea*, among others (Fidalgo et al. 2016; Mesa et al. 2015b; Qin et al. 2014; Sgroy et al. 2009; Zhao et al. 2016). However, the direct effect of these bacteria on plant growth under saline stress has been poorly explored. ACC deaminase-producing endophytes isolated from *L. sinense* belonging to genera *Arthrobacter*, *Bacillus*, *Isoptericola*, and *Streptomyces* showed the ability to promote plant growth under salt stress (Qin et al. 2014). Interestingly, the most promising strains showed no or very low levels of IAA and were not able to solubilize phosphate, suggesting that the observed capacity to alleviate salt stress may be related to their capacity to produce ACC deaminase (Qin et al. 2014). Similarly, endophytes isolated from *S. europaea* that belonged to the genera *Arthrobacter*, *Bacillus*, *Planococcus*, and *Variovorax* with ACC deaminase activity enhanced in vitro seedlings growth in

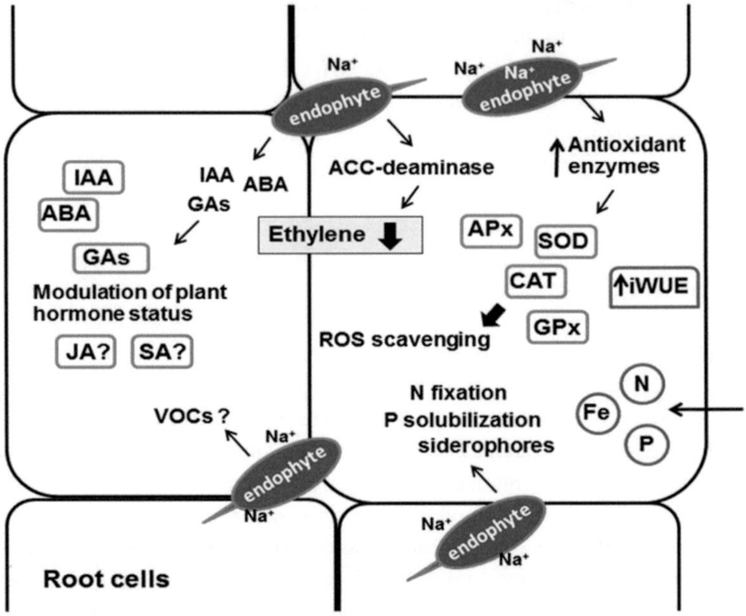

Fig. 6.1 Overview of the mechanisms by which bacterial endophytes isolated from halophyte microbiome could help to ameliorate plant growth in conditions of elevated salinity according to published results. Endophytes can produce ACC deaminase and phytohormones to modulate plant hormone status. ACC deaminase-producing endophytes reduce the excessive ethylene production caused by salt stress. Endophytes are capable of increasing the antioxidative system in plants for ROS scavenging. Endophytes can enhance plant nutrient uptake by several mechanism and could accumulate or bind Na^+. *ABA* abscisic acid, *APx* ascorbate peroxidase, *CAT* catalase, *GAs* gibberellic acids, *GPx* guaiacol peroxidase, *IAA* índole-3-acetic acid, *iWUE* water use efficiency, *JA* jasmonic acid, *ROS* reactive oxygen species, *SA* salicylic acid, *SOD* superoxide dismutase, *VOCs* volatile organic compounds

the presence of increasing concentrations of NaCl; but these strains were also able to solubilize phosphate and/or produce IAA, which could contribute to the observed *S. europaea* growth promotion (Zhao et al. 2016).

Secretion of phytohormones and particularly IAA in vitro is probably the most common trait in PGPB (Duca et al. 2014). IAA is directly involved in root generation and growth (Birnbaum 2016), so IAA-producing endophytes could increase plant salt tolerance by stimulating root proliferation. IAA-secreting endophytes have been isolated from several halophytes, including *P. strombulifera*, *A. macrostachyum*, *S. maritima*, *S. europaea*, *H. portulacoides*, *or Psoralea corylifolia* (Fidalgo et al. 2016; Mesa et al. 2015b; Zhao et al. 2016; Navarro-Torre et al. 2017a; Navarro-Torre et al. 2017b; Sgroy et al. 2009; Sorty et al. 2016). IAA-producing endophytes belonging to the genera *Bacillus*, *Marinobacterium*, and *Sinorhizobium* isolated from *P. corylifolia* plants were able to enhance wheat seedling growth under salinity stress when inoculated separately (Sorty et al. 2016). More recently, inoculation with an endophytic bacterial *consortium* isolated from *A. macrostachyum* mitigated the effects of high salinity on plant growth and physiological performance (Navarro-Torre et al. 2017a). In that work, the bacterial *consortium* was composed of IAA-producing endophytes from genera *Bacillus* and *Gracilibacillus*; nevertheless, the increase in plant growth and salt tolerance cannot be directly attributed to the production of IAA due to the presence in the endophytes of other PGP properties, such as phosphate solubilization and siderophore production.

Modulation of plant abscisic acid (ABA), an important plant stress hormone, by ABA-producing and/or ABA-metabolizing PGPB may also contribute to mitigate the effects of salt stress in plants (Gerhardt et al. 2017). Nevertheless, the ability to produce or metabolize ABA in endophytes isolated from halophytes has been poorly investigated. Sgroy et al. (2009) reported the isolation of seven endophytic bacterial strains from the halophyte *P. strombulifera* with capacity to produce ABA in vitro. These strains belonged to the genera *Achromobacter*, *Bacillus*, *Brevibacterium*, *Lysinibacillus*, and *Pseudomonas*. Most of these strains were also able to produce gibberellic acids (GAs), a plant hormone that positively regulates cell division and elongation, hypocotyl and stem growth, and leaf and root-meristem size (Martínez et al. 2016; Wang et al. 2015). Although GA signaling is a key factor in the inhibition of plant growth under stress (Martínez et al. 2016), production of GAs by endophytes is not frequently assayed. Other hormones that are also involved in abiotic stress resistance in plants, such as jasmonic and salicylic acids (Ahmad et al. 2016), are also synthesized by several endophytes (Chen et al. 2014; Forchetti et al. 2007), but their production by endophytes isolated from halophytes has not been explored until date.

6.4.2 Endophytes and Plant Nutrient Uptake

PGPB can enhance the uptake of nutrients in plants. Endophytes with the ability to solubilize insoluble phosphate, potassium, and zinc, fix atmospheric nitrogen asymbiotically and/or release siderophores to scavenge iron which could promote plant growth under abiotic stress (Santoyo et al. 2016). These bacterial traits are often investigated in endophytes isolated from halophytes, but plant inoculation with selected bacteria under saline stress is not so frequent (Sgroy et al. 2009). *Bacillus* and *Gracilibacillus* endophytes isolated from *A. macrostachyum* showed the ability to solubilize phosphate, fix atmospheric nitrogen, and produce siderophores (Navarro-Torre et al. 2017a). A bacterial *consortium* containing three of these endophytes was able to promote plant growth at high saline concentrations and also to increase the level of NaCl accumulated by the plant (Navarro-Torre et al. 2017a). Mesa et al. (2015b) also reported how a bacterial endophytic *consortium*, including genera *Micrococcus*, *Salinicola*, and *Vibrio*, helped to alleviate the effect of heavy metals and NaCl on *S. maritima* growth in contaminated salt marsh soils. These endophytes combined in vitro IAA and siderophore production, phosphate solubilization, and nitrogen fixation. Interestingly, inoculation of *S. maritima* plants with the endophytes increased (15%) water use efficiency (WUE), an important indicator of how plants manage water under stress conditions (Tardieu 2012). Similar results were obtained using an endophytic bacterial *consortium* isolated from *A. macrostachyum* as inoculants (Navarro-Torre et al. 2017b). In this work, endophytes belonging to genera *Bacillus*, *Halomonas*, *Kushneria*, and *Micrococcus* ameliorated *A. macrostachyum* growth in salt marsh soils that were also contaminated with heavy metals. This *consortium* was able to produce IAA and siderophores, solubilize phosphate, and fix nitrogen asymbiotically.

6.4.3 Endophytes and Levels of Reactive Oxygen Species (ROS)

Increased production of ROS is commonly observed in plants growing under high salinity (Miller et al. 2010). Endophytes may increase plant tolerance to salt and other abiotic stresses modulating the production of ROS. An endophytic bacterial *consortium* isolated from *A. macrostachyum* composed by genera *Bacillus*, *Halomonas*, *Kushneria*, and *Micrococcus* induced changes in antioxidative enzyme activities, such as ascorbate peroxidase (APx), catalase (CAT), guaiacol peroxidase (GPx), and superoxide dismutase (SOD) (Navarro-Torre et al. 2017b). These enzymes contribute to maintain the homeostasis against the high levels of ROS (Mesnoua et al. 2016). *A. macrostachyum* plants inoculated with the endophytic *consortium* showed enhanced APx, GPx, and SOD activities in saline marsh soils contaminated with heavy metals, suggesting that endophytes could contribute to reduce the oxidative stress derived from the amount of metal and NaCl accumulated by the plant. Unfortunately, assays using endophytes and NaCl as single abiotic stress are not found in the literature.

6.4.4 Endophytes and Other Mechanism of Salt Tolerance

Other mechanisms that could be relevant to PGPB-conferred salt tolerance have been partially addressed using rhizobacteria. For example, inoculation with rhizobacteria has shown upregulation of aquaporin gene expression (Gond et al. 2015; Qin et al. 2016; Vurukonda et al. 2016), directly related with root hydraulic conductance that determines root water uptake capacity (Moshelion et al. 2015). Some rhizospheric PGPB release volatile organic compounds (VOCs) that could induce physiological changes in plants (Liu and Zhang 2015). Nevertheless, the involvement of these mechanisms in plant salt tolerance elicited by endophytes has not been explored until date.

Endophytes inhabiting halophytes have to be adapted to the salt concentrations found within plant tissues. As previously indicated, most of the endophytes isolated from halophytes mentioned in this review are halophilic or halotolerants. These bacteria are able to grow in the presence of high concentrations of NaCl by different mechanisms generally involving the synthesis or accumulation of organic solutes to provide osmotic equilibrium of the cytoplasm with the surrounding medium (Ventosa et al. 1998). Since halophilic and halotolerant bacteria are able to accumulate Na^+ within their cytoplasm or in the cell surface, it can be speculated that sodium sequestration and/or accumulation by endophytes could also contribute to plant tolerance and growth in saline environments.

6.5 Conclusion and Perspectives

Bacterial endophytes inhabiting halophytes have to be adapted to salinity and contribute to plant adaptation and growth in salty soils. There are a growing number of manuscripts describing the isolation and characterization of bacterial endophytes from halophyte plants with PGP properties in vitro. Although endophytic bacterial traits related with plant hormone regulation, nutrient uptake, and ROS production, among others, might have a collaborative effect in eliciting plant salt tolerance, direct demonstration of the concrete molecular mechanisms involved in such induced tolerance has not been explored until date. Studies of complete microbiomes of halophytes and comparison of recently sequenced genomes of endophytes isolated from these plants could provide the genomic tools to decipher the influence of the different PGP properties in plant tolerance. For example, the construction of knockout mutants affected in the production of concrete traits and the inoculation of plants with such mutants will help to establish the influence of each trait or property in conferring salt tolerance. These studies could provide useful information to select and design bioinoculants aimed to improve crop production in salt-affected soils.

Acknowledgment We wish to thank the University of Sevilla for providing the necessary facilities for this study.

References

Ahmad P, Rasool S, Gul A, Sheikh SA, Akram NA, Ashraf M, Kazi AM, Gucel S (2016) Jasmonates: multifunctional roles in stress tolerance. Front Plant Sci 7:813

Al-Mailem D, Sorkhoh N, Marafie M, Al-Awadhi H, Eliyas M, Radwan S (2010) Oil phytoremediation potential of hypersaline coasts of the Arabian Gulf using rhizosphere technology. Bioresour Technol 101:5786–5792

Arora S, Patel PN, Vanza MJ, Rao GG (2014) Isolation and characterization of endophytic bacteria colonizing halophyte and other salt tolerant plant species from coastal Gujarat. Afr J Microbiol Res 8:1779–1788

Bai Y, Müller DB, Srinivas G, Garrido-Oter R, Potthoff E, Rott M, Dombrowski N, Münch PC, Spaepen S, Remus-Emsermann M, Hüttel B, McHardy AC, Vorholt JA, Schulze-Lefert P (2015) Functional overlap of the *Arabidopsis* leaf and root microbiota. Nat Res 528:364–369

Bhattacharyya PN, Jha DK (2012) Plant growth-promoting rhizobacteria (PGPR): emergence in agriculture. World J Microbiol Biotechnol 28:1327–1350

Birnbaum KD (2016) How many ways are there to make a root? Curr Opin Plant Biol 34:61–67

Bui EN, Thornhill A, Miller JT (2014) Salt- and alkaline-tolerance are linked in *Acacia*. Biol Lett 10:20140278

Camacho M, Montero-Calasanz M, Redondo-Gómez S, Rodríguez-Llorente I, Schumann P, Klenk HP (2016a) *Microbulbifer rhizosphaerae* sp. nov., isolated from the rhizosphere of the halophyte *Arthrocnemum macrostachyum*. Int J Syst Evol Microbiol 66:1844–1850

Camacho M, Redondo-Gómez S, Rodríguez-Llorente I, Rohde M, Spröer C, Schumann P, Klenk HP, Montero-Calasanz MD (2016b) *Labrenzia salina* sp. nov., isolated from the rhizosphere of the halophyte *Arthrocnemum macrostachyum*. Int J Syst Evol Microbiol 66:5173–5180

Camacho M, Redondo-Gomez S, Rodríguez-Llorente I, Rohde M, Spröer C, Schumann P, Klenk HP, Montero-Calasanz MDC (2017) *Kocuria salina* sp. nov., an actinobacterium isolated from the rhizosphere of the halophyte *Arthrocnemum macrostachyum* and emended description of *Kocuria turfanensis*. Int J Syst Evol Microbiol 67:5006–5012

Chen Y, Fan J-B, Du L, Xu H, Zhang Q-H, He Y-Q (2014) The application of phosphate solubilizing endophyte *Pantoea dispersa* triggers the microbial community in red acidic soil. Appl Soil Ecol 84:235–244

Chung EJ, Hwang JM, Kim KH, Jeon CO, Chung YR (2016) *Martelella suaedae* sp. nov. and *Martelella limonii* sp. nov., isolated from the root of halophytes. Int J Syst Evol Microbiol 66:3917–3922

Coleman-Derr D, Tringe SG (2014) Building the crops of tomorrow: advantages of symbiont-based approaches to improving abiotic stress tolerance. Front Microbiol 5:283

Deinlein U, Stephan AB, Horie T, Luo W, Xu G, Schroeder JI (2014) Plant salt tolerance mechanisms. Trends Plant Sci 19:371–379

Duca D, Lorv J, Patten CL, Rose D, Glick BR (2014) Indole-3-acetic acid in plant–microbe interactions. Ant Leeuw 106:85–125

El-Awady MAM, Hassan MM, Al-Sodany YM (2015) Isolation and characterization of salt tolerant endophytic and rhizospheric plant growth promoting bacteria (PGPB) associated with the halophyte plant (*Sesuvium verrucosum*) grown in KSA. Int J Appl Sci Biotechnol 33:552–560

Etesami H, Beattie GA (2018) Mining halophytes for plant growth-promoting halotolerant bacteria to enhance the salinity tolerance of non-halophytic crops. Front Microbiol 9:148

Farag MA, Zhang H, Ryu CM (2013) Dynamic chemical communication between plants and bacteria through airborne signals: induced resistance by bacterial volatiles. J Chem Ecol 39:1007–1018

Fedoroff NV, Battisti DS, Beachy RN, Cooper PJ, Fischhoff DA, Hodges CN, Knauf VC, Lobell D, Mazur BJ, Molden D, Reynolds MP, Ronald PC, Rosegrant MW, Sanchez PA, Vonshak A, Zhu J-K (2010) Radically rethinking agriculture for the 21st century. Science 327:833–834

Feng WW, Wang TT, Bai JL, Ding P, Xing K, Jiang JH, Peng X, Qin S (2017) *Glutamicibacter halophytocola* sp. nov., an endophytic actinomycete isolated from the roots of a coastal halophyte, *Limonium sinense*. Int J Syst Evol Microbiol 67:1120–1125

Fidalgo C, Henriques I, Rocha J, Tacao M, Alves A (2016) Culturable endophytic bacteria from the salt marsh plant *Halimione portulacoides*: phylogenetic diversity, functional characterization, and influence of metal(loid) contamination. Environ Sci Pollut Res 23:10200–11024

Fidalgo C, Martins R, Proença DN, Morais PV, Alves A, Henriques I (2017a) *Zunongwangia endophytica* sp. nov., an endophyte isolated from the salt marsh plant, *Halimione portulacoides*, and emended description of the genus *Zunongwangia*. Int J Syst Evol Microbiol 67:3004–3009

Fidalgo C, Rocha J, Martins R, Proença DN, Morais PV, Henriques I, Alves A (2017b) *Altererythrobacter halimionae* sp. nov. and *Altererythrobacter endophyticus* sp. nov., two endophytes from the salt marsh plant *Halimione portulacoides*. Int J Syst Evol Microbiol 67:3057–3062

Fidalgo C, Rocha J, Proença DN, Morais PV, Alves A, Henriques I (2017c) *Saccharospirillum correiae* sp. nov., an endophytic bacterium isolated from the halophyte *Halimione portulacoides*. Int J Syst Evol Microbiol 67:2026–2030

Flowers TJ, Colmer TD (2008) Salinity tolerance in halophytes. New Phytol 179:945–963

Forchetti G, Masciarelli O, Alemano S, Alvarez D, Abdala G (2007) Endophytic bacteria in sunflower (*Helianthus annuus* L.): isolation, characterization, and production of jasmonates and abscisic acid in culture medium. Appl Microbiol Biotechnol 76:1145–1152

Gerhardt KE, Macneill GJ, Gerwing PD, Greenberg BM (2017) Phytoremediation of salt-impacted soils and use of plant growth-promoting rhizobacteria (PGPR) to enhance phytoremediation. Phytoremediation. https://doi.org/10.1007/978-3-319-52381-1

Gond SK, Torres MS, Bergen MS, Helsel Z, White JF (2015) Induction of salt tolerance and up-regulation of aquaporin genes in tropical corn by rhizobacterium *Pantoea agglomerans*. Lett Appl Microbiol 60:392–399

Gupta B, Huang B (2014) Mechanism of salinity tolerance in plants: physiological, biochemical, and molecular characterization. Int J Genomics 2014:701596

Gururani MA, Upadhyaya CP, Baskar V, Venkatesh J, Nookaraju A, Park SW (2013) Plant growth-promoting rhizobacteria enhance abiotic stress tolerance in *Solanum tuberosum* through inducing changes in the expression of ROS-scavenging enzymes and improved photosynthetic performance. J Plant Growth Regul 32:245–258

Hardoim PR, Overbeek LSV, Elsas JDV (2008) Properties of bacterial endophytes and their proposed role in plant growth. Trends Microbiol 16:463–471

Ismail AM, Horie T (2017) Genomics, physiology, and molecular breeding approaches for improving salt tolerance. Annu Rev Plant Biol 68:405–434

Jesús JM, Danko AS, Fiúza A, Borges M-T (2015) Phytoremediation of salt-affected soils: a review of processes, applicability, and the impact of climate change. Environ Sci Pollut Res 22:6511–6525

Jewell MC, Campbell BC, Godwin ID (2010) Transgenic plants for abiotic stress resistance. In: Kole C, Michler C, Abbott AG, Hall TC (eds) Transgenic crop plants. Springer, Berlin, pp 67–132

Kataoka R, Güneri E, Turgay OC, Yaprak AE, Sevilir B, Başköse I (2017) Sodium-resistant plant growth-promoting rhizobacteria isolated from a halophyte, *Salsola grandis*, in saline-alkaline soils of Turkey. Eur J Soil Sci 6:216–225

Kumar M, Choi JY, Kumari N, Pareek A, Kim SR (2015) Molecular breeding in *Brassica* for salt tolerance: importance of microsatellite (SSR) markers for molecular breeding in *Brassica*. Front Plant Sci 6:688

Kumari S, Vaishnav A, Jain S, Varma A, Choudhary DK (2015a) Bacterial-mediated induction of systemic tolerance to salinity with expression of stress alleviating enzymes in soybean (*Glycine max* L. Merrill). J Plant Growth Regul 34:558–573

Kumari A, Das P, Parida AK, Agarwal PK (2015b) Proteomics, metabolomics, and ionomics perspectives of salinity tolerance in halophytes. Front Plant Sci 6:537

Liu XM, Zhang H (2015) The effects of bacterial volatile emissions on plant abiotic stress tolerance. Front Plant Sci 6:774

Liu X, Wang Z, Zhang X, Wang J, Xu G, Cao Z, Zhong C, Su P (2011) Degradation of diesel-originated pollutants in wetlands by Scirpus triqueter and microorganisms. Ecotoxcol Environ Safe 74:1967–1972

Liu BB, Wang HF, Li QL, Zhou XK, Zhang YG, Xiao M, Li QQ, Zhang W, Li WJ (2016) *Aurantimonas endophytica* sp. nov., a novel endophytic bacterium isolated from roots of *Anabasis elatior* (C. A. Mey.) Schischk. Int J Syst Evol Microbiol 66:4112–4117

Lucena T, Mesa J, Rodriguez-Llorente ID, Pajuelo E, Caviedes MÁ, Ruvira MA, Pujalte MJ (2016) *Marinomonas spartinae* sp. nov., a novel species with plant-beneficial properties. Int J Syst Evol Microbiol 66:1686–1691

Lucena T, Arahal DR, Ruvira MA, Navarro-Torre S, Mesa J, Pajuelo E, Rodriguez-Llorente ID, Rodrigo-Torres L, Piñar MJ, Pujalte MJ (2017) *Vibrio palustris* sp. nov. and *Vibrio spartinae* sp. nov., two novel members of the Gazogenes clade, isolated from salt-marsh plants (*Arthrocnemum macrostachyum* and *Spartina maritima*). Int J Syst Evol Microbiol 67:3506–3512

Mackova M, Dowling D, Macek T (2006) Phytoremediation and rhizoremediation. Springer, Dordrecht. (Focus on Biotechnology). Available at: https://books.google.es/books?id=mph14ko78hYC

Mahajan S, Tuteja N (2005) Cold, salinity and drought stresses: an overview. Arch Biochem Biophys 444:139–158

Martínez C, Espinosa-Ruiz A, Prat S (2016) Gibberellins and plant vegetative growth. Annu Plant Rev 49:285–322

Mateos-Naranjo E, Mesa J, Pajuelo E, Perez-Martin A, Caviedes MA, Rodríguez-Llorente ID (2015) Deciphering the role of plant growth-promoting rhizobacteria in the tolerance of the invasive cordgrass *Spartina densiflora* to physicochemical properties of salt-marsh soils. Plant Soil 394:45–55

Mesa J, Mateos-Naranjo E, Caviedes MA, Redondo-Gómez S, Pajuelo E, Rodríguez-Llorente ID (2015a) Scouting contaminated estuaries: heavy metal resistant and plant growth promoting rhizobacteria in the native metal rhizoaccumulator *Spartina maritima*. Mar Pollut Bull 90:150–159

Mesa J, Mateos-Naranjo E, Caviedes MA, Redondo-Gomez S, Pajuelo E, Rodriguez-Llorente ID (2015b) Endophytic cultivable bacteria of the metal bioaccumulator *Spartina maritima* improve plant growth but not metal uptake in polluted marshes soils. Front Microbiol 6:1450

Mesa J, Rodriguez-Llorente ID, Pajuelo E, Piedras JM, Caviedes MA, Redondo-Gómez S, Mateos-Naranjo E (2015c) Moving closer towards restoration of contaminated estuaries: bioaugmentation with autochthonous rhizobacteria improves metal rhizoaccumulation in native *Spartina maritima*. J Hazard Mater 300:263–271

Mesnoua M, Mateos-Naranjo E, Barcia-Piedras JM, Perez-Romero JA, Lotmani B, Redondo-Gomez S (2016) Physiological and biochemical mechanisms preventing Cd-toxicity in the hyperaccumulator *Atriplex halimus* L. Plant Physiol Biochem 106:30–38

Mickelbart MV, Hasegawa PM, Bailey-Serres J (2015) Genetic mechanisms of abiotic stress tolerance that translate to crop yield stability. Nat Rev Genet 16:237–251

Miller G, Suzuki N, Ciftci-Yilmaz S, Mittler R (2010) Reactive oxygen species homeostasis and signalling during drought and salinity stresses. Plant Cell Environ 33:453–467

Mittler R, Vanderauwera S, Gollery M, Van Breusegem F (2004) Reactive oxygen gene network of plants. Trends Plant Sci 9:490–498

Moshelion M, Halperin O, Wallach R, Oren R, Way DA (2015) Role of aquaporins in determining transpiration and photosynthesis in water-stressed plants: crop water use efficiency, growth and yield. Plant Cell Environ 38:1785–1793

Munns R, Gilliham M (2015) Salinity tolerance of crops – what is the cost? New Phytol 208:668–673

Nadeem SM, Ahmad M, Naveed M, Imran M, Zahir ZA, Crowley DE (2016) Relationship between in vitro characterization and comparative efficacy of plant growth-promoting rhizobacteria for improving cucumber salt tolerance. Arch Microbiol 198:379–387

Navarro-Torre S, Mateos-Naranjo E, Caviedes MA, Pajuelo E, Rodríguez-Llorente ID (2016) Isolation of plant-growth-promoting and metal-resistant cultivable bacteria from *Arthrocnemum macrostachyum* in the Odiel marshes with potential use in phytoremediation. Mar Pollut Bull 110:133–142

Navarro-Torre S, Barcia-Piedras JM, Mateos-Naranjo E, Redondo-Gómez S, Camacho M, Caviedes MA, Pajuelo E, Rodríguez-Llorente ID (2017a) Assessing the role of endophytic bacteria in the halophyte *Arthrocnemum macrostachyum* salt tolerance. Plant Biol 19:249–256

Navarro-Torre S, Barcia-Piedras JM, Caviedes MA, Pajuelo E, Redondo-Gómez S, Rodríguez-Llorente ID, Mateos-Naranjo E (2017b) Bioaugmentation with bacteria selected from the microbiome enhances *Arthrocnemum macrostachyum* metal accumulation and tolerance. Mar Pollut Bull 117:340–347

Paredes-Páliz KI, Mateos-Naranjo E, Doukkali B, Caviedes MA, Redondo-Gómez S, Rodríguez-Llorente ID, Pajuelo E (2017) Modulation of *Spartina densiflora* plant growth and metal accumulation upon selective inoculation treatments: a comparison of gram negative and gram positive rhizobacteria. Mar Pollut Bull 125:77–85

Pitman MG, Läuchli A (2002) Global impacts of salinity and agricultural ecosystem. In: Läuchli A, Lüttge U (eds) Salinity: environment-plants-molecules. Kluwer Academic, Dordrecht, pp 3–20

Qin S, Zhang Y-J, Yuan B, Xu P-Y, Xing K, Wang J, Jiang J-H (2014) Isolation of ACC deaminase-producing habitat-adapted symbiotic bacteria associated with halophyte *Limonium sinense* (Girard) Kuntze and evaluating their plant growth-promoting activity under salt stress. Plant Soil 374:753–766

Qin Y, Druzhinina IS, Pan X, Yuan Z (2016) Microbially mediated plant salt tolerance and microbiome-based solutions for saline agriculture. Biotechnol Adv 34:1245–1259

Redondo-Gómez S (2013) Bioaccumulation of heavy metals in *Spartina*. Funct Plant Biol 40:913–921

Redondo-Gómez S, Mateos-Naranjo E, Andrades-Moreno L (2010) Accumulation and tolerance characteristics of cadmium in a halophytic Cd-hyperaccumulator, *Arthrocnemum macrostachyum*. J Hazard Mater 184:299–307

Redondo-Gómez S, Andrades-Moreno L, Parra R, Valera-Burgos J, Real M, Mateos-Naranjo E, Cox L, Cornejo J (2011) *Spartina densiflora* demonstrates high tolerance to phenanthrene in soil and reduces it concentration. Mar Pollut Bull 62:1800–1808

Roy SJ, Negrão S, Tester M (2014) Salt resistant crop plant. Curr Opin Biotechnol 26:115–124

Ruppel S, Franken P, Witzel K (2013) Properties of the halophyte microbiome and their implications for plant salt tolerance. Funct Plant Biol 40:940–951

Santoyo G, Moreno-Hagelsieb G, Orozco-Mosqueda MC, Glick B (2016) Plant growth-promoting bacterial endophytes. Microbiol Res 183:92–99

Senthilkumar M, Swarnalakshmi K, Govindasamy V, Lee YK, Annapurna K (2009) Biocontrol potential of soybean bacterial endophytes against charcoal root fungus, *Rhizoctonia bataticola*. Curr Microbiol 58:288–293

Sgroy V, Cassán F, Masciarelli O, Del Papa MF, Lagares A, Luna V (2009) Isolation and characterization of endophytic plant growth-promoting (PGPB) or stress homeostasis-regulating (PSHB) bacteria associated to the halophyte *Prosopis strombulifera*. Appl Microbiol Biotechnol 85:371–381

Shabala S (2013) Learning from halophytes: physiological basis and strategies to improve abiotic stress tolerance in crops. Ann Bot 112:1209–1221

Sorty AM, Meena KK, Choudhary K, Bitla UM, Minhas PS, Krishnani KK (2016) Effect of plant growth promoting bacteria associated with halophytic weed (*Psoralea corylifolia* L) on germination and seedling growth of wheat under saline conditions. Appl Biochem Biotechnol 180:872–882

Suarez C, Ratering S, Schäfer J, Schnell S (2017) *Ancylobacter pratisalsi* sp. nov. with plant growth promotion abilities from the rhizosphere of *Plantago winteri* Wirtg. Int J Syst Evol Microbiol 67:4500–4506

Syranidou E, Christofilopoulos S, Politi M, Weyens N, Venieri D, Vangronsveld J, Kalogerakis N (2017) Bisphenol-a removal by the halophyte *Juncus acutus* in a phytoremediation pilot: characterization and potential role of the endophytic community. J Hazard Mater 32:350–358

Tardieu F (2012) Any trait or trait related allele can confer drought tolerance: just design the right drought scenario. J Exp Bot 63:25–31

Tkacz A, Poole P (2015) Role of root microbiota in plant productivity. J Exp Bot 66:2167–2175

Ventosa A, Nieto JJ, Oren A (1998) Biology of moderately halophilic aerobic bacteria. Microbiol Mol Biol Rev 62:504–544

Vurukonda SSKP, Vardharajula S, Shrivastava M, SkZ A (2016) Enhancement of drought stress tolerance in crops by plant growth promoting rhizobacteria. Microbiol Res 184:13–24

Wang G-L, Que F, Xu Z-S, Wang F, Xiong A-S (2015) Exogenous gibberellin altered morphology, anatomic and transcriptional regulatory networks of hormones in carrot root and shoot. BMC Plant Biol 15:290

Yang J, Kloepper JW, Ryu CM (2009) Rhizosphere bacteria help plants tolerate abiotic stress. Trends Plant Sci 14:1–4

Yensen NP, Biel KY (2006) Soil remediation via salt-conduction and the hypotheses of halosynthesis and photoprotection ecophysiology of high salinity tolerant plants. In: Khan MA, Weber DJ (eds) Task for vegetation science, vol 34. Springer, Dordrecht, pp 313–344

Yuan Z, Druzhinina IS, Labbé J, Redman R, Qin Y, Rodriguez R, Zhang C, Tuskan GA, Lin F (2016) Specialized microbiome of a halophyte and its role in helping non-host plants to withstand salinity. Sci Rep 6:32467

Zhao S, Zhou N, Wang L, Tian C-Y (2013) Halophyte-endophyte coupling: a promising bioremediation system for oil-contaminated soil in Northwest China. Environ Sci Technol 47:11938–11939

Zhao S, Zhou N, Zhao Z-Y, Zhang K, Wu G-H, Tian C-Y (2016) Isolation of endophytic plant growth-promoting bacteria associated with the halophyte *Salicornia europaea* and evaluation of their promoting activity under salt stress. Curr Microbiol 73:574–581

Zhao GY, Zhao LY, Xia ZJ, Zhu JL, Liu D, Liu CY, Chen XL, Zhang YZ, Zhang XY, Dai MX (2017) *Salinicola tamaricis* sp. nov., a heavy-metal-tolerant, endophytic bacterium isolated from the halophyte *Tamarix chinensis* Lour. Int J Syst Evol Microbiol 67:1813–1819

Zhu J-K (2016) Abiotic stress signaling and responses in plants. Cell 167:313–324

Effects of Halophilic Bacteria on Biochemical Characteristics of Rice Cultivars Under Salinity Stress Conditions

7

Mehvish Riaz Khattak, Sami Ullah Jan, Ijaz Malook, Sehrish Riaz Khattak, Nazneen Akhtar, Sehresh Khan, and Muhammad Jamil

Abstract

Soil salinization is a serious ecological problem affecting cereal crops including rice. Different bacterial strains have been identified and used as salt remediator in salt-affected soil, while there is confined use of halophilic bacteria. Present study was aimed to find the growth stimulatory effects of two salt-resistant *Bacillus* strains (NCCP-71 and NCCP-77) on biochemical changes in rice cultivars (NIAB-IR-9 and KSK-282) under control (0 mM) and different salt stress treatments (50, 100, and 150 Mm NaCl). High concentration of salinity increased the Na^+ ion content and lowered the K^+ and Ca^+ ion contents in rice. The rice seedlings inoculated with *Bacillus* strains showed a decrease in Na^+ content followed by an increase in K^+ ion content. Calcium ions were increased by NCCP-71 in both varieties; however NCCP-77 inoculation decreased calcium ion content in KSK-282 and in NIAB-IR-9 at 50 mM NaCl. Salinity gradually lowered the contents of photosynthetic pigments; however inoculation with NCCP-71 and NCCP-77 caused a prominent increase in photosynthetic content (Chl a, Chl b, and carotenoids) under salt stress. On the other side, significant reduction in protein and nitrogen occurred with increasing salt stress. Inoculation with bacterial strains NCCP-71 enhanced protein and nitrogen content in both varieties, while inoculation with NCCP-77 enhanced nitrogen and protein content at 50 mM NaCl concentration. Both halophilic bacteria had a great impact on the biochemical characteristics of inoculated rice cultivar by modifying different biochemical processes under saline conditions.

Keywords

Halophilic bacteria · Biochemical characteristics · Rice · Salt stress

M. R. Khattak · S. U. Jan · I. Malook · S. R. Khattak · N. Akhtar · S. Khan · M. Jamil (✉)
Department of Biotechnology and Genetic Engineering, Kohat University of Science and Technology, Kohat, Pakistan
e-mail: dr.jamil@kust.edu.pk

M. Kumar et al. (eds.), *Saline Soil-based Agriculture by Halotolerant Microorganisms*, https://doi.org/10.1007/978-981-13-8335-9_7

7.1 Introduction

Plants are most susceptible to the climatic changes. The production of major crops has been mainly affected by different stresses (Grover et al. 2011). Salinity is one of the major obstacles to the agricultural practices today. The salt content in the cultivated soil has been increasing due to irrigated water, desertification, and extreme fertilization (Bacilio et al. 2004). Salinity affects the plants by inhibiting cell division, elongation growth due to changes in homeostasis, ionic imbalance, and osmotic potential (Principe et al. 2007). Salinity affects the physiological parameters of plant growth by lowering the germination rate and percentage. Abiotic stresses caused the alteration of growth parameters in plants (Kaymak et al. 2009). Accumulation of toxic ions leads to reduced water uptake from the soil and plants that rely on its carbohydrates to adjust its osmotic potential. This causes reduction of the photosynthesis rate and closure of stomata (Yildirim et al. 2011).

Different techniques were streamlined by researchers in which most are using sucrose (Siringam et al. 2012), phosphorus (Naheed et al. 2007), hormones (Rafique et al. 2011; Afzal et al. 2006), priming with plant-derived smoke solution (Jamil et al. 2013; Malook et al. 2014; Malook et al. 2017), and plant growth-promoting rhizobacteria (PGPR) (Bhattacharyya and Jha 2012; Ashrafuzzaman et al. 2009; Kaymak et al. 2008; Shah et al. 2017; Khan et al. 2017).

The free-living microbes that reside in plant roots and promote its germination are called PGPR (Pallai 2005). PGPR uses two types of mechanism for increase of growth by stimulating the maximum phosphorus uptake and solubilization, biosynthesis and stimulating of plant growth hormones (indol acetic acid (IAA) auxin, cytokinins and gibberellins), increased nitrogen and iron utilized by plants roots through the production of iron rich siderophore etc (Mallesh 2008). Indirect mechanisms involved the production of antibiotics against pathogenic bacterial strains, biosynthesis of fungal cell disrupting enzymes, and formation of proline content and glycine betaine which enhanced the overall systematic resistance to pathogens and biotic stresses (Pallai 2005). Rhizobacteria are also termed as biopesticides, biofertilizer, phytostimulator, and rhizomediators (Antoun and Prévost 2005). Many bacterial strains were reported to have increases in plant growth such as *Pseudomonas*, *Azospirillum*, *Azotobacter*, *Enterobacter*, *Klebsiella*, *Alcaligenes*, *Burkholderia*, *Arthrobacter*, *Serratia*, *Rhizobium*, and *Bacillus* (Saharan and Nehra 2011). *Bacillus* having a PGP activity especially has resistant activity against plant pathogens. These bacteria form endospores and have many beneficial compounds for agronomic purposes (Nihorimbere et al. 2010). *Bacillus* spp. increase plant growth by increasing beneficial compounds such as butanediol and acetoin in a diverse of abiotic stress conditions including salt and heavy metal stresses (Ryu et al. 2003). The most frequently PGPR isolated from rhizosphere are of *Bacillus* spp. Research showed an increase in yield of wheat by inoculation of *Bacillus*. Soybean seeds when inoculated with *Bacillus* strains showed enhanced nodulation and growth (Gururani et al. 2013).

Soil salinization inhibits plant growth such as rice that is used as staple food. Different techniques were used to minimize or remediate salt-affected soil, but these methods are costly and also abolishing the soil structure and texture. Therefore, different environment-friendly and economical methods have been utilized to overcome this problem. Different bacterial species have been used to increase the organic or inorganic contaminants in soils. The use of plant growth-promoting bacterial strains isolated from rhizosphere has the potential to remediate soil contamination. There is little information on the halophilic bacteria used for the remediation of salt-affected soil. Therefore, the current study is to explore the role of halophilic *Bacillus* strains (NCCP-71 and NCCP-77) on biochemical characteristics of rice cultivars under different saline conditions.

7.2 Plant Materials

Rice cultivar seeds (cvs. NIAB-IR-9 and KSK-282) were obtained from the National Agricultural Research Centre (NARC), Islamabad, Pakistan. Seeds were sterilized with sodium hypochlorite solution 3.5% *(*v/v*)* and rinsed with distilled water.

7.3 Bacterial Strains

Halophilic bacterial strains were taken from Plant and Microbial Biotechnology lab, Kohat University of Science and Technology (KUST), Kohat, Pakistan, already collected from Karak salt mines (Roohi et al. 2014). Halophilic bacterial strains *Bacillus* spp. NCCP-71 and NCCP-77 were cultured according to Roohi et al. (2012). The bacterial strains were grown at 37 °C for 24 h with continuous mixing in shaking growth incubator (Wise Cube, WIS-20R). Colony-forming unit (CFU) was calculated to obtain 10^8 CFU/ml before inoculation of bacteria according to the standard protocol of Rani and Arundhathi Reddy (2012) using TSA agar supplemented with 5% (w/v) sodium content. The CFU was calculated from the following equation:

$$\text{Colony Forming Unit} = \frac{\text{Number of Colonies} \times \text{Dilution Factor}}{\text{Volume of Inoculum}}$$

7.3.1 Sand Culture Experiment

The 10-day-old germinated seedlings of both varieties NIAB-IR-9 and KS-282 were transferred to plastic pots (having a hole in the bottom, covered with cloth) containing about 1 kg of well-washed dried sand in the greenhouse. The seedlings were sown 1 inch deep at a relative distance of a few cm apart from each other. All the pots were germinated in Hoagland's solution (Hoagland and Arnon 1950) having pH of 5–6 and were allowed growing for 24 days.

7.3.2 Salt Stress and PGPR Treatment

After 24 days, stress of different salt concentrations 50, 100, and 150 mM NaCl including control (0 mM NaCl) was given to plants mixed with Hoagland's solution for 1 week. To find the growth of halophilic strains *Bacillus* spp. NCCP-71 and *Bacillus* spp. NCCP-77 under NaCl stress conditions, 10 ml of broth (10^8 CFU/ml) of each strain was also given along with Hoagland's solution (Guo et al. 2011; Nabti et al. 2010). Fresh leaves randomly taken from three replicates were crushed and placed at -40 °C in Eppendorf tubes for molecular determination. The rest of the materials was dried in an incubator at 80 °C for 48 h and was powdered for different biochemical tests.

7.3.3 Biochemical Parameters

7.3.3.1 Ion Analysis (Na^+, K^+, and Ca^{+2})

Ions were determined by using Flame Photometer (Jenway PFP7) according to the standard protocol of Awan and Salim (1997). The plant material was oven-dried and digested with to H_2SO_4 and H_2O_2 at 2:1 (v/v) on a hot plate. The solution was then diluted with 20 ml of distilled water, filtered with Whatman filter paper no. 41, and was processed for ion analysis.

7.3.3.2 Determination of Photosynthetic Pigments

Photosynthetic pigments (Chl a, Chl b, and carotenoids) were analyzed by taking the 25 mg plant material along with 25 mg of MgO in 10 ml falcon tubes. The mixture was supplemented with 5 ml of methanol and allowed to vortex at room temperature. The homogenized mixture was kept on a shaker for 3 h at 250 rpm. After that, all the samples were centrifuged for 7 min at 6000 rpm at ambient temperature. All the plant extracts were noted on UV-visible spectrophotometer at 470, 653, and 666 nm using methanol as blank. The concentration of pigments was calculated from the formulas mentioned by Lichtenthaler and Wellburn (1985):

$$(C_a = 15.65\ A_{666} - 7.340\ A_{653})$$

$$(C_b = 27.05\ A_{653} - 11.21\ A_{666})$$

$$(C_{x+c} = 1000\ A_{470} - 2.860\ C_a - 129.2\ C_b/245)$$

7.3.3.3 Nitrogen and Protein Content Determination

In digestion flasks, dried plant material was digested with H_2SO_4 at 400 °C until the solution becomes pale yellow. The mixture was diluted to 20 ml of distilled water followed by the rapid drainage of 50% (w/v) of NaOH solution for 4 s. In 250 ml volumetric flask, 50 ml solution of 4% boric acid was taken with methyl red as indicator. The distillation was terminated when the color of solution changed into colorless. The mixture was titrated against 0.1 N H_2SO_4 and until the pink color

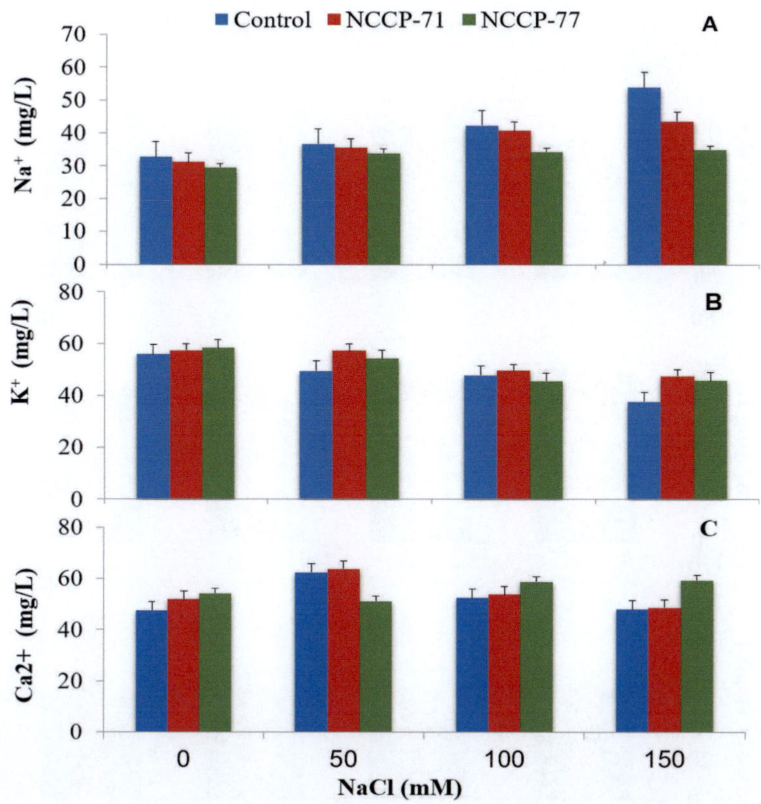

Fig. 7.1 Effect of plant growth-promoting halophilic bacteria on (**a**) Na^{+}, (**b**) K^{+}, and (**c**) Ca^{+2} ion contents of NIAB-IR-9 under different saline conditions

appeared (Peter and Young 1980). Total protein and organic nitrogen contents were find by

$$\text{Total Protein}\left(\frac{g}{g}\right) = \frac{\text{Sample Volume} - \text{Blank Volume} \times 0.1\text{N}}{\text{Dry Weight of Sample}} \times 1.4007$$

$$\text{Total Organic Nitrogen}\left(\frac{g}{g}\right) = \frac{\text{Sample Volume} - \text{Blank Volume} \times 0.1\text{N}}{\text{Dry Weight of Sample}} \times 6.25$$

7.3.4 Results Ionic Content

The sodium ion content was increased with increase in salt concentration followed by decrease in potassium and calcium ion concentrations in both varieties (Figs. 7.1 and 7.2). The inoculation of halophilic strains showed variations in ionic contents of

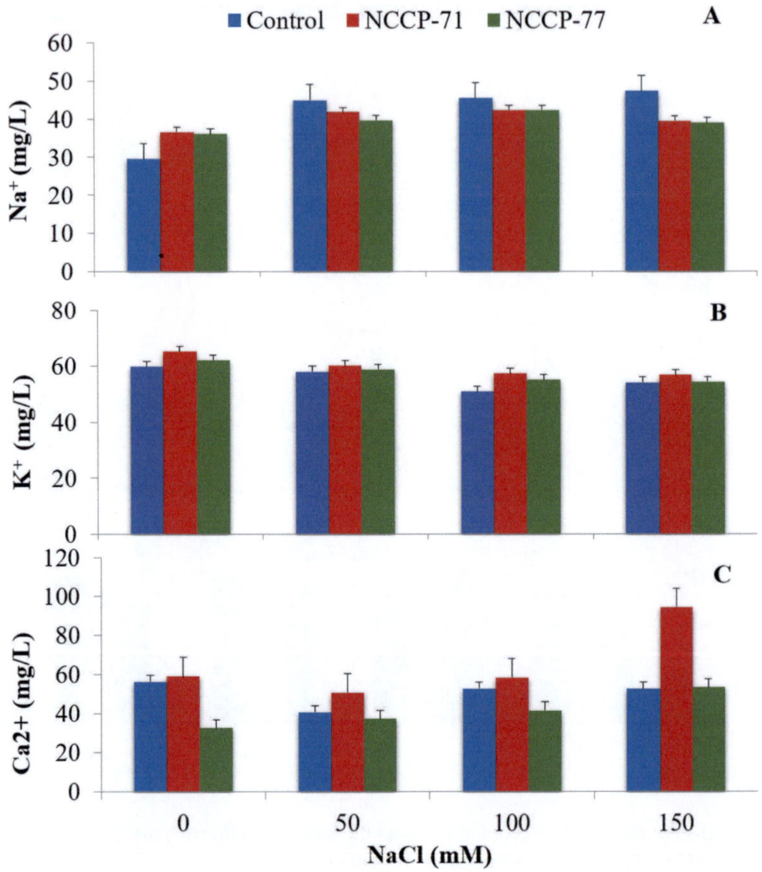

Fig. 7.2 Effect of plant growth-promoting halophilic bacteria on (**a**) Na^+, (**b**) K^+, and (**c**) Ca^{+2} ion contents of KSK-28 under different saline conditions

rice plants. The inoculation of plant samples with halophilic strains NCCP-71 and NCCP-77 showed a decrease in Na^+ contents and an increase in K^+ ion contents in both varieties (Figs. 7.1 and 7.2). Calcium ion content was increased by inoculation of NCCP-71 in both varieties (Figs. 7.1 and 7.2). NIAB-IR-9 showed an increase in calcium ion content by NCCP-77 inoculation except at 100 mM NaCl (Fig. 7.1c). KSK-282 showed a decrease in calcium ion contents by the inoculation of NCCP-77 (Fig. 7.2c). This indicated that NCCP-71 has more potential for stress alleviation.

7.3.5 Photosynthetic Pigments

Concentration of Chl a, Chl b, and carotenoids lowered under various salt concentrations in both varieties (Figs. 7.3 and 7.4). The use of bacteria induced significant changes in concentration of photosynthetic pigments (Chl a, Chl b,

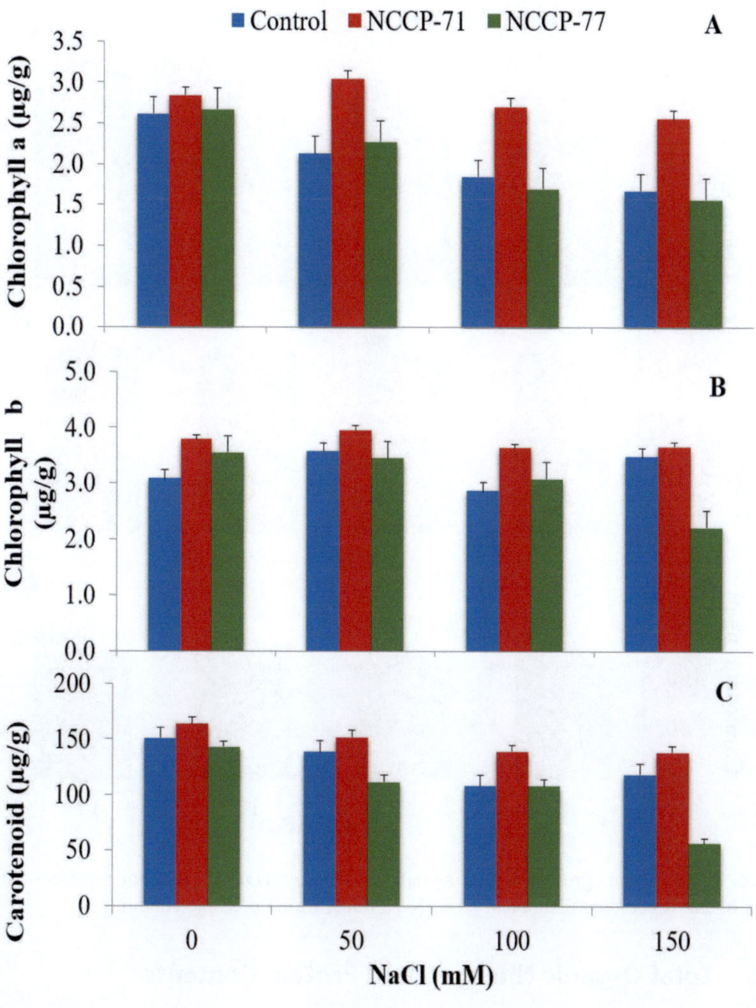

Fig. 7.3 Effect of plant growth-promoting halophilic bacteria on photosyntetic pigments (**a**) Chl a, (**b**) Chl b, and (**c**) carotenoid content of NIAB-IR-9 under different saline conditions

and carotenoids) in both rice varieties under saline conditions. The inoculation of NCCP-71 led to an increase in pigment concentration in NIAB-IR-9 (Fig. 7.3a–c). NCCP-77 inoculation showed an increase in Chl a and Chl b concentration at all salt concentration except 150 mM NaCl in NIAB-IR-9 (Fig. 7.3). NCCP-77 had no growth-promoting effect on carotenoid contents in NIAB-IR-9 (Fig. 7.3c). KSK-282 showed an increase in concentration of Chl a, Chl b, and carotenoids in control and 100 mM NaCl by inoculation of both strains NCCP-71 and NCCP-77 (Fig. 7.4).

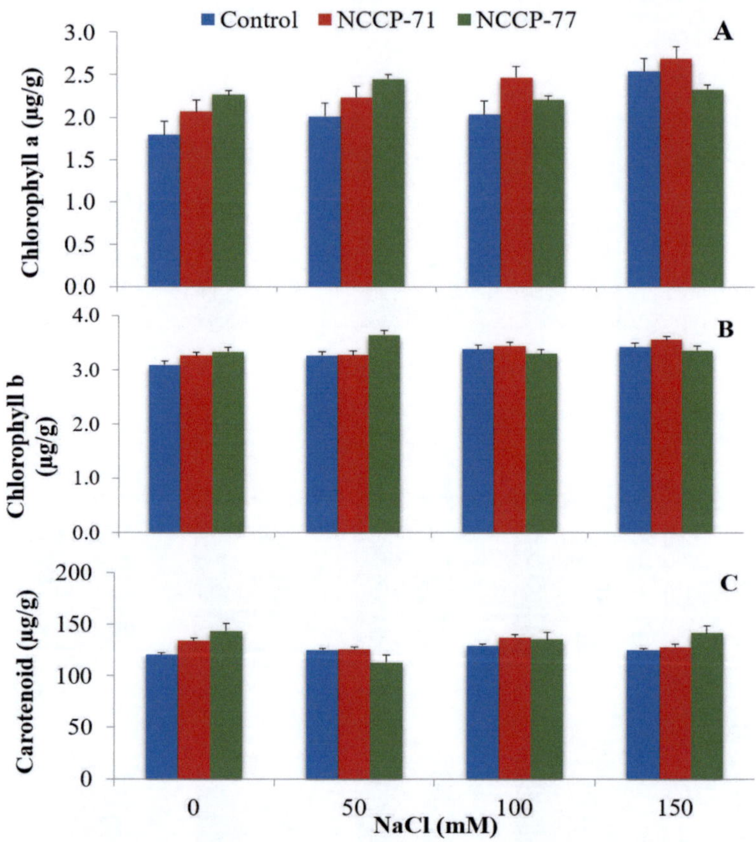

Fig. 7.4 Effect of plant growth-promoting halophilic bacteria on photosyntetic pigments (**a**) Chl a, (**b**) Chl b, and (**c**) carotenoid content of KSK-28 under different saline conditions

7.3.6 Total Organic Nitrogen and Protein Contents

Increase in salt concentration led to a decrease in protein and nitrogen contents (Fig. 7.5). The inoculation of halophilic bacteria also induced changes in protein and nitrogen content. NIAB-IR-9 plants inoculated with NCCP-71 had more protein contents under all saline conditions. NCCP-77 inoculation increased protein contents at 0 mM and 50 mM NaCl (Fig. 7.5a). The inoculation of both strains increased nitrogen content under 0 mM and 100 mM NaCl concentration in NIAB-IR-9 (Fig. 7.5b). NCCP-71 showed better results as compared to NCCP-77. KSK-282 also showed a decrease in total organic nitrogen and protein contents by increasing the concentration of salt (Fig. 7.6). KSK-282 plants inoculated with NCCP-71 led to an increase in protein contents under all saline conditions. NCCP-77 inoculation increased protein contents at 0 and 50 mM NaCl (Fig. 7.6a). NCCP-71-inoculated plants showed an increase in nitrogen contents under all salt

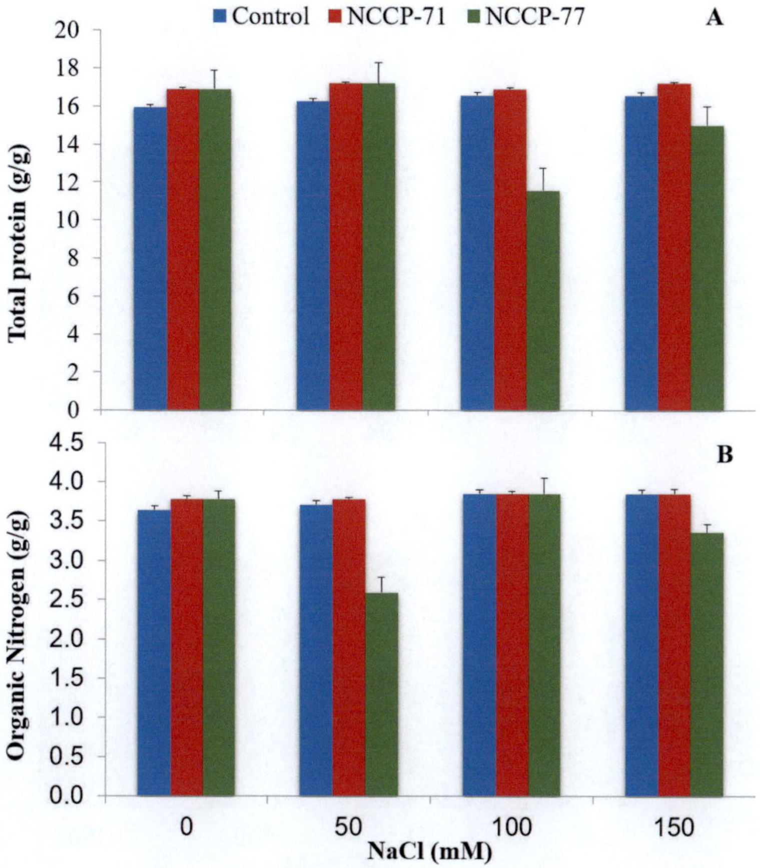

Fig. 7.5 Effect of plant growth-promoting halophilic bacteria on protein (**a**) and total organic nitrogen (**b**) content of NIAB-IR-9 under different saline conditions

concentrations. NCCP-77 inoculation increased nitrogen contents at 0 and 50 mM NaCl (Fig. 7.6b). NCCP-71 yielded more increase in nitrogen and protein in comparison to NCCP-77.

7.4 Discussion

Microbes isolated from rhizosphere has been widely used for the remediation of inorganic metals, organic petroleum waste, and pesticides and for reclamation of saline soils (Bose et al. 2008; Zeeb et al. 2006; Huang et al. 2004, 2005; Olson et al. 2008; Lunney et al. 2004; Qadir et al. 2007; Su et al. 2008). Bacteria isolated from salt habitat improve the effectiveness of the process. Therefore, the current findings were aimed to assess the role of halophilic *Bacillus* strains (NCCP-71 and

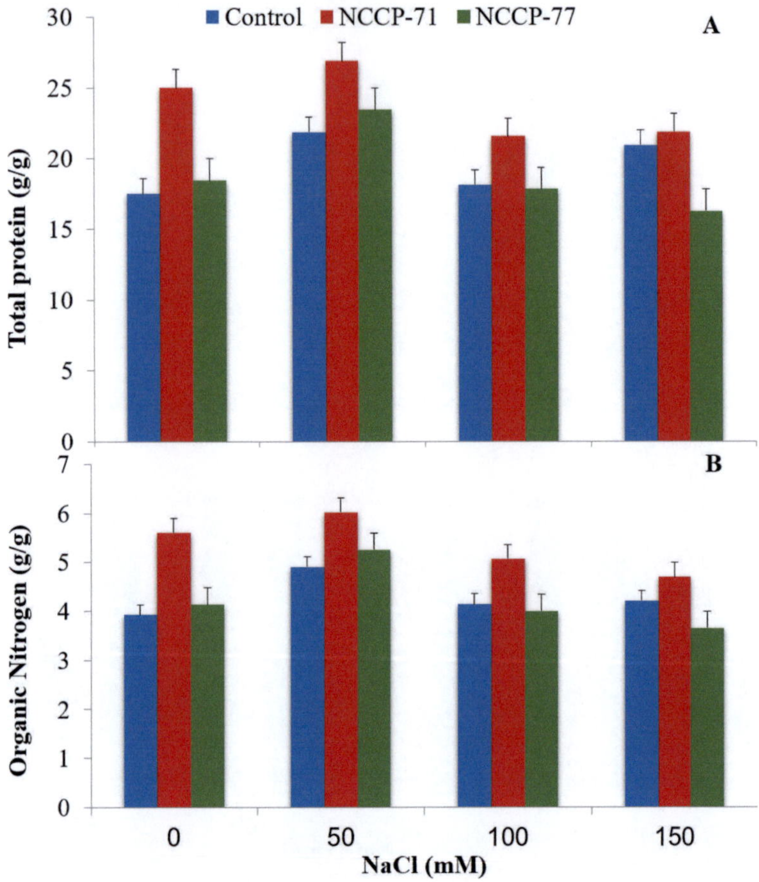

Fig. 7.6 Effect of plant growth-promoting halophilic bacteria on protein (**a**) and total organic nitrogen (**b**) content of KSK-28 under different saline conditions

NCCP-77) on biochemical characteristics of rice grown in salt-affected soil. Increasing concentration of salt in soil increased Na^+ ions, while K^+ and Ca^+ ion contents decreased in the plant under saline condition (Figs. 7.1 and 7.2). This high concentration in Na^+ and lower in K^+ and Ca^{+2} content may be due to the ionic pool available in direct contact to plant cells with the outer environment under stress. Plants exposed to the high salt concentration induce ionic and osmotic stress (Niu et al. 1995; Serrano et al. 1999; Zhu et al. 1998). It is also known that the hydrophobic-electrostatic imbalance occurs at the cellular level. The regulation in ionic concentration is primarily vital for optimal functions of cellular enzymes which are very crucial in maintaining the cellular protein (Wyri Jones and Pollard 1983). Wang et al. (2012)

reported high contents of sodium ions under saline stress condition as the NaCl concentration increased. Similar results were reported by Jamil et al. (2012) who studied the effect of salt stress on physiology and biochemistry of three rice varieties including NIAB-IR-9. The inoculation of rice seedling with NCCP-71 and NCCP-77 reduced the sodium ions and increased the calcium and potassium ions (Figs. 7.1 and 7.2). Both strains used were potent enough to reduce the Na^+ concentration while they were able to keep the K^+ and Ca^+ concentration above the untreated plants. Different research studies indicated that plant growth-promoting bacteria inoculation reduced sodium ion content and increased potassium ion content in plants (Nadeem et al. 2006; Zahir et al. 2009, Mishra et al. 2011; Yildirim et al. 2011).

Photosynthetic pigments are very significant traits that link with photosynthetic capability in rice (Teng et al. 2004). Photosynthetic pigments decreased with increasing salt concentrations (Figs. 7.3 and 7.4). Salt stress decreased chlorophyll a more than chlorophyll b (Figs. 7.3 and 7.4), as they are mostly affected by high salts (Daiz et al. 2002; Santos 2004). The decreased ratio is the cause for lowering the occurrence of photosynthetic contents in rice (Moradi and Ismail 2007; Zhen-Hua et al. 2012). Inoculation with NCCP-71 significantly increased pigment contents under saline conditions, while NCCP-77 increased pigment contents at 100 mM salt concentration in both varieties (Figs. 7.3 and 7.4). This increase in contents may be due to important compounds such as butanediol and acetoin formed by bacteria to increase the growth of a wide range of plants (Ryu et al. 2003). Priming with plant growth-promoting bacteria enhanced the pigment content as reported by different studies (Zahir et al. 2009; Mishra et al. 2011; Yildirim et al. 2011; Nadeem et al. 2006).

Increase in salt concentration also decreased the protein and nitrogen content (Figs. 7.5 and 7.6). High salt content decreased the protein content and increased the polyribosomes (Jones 1996). Lower concentration of protein content with increase in NaCl was also reported by other studies (Khan 1998; Azooz et al. 2004; Dagar et al. 2004). PGPR inoculation led to an increase in organic nitrogen and protein content (Nadeem et al. 2006; Mishra et al. 2011). In this study, protein and nitrogen contents were enhanced in both varieties by NCCP-71 inoculation, while NCCP-77 yielded enhanced nitrogen and protein content at 50 mM salt concentration (Figs. 7.5 and 7.6). PGPR also stimulated increased nitrogen and iron by plant which increased the protein contents in inoculated plants (Mallesh 2008).

7.5 Conclusion

The halophilic strains *Bacillus* spp. NCCP-71 and *Bacillus* spp. NCCP-77 isolated from salt mines of Karak had an efficient ability to overcome salt stress by inoculating with plant growth-promoting bacteria strains. These strains yielded an increase in ionic contents, photosynthetic pigments, and nitrogen and protein content. However, NCCP-77 showed sensitivity at some salt concentration.

References

Afzal I, Basra SM, Farooq M, Nawaz A (2006) Alleviation of salinity stress in spring wheat by hormonal priming with ABA, salicylic acid and ascorbic acid. Int J Agric Biol 8:23–28

Antoun H, Prévost D (2005) Ecology of plant growth promoting rhizobacteria. In: PGPR: biocontrol and biofertilization. Springer, pp 1–38

Ashrafuzzaman M, Hossen FA, Ismail MR, Hoque A, Islam MZ, Shahidullah SM, Meon S (2009) Efficiency of plant growth-promoting rhizobacteria (PGPR) for the enhancement of rice growth Afr J Biotechnol 8

Awan JA, Salim UR (1997) Food analysis manual. Pak Vet J Publ 5:2–7

Azooz MM, Shaddad MA, Abdel-Latef AA (2004) The accumulation and compartmentation of proline in relation to salt tolerance of three sorghum cultivars. Indian J Plant Physiol 9:1–8

Bacilio M, Rodriguez H, Moreno M, Hernandez J-P, Bashan Y (2004) Mitigation of salt stress in wheat seedlings by a gfp-tagged Azospirillum lipoferum. Biol Fertil Soils 40:188–193

Bhattacharyya PN, Jha DK (2012) Plant growth-promoting rhizobacteria (PGPR): emergence in agriculture. World J Microbiol Biotechnol 28:1327–1350

Bose S, Vedamati J, Rai V, Ramanathan AL (2008) Metal uptake and transport by *Typha angustata* L. grown on metal contaminated waste amended soil: an implication of phytoremediation. Geoderma 145:136–142

Dagar JC, Bhagwan H, Kumar Y (2004) Effect on growth performance and biochemical contents of Salvadora persica when irrigated with water of different salinity. Indian J Plant Physiol 9:234–238

Daiz MGQ, Bebing NN, Villegas MJ, Manuel MC (2002) Cell biology, laboratory manual (fifth edition): genetics and molecular biology division. The University of the Philippines, Philippines, p 47

Grover M, Ali SZ, Sandhya V, Rasul A, Venkateswarlu B (2011) Role of microorganisms in adaptation of agriculture crops to abiotic stresses. World J Microbiol Biotechnol 27:1231–1240

Guo J, Tang S, Ju X, Ding Y, Liao S, Song N (2011) Effects of inoculation of a plant growth promoting rhizobacterium Burkholderia sp. D54 on plant growth and metal uptake by a hyperaccumulator Sedum alfredii Hance grown on multiple metal contaminated soil. World J Microbiol Biotechnol 27:2835–2844

Gururani MA, Upadhyaya CP, Baskar V, Venkatesh J, Nookaraju A, Park SW (2013) Plant growth-promoting rhizobacteria enhance abiotic stress tolerance in *Solanum tuberosum* through inducing changes in the expression of ROS-scavenging enzymes and improved photosynthetic performance. J Plant Growth Regul 32:245–258

Hoagland DR, Arnon DI (1950) The water-culture method for growing plants without soil. Circ Calif Agric Exp Station 347:32

Huang XD, El-Alawi Y, Penrose DM, Glick BR, Greenberg BM (2004) A multi-process phytoremediation system for removal of polycyclic aromatic hydrocarbons from contaminated soils. Environ Pollution 130(3):465–476

Huang XD, El-Alawi Y, Gurska J, Glick BR, Greenberg BM (2005) A multi-process phytoremediation system for decontamination of persistent total petroleum hydrocarbons (TPHs) from soils. Microchem J 81(1):139–147

Jamil M, Bashir S, Anwar S, Bibi S, Bangash A, Ullah F, Rha ES (2012) Effect of salinity on physiological and biochemical characteristics of different varieties of rice. Pak J Bot 44:7–13

Jamil M, Malook I, Parveen S, Naz T, Ali A, Jan SU, Rehman SU (2013) Smoke priming, a potent protective agent against salinity: effect on proline accumulation, elemental uptake, pigmental attributes and protein banding patterns of rice (*Oryza Sativa*). J Stress Physiol & Biochem 9(1):169–183

Jones GH (1996) Plants and microclimate, 2nd edn. Cambridge, MA, pp 72–108

Kaymak HC, Yarali F, Guvenc I, Donmez MF (2008) The effect of inoculation with plant growth rhizobacteria (PGPR) on root formation of mint (Mentha piperita L.) cuttings. Afr J Biotechnol 7:4479–4483

Kaymak HÇ, Güvenç IS, Yarali F, Dönmez MF (2009) The effects of bio-priming with PGPR on germination of radish (*Raphanus sativus* L.) seeds under saline conditions. Turk J Agric For 33:173–179

Khan GS (1998) Soil salinity/sodicity status in Pakistan. Soil Survey of Pakistan, Lahore 59

Khan N, Bano A, Babar MA (2017) The root growth of wheat plants, the water conservation and fertility status of sandy soils influenced by plant growth promoting rhizobacteria. Symbiosis 72:195–205

Lichtenthaler HK, Wellburn AR (1985) Determination of total carotenoids and chlorophyll a and B of leaf in different solvents. Biochem Soc Trans 11:591–592

Lunney AI, Zeeb BA, Reimer KJ (2004) Uptake of weathered DDT in vascular plants: potential for phytoremediation. Environ Sci Technol 38(22):6147–6154

Mallesh SB (2008) Plant growth promoting rhizobacteria, their characterization and mechanisms in the suppression of soil bore pathogens of Coleus and Ashwagandha, P.HD thesis, University of agricultural sciences, Dharwad

Malook I, Atlas A, Rehman SU, Wang W, Jamil M (2014) Smoke alleviates adverse effects induced by stress on rice. Toxicol Environ Chem 96:755–767

Malook I, Shah G, Jan M, Shinwari KI, Aslam MM, Rehman S, Jamil M (2017) Smoke priming regulates growth and the expression of myeloblastosis and zinc-finger genes in rice under salt stress. Arab J Sci Eng 42:2207–2215. https://doi.org/10.1007/s13369-016-2378-x

Mishra PK et al (2011) Alleviation of cold stress in inoculated wheat (*Triticum aestivum* L.) seedlings with psychrotolerant pseudomonads from NW Himalayas. Arch Microbiol 193:497–513

Moradi F, Ismail AM (2007) Responses of photosynthesis, chlorophyll fluorescence and ROS-scavenging systems to salt stress during seedling and reproductive stages in rice. Ann Bot 99:1161–1173. https://doi.org/10.1093/aob/mcm052

Nabti E, Sahnoune M, Ghoul M, Fischer D, Hofmann A, Rothballer M, Schmid M, Hartmann A (2010) Restoration of growth of durum wheat (*Triticum durum* var. *waha*) under saline conditions due to inoculation with the rhizosphere bacterium *Azospirillum brasilense* NH and extracts of the marine alga *Ulva lactuca*. J Plant Growth Regul 29:6–22

Nadeem S, Zahir Z, Naveed M, Arshad M, Shahzad S (2006) Variation in growth and ion uptake of maize due to inoculation with plant growth promoting rhizobacteria under salt stress. Soil Environ 25:78–84

Naheed G, Shahbaz M, Latif T, Rha ES (2007) Alleviation of the adverse effects of salt stress on rice (*Oryza sativa* L.) by phosphorus applied through rooting medium: growth and gas exchange .... Pak J Bot 39:729–737

Nihorimbere V, Ongena M, Cawoy H, Brostaux Y, Kakana P, Jourdan E, Thonart P (2010) Beneficial effects of bacillus subtilis on field-grown tomato in Burundi: reduction of local fusarium disease and growth promotion. Afr J Microbiol Res 4:1135–1142

Niu X, Bressan RA, Hasegawa PM, Pardo JM (1995) Ion homeostasis in NaCl. Stress Environ Plant Physiol 109:735

Olson PE, Castro A, Joern M, DuTeau NM, Pilon-Smits E, Reardon KF (2008) Effects of agronomic practices on phytoremediation of an aged PAH-contaminated soil. J Environ Quality 37(4):1439–1446

Pallai R (2005) Effect of plant growth-promoting rhizobacteria on Canola (Brassica napus. L) and lentil (*Lens culinaris*. Medik) plants. MSc thesis, University of Saskatchewan

Peter LP, Young VR (1980) Nutritional evaluation of proteins foods. United Nation University, Japan, p 8

Príncipe A, Alvarez F, Castro MG, Zachi L, Fischer SE, Mori GB, Jofré E (2007) Biocontrol and PGPR features in native strains isolated from saline. Soils Argent Curr Microbiol 55:314–322. https://doi.org/10.1007/s00284-006-0654-9

Qadir M, Oster JD, Schubert S, Noble AD, Sahrawat KL (2007) Phytoremediation of sodic and saline-sodic soils. In: Sparks DL (ed) Advan Agronomy, vol 96, pp 197–247

Rafique N, Raza SH, Qasim M, Iqbal N (2011) Pre-sowing application of ascorbic acid and salicylic acid to seed of pumpkin and seedling response to salt. Pak J Bot 43(6):2677–2682

Rani UM, Arundhathi Reddy G (2012) Screening of rhizobacteria containing plant growth promoting (PGPR) traits in rhizosphere soils and their role in enhancing growth of pigeon pea. Afr J Biotechnol 11(32):8085–8091

Roohi A, Ahmed I, Iqbal M, Jamil M (2012) Preliminary isolation and characterization of halotolerant and halophilic bacteria from salt mines of Karak, Pakistan. Pak J Bot 44:365–370

Roohi A, Ahmed I, Khalid N, Iqbal M, Jamil M (2014) Isolation and phylogenetic identification of halotolerant/halophilic bacteria from the salt mines of Karak, Pakistan. Int J Agric Biol 16:564–570

Ryu CM, Farag MA, Hu CH, Reddy MS, Wei HX, Pare PW, Kloepper JW (2003) Bacterial volatiles promote growth in *Arabidopsis*. Proc Natl Acad Sci U S A 100:4927–4932

Saharan BS, Nehra V (2011) Plant growth promoting rhizobacteria: a critical review. Life Sci Med Res 21:1–30. LSMR-21

Santos CV (2004) Regulation of chlorophyll biosynthesis and degradation by salt stress in sunflower leaves. Sci Hortic 103:93–99

Serrano R et al (1999) A glimpse of the mechanisms of ion homeostasis during salt stress. J Exp Bot 50:1023–1036

Shah G et al (2017) Halophilic bacteria mediated phytoremediation of salt-affected soils cultivated with rice. J Geochem Explor 174:59–65. https://doi.org/10.1016/j.gexplo.2016.03.011

Siringam K, Juntawong N, Cha-um S, Boriboonkaset T, Kirdmanee C (2012) Salt tolerance enhancement in indica rice ('*Oryza sativa*' L. spp. indica) seedlings using exogenous sucrose supplementation. Plant Omics 5:52

Su Y, Han FX, Chen J, Sridhar BBM, Monts DL (2008) Phytoextraction and accumulation of mercury in three plant species: Indian mustard (*Brassica juncea*), beard grass (Polypogon monospeliensis), and Chinese brake fern (*Pteris vittata*). Int J Phytoremediation 10(6):547–560

Teng S, Qian Q, Zeng D, Kunihiro Y, Fujimoto K, Huang D, Zhu L (2004) QTL analysis of leaf photosynthetic rate and related physiological traits in rice (*Oryza sativa* L.). Euphytica 135:1–7

Wang H, Wu Z, Zhou Y, Han J, Shi D (2012) Effects of salt stress on ion balance and nitrogen metabolism in rice. Plant Soil Environ 58:62–67

Wyn Jones R, Pollard A (1983) Proteins, enzymes and inorganic ions. In: Encyclopedia of plant physiology new series

Yildirim E, Turan M, Ekinci M, Dursun A, Cakmakci R (2011) Plant growth promoting rhizobacteria ameliorate deleterious effect of salt stress on lettuce. Sci Res Essays 6:4389–4396

Zahir ZA, Ghani U, Naveed M, Nadeem SM, Asghar HN (2009) Comparative effectiveness of Pseudomonas and Serratia sp. containing ACC-deaminase for improving growth and yield of wheat (*Triticum aestivum* L.) under salt-stressed conditions. Arch Microbiol 191:415–424

Zeeb BA, Amphlett JS, Rutter A, Reimer KJ (2006) Potential for phytoremediation of polychlorinated biphenyl-(PCB-) contaminated soil. Int J Phytoremediation 8(3):199–221

Zhen-hua Z, Qiang LIU, Hai-xing S, Xiang-min R, Ismail AM (2012) Responses of different rice (*Oryza sativa* L.) genotypes to salt stress and relation to carbohydrate metabolism and chlorophyll content African. J Agric Res 7:19–27

Zhu J-K, Liu J, Xiong L (1998) Genetic analysis of salt tolerance in Arabidopsis: evidence for a critical role of potassium nutrition. Plant Cell 10:1181–1191

Production of Rhamnolipids by a Halotolerant Bacterial Strain with Seawater and Its Application in the Treatment of Powdery Mildew of Strawberry (*Fragaria ananassa*)

8

Xiangsheng Zhang and Boping Tang

Abstract

Biosurfactants are amphipathic compounds, a kind of natural agricultural chemical, excreted by microorganisms that exhibit surface activity. Biosurfactants have advantages over their chemical counterparts in biodegradability, low toxicity, and ecological acceptability and effectiveness at extreme temperature and pH. Fermentation water and mineral salt costs are among the production costs. The seawater is usually rich in various mineral salts. Biosurfactant production with seawater and waste vegetable oil with simple facilities will lower the cost greatly. In our study, strain screening, shaking flask fermentation, and 5 L liquid fermentor fermentation were carried out consequently, and the fermentation yield reached above 10 $g \cdot L^{-1}$, confirming this kind of low-cost production is feasible and practicable. And the application of rhamnolipids in plant protection in saline soils was also conducted. The fermentation broth was diluted to 1 $g \cdot L^{-1}$ and 0.5 $g \cdot L^{-1}$ separately and was directly used to treat the downy mildew of strawberry. The results showed that the fermentation broth could be used as an efficient kind of fungicide. The pot experiments showed that control efficiency could be reached over 90.8% and 87.6%, respectively, compared to blank group, performing better than the tested dominant chemical fungicides. Furthermore, the rhamnolipids fermentation broth could also enhance the development of root and shoot of strawberry.

X. Zhang · B. Tang (✉)
Jiangsu Key Laboratory for Bioresources of Saline Soil & Jiangsu Provincial Key Laboratory of Coastal Wetland Bioresources and Environmental Protection, Yancheng Teachers University, Yancheng, Jiangsu, China

Jiangsu Coastal Biological Agriculture Synthetic Innovation Center, Yancheng Teachers University, Yancheng, Jiangsu, China

M. Kumar et al. (eds.), *Saline Soil-based Agriculture by Halotolerant Microorganisms*, https://doi.org/10.1007/978-981-13-8335-9_8

Keywords
Seawater fermentation · Halotolerant biosurfactant producer · Rhamnolipids · Fungicide · Powdery mildew · *Fragaria ananassa*

8.1 Research Background

Biosurfactants are amphipathic compounds excreted by microorganisms that exhibit surface activity (Zhang et al. 2012a, b). They have been found to be produced mainly by aerobically growing microorganisms in aqueous media from a carbon source feedstock, e.g., carbohydrates, hydrocarbons, oils, and other lipids or mixtures thereof (Zhang et al. 2012b). Producers of surfactants include *Bacillus*, *Pseudomonas*, *Acinetobacter*, *Achromobacter*, *Arthrobacter*, *Brevibacterium*, *Corynebacterium*, *Candida*, and *Rhodotorula* (Pirog et al. 2012). Five major classes of biosurfactants are known, including glyco- and phospholipids, lipopeptides, polymeric, and particulate surfactants (Shen et al. 2011; Zhang and Xiang 2010a).

Compared to the pure chemical surfactants, biosurfactants have many advantages, such as being biodegradable, nontoxic, and eco-friendly and having highly efficient surface activity, high tolerance to extreme environmental conditions, and much lower critical micelle concentration (CMC) (Silva et al. 2017; Zhang and Xiang 2010a). Therefore, biosurfactants have been used in microbial enhanced oil recovery (MEOR) (Bordoloi and Konwar 2008; Zhang and Xiang 2010b; Zhao and Jiang 2004), petroleum chemical engineering, environmental biological remediation (Li et al. 2000; Zhang 2013), agriculture, cosmetic industry, food industry, etc. (Banat et al. 2000).

Glycolipids are among the most often bacterially produced biosurfactants (Hatha et al. 2007; Zhang et al. 2012a), while the rhamnolipids are well-known with many potential applications and are amidst the most widely studied glycolipids (Henkel et al. 2012; Huang et al. 2009; Sodagari et al. 2017; Zhang and Lu 2013). In this chapter, relevant fields of biosurfactants such as the screening, fermentation, and application will be focused on rhamnolipid.

Furthermore, biosurfactants (in this paper, rhamnolipids) can be also used as pesticides, bactericides, or fungicide. Ji Ye Ahn et al. once reported that one isolate named SG3 could produce rhamnolipids, whose fermentation broth inhibited mycelial growth of all of the tested fungal pathogens and effectively suppressed the development of various plant diseases including rice blast, tomato gray mold, tomato late blight, wheat leaf rust, barley powdery mildew, and red pepper anthracnose. But presently other similar researches are not reported very often.

The powdery mildew in strawberry (*F. ananassa*) in greenhouse often brings about a huge loss in yield and qualities. In this study, an exploratory research on fermentation of rhamnolipids by PW in laboratory scale and its application in treatment of powdery mildew of strawberry in field scale was implemented with the aim of clarification of the effectiveness of PW fermented rhamnolipids in the treatment of powdery mildew in strawberry treatment.

8.2 Low-Cost Production of Biosurfactants

Theoretically, production of biosurfactants can carried out both by chemical synthesis and microbial fermentation. But due to the complexity of biosurfactants' structures, the former production is very hard to implement. Presently the fermentation method is the dominant one in biosurfactant production, including living cell method, metabolically controlled cell method, static cell method, and precursor addition method (Sarachat et al. 2010; Wu et al. 2007). Presently, free-living cell fermentation is the simplest and the most common production method.

Although microbial biosurfactants have many advantages, their application in abovementioned fields, especially in agriculture, was restrained due to the high cost in production. So, lowering the production cost is among the research focuses of microbiologists. The cost of the substrate will greatly influence the spread of the biosurfactants. To date the low-cost production studies are mainly attracted by using low-cost materials as carbon sources (Jain et al. 2013; Mukherjee et al. 2006), such as waste oil (Nitschke et al. 2005), slaughterhouse waste (Ramani et al. 2012), distillery and curd whey wastes (Dubey and Juwarkar 2001), sugarcane bagasse (Neto et al. 2009), potato process effluent (Noah et al. 2005), etc. It was also reported that the production cost was also due to the severe foaming tendency in aerobic fermentation and the relatively low productivity and yield. The optimized fermentation produced 42 g L^{-1} rhamnolipids (Sodagari et al. 2017). This might the highest yield once reported.

Waste oils such as thermally abused frying oils and oil phases accumulating in communal sewage treatment plants are the most used carbon source to reduce the production cost (Nitschke et al. 2005; Raza et al. 2006). They are largely not recovered as a secondary resource, causing environmental pollution if disposed of improperly or causing damage to public health if resold illegally as edible oils. Waste oils, however, can be utilized to produce biosurfactants, such as rhamnolipid used as emulsifier or demulsifier (Nitschke et al. 2005; Yao and Min 2010) as well as biodiesel (Xiong et al. 2007).

Although a mass of researches on halotolerant or halophilic biosurfactant producers were carried out, production of biosurfactant from natural seawater without or with adding mineral salt was rarely reported. The author of this chapter put forward a new idea of fermentation with seawater to produce biosurfactants (rhamnolipids) and put it into practices, achieving satisfactory results.

8.3 Screening of Biosurfactant Strains

Fermentation water and mineral salt costs are among the production costs. The seawater is rich in various mineral salts, while presently seawater desalination is expensive. Biosurfactant production with seawater and waste vegetable oil with simple facilities will lower greatly the cost greatly. Fermentation with high salinity (no lower than 30%) seawater is somewhat similar to fermentation under extreme environments; then the screening of seawater-tolerant biosurfactant-producing strains that can make use of seawater for fermentation is meaningful in coastal areas. These strains also have a potential application in bioremediation of hydrophobic contaminants in the sea. Usually screening of seawater fermentation strains is the following (Miao et al. 2013; Zhang 2014):

8.3.1 Culture Media

High salt nutrient broth (NB) or seed medium was composed of, in g/L de-ionized water, peptone 10, beef extract 5, and NaCl 35. For the preparation of nutrient agar (NA) plates or slants, 15.0 $g \cdot L^{-1}$agar (strength 1300) was added. Blood NA medium (3.5% NaCl) was all prepared as reference. Seawater fermentation broth was composed of (g/L filtered seawater) yeast extract 1.0; pH was set as natural value; 3% (V/V) frying oil (as the sole carbon source) was added to individual shaking flasks. All the media were autoclaved at 121 °C for 20 min.

8.3.2 Isolation and Screening of Halophilic Biosurfactant-Producing Strains

About 5 g coastal soil samples or 5 mL seawater samples were aseptically added to 45 mL sterilized seawater and streaked on blood NA plates. Colonies with big and clear hemolytic circles were selected, incubated on slant media, incubated at 30 °C, and stored at 4 °C until further use. One loopful of culture was inoculated in 20 mL of seed media in a 100 mL flask and incubated on a rotary shaker at 30 °C and 180 rpm for 10 h. Then 2.5 mL (5%) inoculum was transferred to 50 mL seawater fermentation broth in 250 mL shaking flask for 3 days. Then the broth supernatant was centrifuged at 9000 rpm, and the biosurfactant-producing capacity was measured. The strains that performed best were selected.

8.3.3 Measurement of Biosurfactant-Producing Capacity

In this section, all the protocols below are adopted as the author's laboratory conditions (Zhang et al. 2012b).

8.3.3.1 Oil-Spreading Method

About 30 mL distilled water was put into the Petri dishes (90 mm), and then 50 μL diesel was dropped on the water surface to form an oil film, in the midst of which 5 uL fermentation broth, diluted by a factor of 10 with deionized water, was dropped. The diameter of the oil expelling circles was measured by slide caliber (with a degree of accuracy of 0.02 mm). The strains with big circle diameters were selected for surface tension measurement. This method is usually used in preliminary evaluation of fermentation broth surfactant production.

8.3.3.2 Surface Tension Measurement

The surface tension of the culture broth supernatant (diluted by a factor of 100) was measured with a JYW-200 surface tension instrument.

8.3.3.3 Sulfuric Acid-Phenol Reaction Method

The fermentation broth was diluted by 100 times first, and 2 mL of the diluted broth was transferred into a 15 mL glass test tube, 1 mL phenol and 5 mL sulfuric acid was added and vortexed. Then the tubes were heated for 15 min in a boiled water bath, cooled to room temperature, after which the optical density at 480 nm was measured.

8.4 Fermentation Technology

In the laboratory level, there are three kinds of methods for free-cell fermentation: shaking flask fermentation, small fermentor fermentation (liquid or solid), and simplified fermentation. In shaking flasks, after 3 days fermentation, the waste oil was emulsified obviously (Fig. 8.1). Using seawater media, the rhamnolipid production was 10.5 $g{\cdot}L^{-1}$ in shaking flasks.

The appearance of fermentation broth in liquid fermentor was shown in Fig. 8.2, showing very good performance of seawater fermentation. The highest yield could reach 33.0 $g{\cdot}L^{-1}$. With the fermentation process, the oil expelling circles were increasing from 6.61 to 14.47 mm, while the surface tensions were decreasing from 78.9 $mN{\cdot}m^{-1}$ to 36.2 $N{\cdot}m^{-1}$. The 5 L fermentor fermentation performed even better than shaking flasks, and this laid a solid foundation for industrialization.

8.5 Application in Strawberry in Saline Soil

8.5.1 Treatment of Powdery Mildew of Strawberry

As shown in Fig. 8.3, the controlling effects of rhamnolipid fermentation broth treatments were both better than the controls (A, mancozeb and B, triadimefon), the control efficiency of 1 $g{\cdot}L^{-1}$ and 0.5 $g{\cdot}L^{-1}$ especially R1 treatment were 90.8% and 87.6%, increasing by 28.6% at most, and the difference was very significant.

8.5.2 Effects on Root Development of Strawberry

As shown in Table 8.1, four treatments enhanced root development, the fresh root weight, dry root weight, root length, root diameter, and root respiratory intensity, especially rhamnolipids fermentation broth treatment, increasing over twofold.

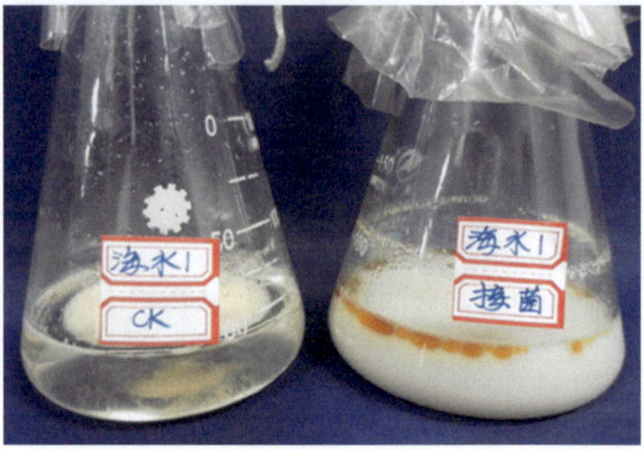

Fig. 8.1 Fermentation by seawater (left, no inoculum; right, inoculated seawater fermentation broth after 3 days incubation at 30 °C)

Fig. 8.2 Rhamnolipid production in 5 L liquid fermentor

8.5.3 Effects on Shoot Development

As shown in Table 8.2, the shoot development was all enhanced after treatments. The fresh shoot weight, dry shoot weight, average leaf area, and chlorophyll content were all enhanced, especially R1 and R2, increasing by 10.8–48.2%.

8.6 Cost Evaluation

The cost evaluation is shown in Table 8.3.

It can be expected that the total cost of materials for fermentation broth is very low.

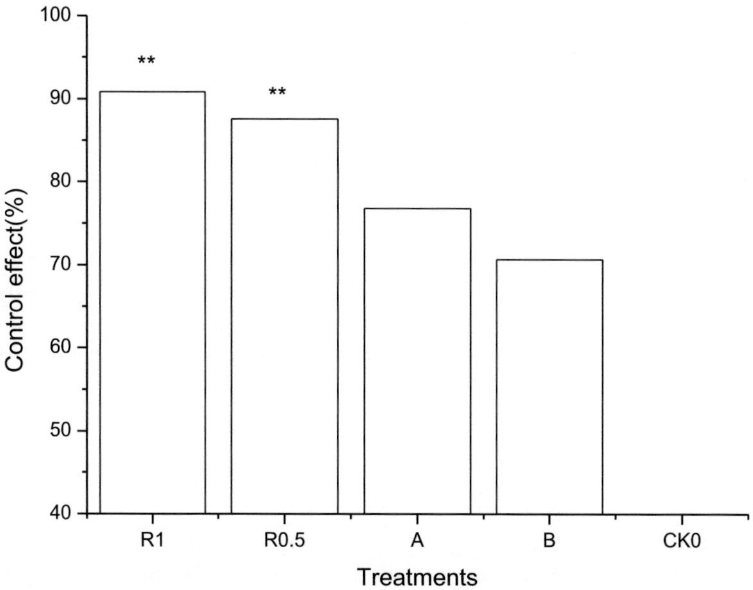

Fig. 8.3 Control efficiency of different treatments (R1, rhamnolipid fermentation broth 1.0 g·L^{-1}; R2, rhamnolipid fermentation broth 0.5 g·L^{-1}; A, mancozeb diluted by 800 times; B, triadimefon diluted by 1000 times; CK0, deionized water; **, means significant at $P < 0.01$)

Table 8.1 Effects on strawberry root development

Treatments	Fresh root weight (g)	Dry root weight (g)	Root length (mm)	Root diameter (mm)	Root respiratory intensity (μmolO_2. min^{-1}. g^{-1}FW)
R1	10.52**	2.95**	352**	0.65**	0.83**
R2	9.37**	2.54**	314**	0.54*	0.70**
A	6.86*	1.33*	229**	0.46	0.60**
B	6.27*	1.29*	201*	0.44	0.52**
Control	4.84	1.18	162	0.42	0.23

Duncan's analyses were used. R1, rhamnolipid fermentation broth 1.0 g·L^{-1}; R2, rhamnolipid fermentation broth 0.5 g·L^{-1}; A, mancozeb diluted by 800 times; B, triadimefon diluted by 1000 times; CK0, deionized water
*significant at $P < 0.05$; **significant at $P < 0.01$

8.6.1 The Principle of Fermentation by Seawater

The seawater is rich in mineral salts (Orban et al. 2007), although some researchers found that cleaning up oil spills with the aid of microorganisms should take into consideration the nutritional deficiencies of sea water (Atlas and Bartha 1972), that is, the seawater is deficient in nutritional elements for microorganisms used for oil spill remediation. There are 11 kinds of the most abundant elements in the seawater, including 5 kinds of cations: Na, Mg, Ca, K, and Sr (Li and Wu 1993). By

Table 8.2 Effects on shoot development

Treatments	Fresh shoot weight (g)	Increasing by (%)	Dry shoot weight (g)	Increasing by (%)	Chlorophyll content (mg. g^{-1}FW)	Increasing by (%)
R1	32.78**	48.1	10.67**	48.2	38.63*	16.4
R2	29.40**	32.9	9.57**	32.9	36.78*	10.8
A	25.54*	15.4	8.31*	15.4	35.32	6.4
B	24.47*	10.6	7.96	10.6	34.96	5.3
Control	22.13	0.0	7.2	0.0	33.2	0.0

Duncan's analyses were used. R1, rhamnolipid fermentation broth 1.0 g L^{-1}; R2, rhamnolipid fermentation broth 0.5 g L^{-1}; A, mancozeb diluted by 800 times; B, triadimefon diluted by 1000 times; CK0, deionized water
*significant at $P < 0.05$; **significant at $P < 0.01$

Table 8.3 Preliminary evaluation of cost only by fermentation materials

	Materials	Demand amount		Price	Total cost
1 t fermentation broth	Seawater	1	t	0	0
	Waste oil	30	L	1 USD·L^{-1}	30 USD
	Yeast extract	1	kg	5 USD·kg^{-1}	5 USD
Sum					**35 USD**

comparison of seawater media fermentation and synthetic media fermentation, it was found that NO_3^-, PO_4^{3-}, and K^+, as nitrogen, phosphorus, and potassium, respectively, had great influence on fermentation yield. While too high concentrations of Ca^{2+} and Mg^{2+} in seawater might suppress biosurfactant yield (Liang et al. 2009). But in this study, fermentation by seawater showed good results, the reasons were the follows: firstly, the other elements in the seawater might compensate for negative effects of too high concentrations of Ca^{2+} and Mg^{2+} ions; secondly, the bacterial strains resisted high concentrations of cations. Therefore, for the industrialization of fermentation by seawater, the screening of optional strains is very important. There is only very limited nitrate-N in the seawater (We et al. 2010). In this study, the concentration of nitrate-N in the seawater was about 5 mg/L, much less than that in synthetic media; therefore, it became one of limiting factors for seawater fermentation, and the research of influence of adding nitrate or urea should be carried out for further improvement of biosurfactant yield.

8.6.2 Potential Application of Seawater Fermented Rhamnolipids in Agriculture

Biosurfactant has a strong commercial potential (Jamal et al. 2014), but its application in different fields is limited by the production cost. These papers mainly deal with the cost relevant problems of the application of biosurfactant by two means. The first is lowering the fermentation cost. In this study, seawater was employed to take place of pure water, lowering the production cost greatly. Secondly, the direct application in plant protection with fermentation broth also decreased the cost of its application. In this paper, the treatment of powdery mildew of strawberry was carried out, with the controlling effects over 87.57%, respectively, compared to blank group, performing better than commonly used chemicals. Furthermore, the rhamnolipid fermentation broth can also be found to stimulate the root and shoot development of strawberry. Rhamnolipids are of great potential as a kind of natural fungicide. As for the application of rhamnolipids in strawberry culture, it is suggested that it should be used at the concentration of 0.5–1 $g{\cdot}L^{-1}$ every 7 days, totally 2–3 times to achieve more stable controlling effects.

8.7 Conclusions

The author first proposed biosurfactant fermentation by halotolerant strain with seawater adding a limited amount of yeast extract (growth factor) and sufficient waste vegetable oil (as carbon source) using the natural mineral salts in the seawater with a bacterial strain isolated from coastal soil, lowering the production cost greatly, while the rhamnolipids yield was higher than fermentation yield with oil-containing wastewater in other references (Liang et al. 2009). In summary, the seawater can be used to ferment biosurfactants including rhamnolipids, and the fermentation broth is effective at treating various plant pathogens, including powdery mildew of strawberry. This study laid a foundation for further application of wastewater fermentation broth in horticultural crop protection.

Acknowledgments This research was supported by the "333" Project of Jiangsu Province, National Spark Program (2015GA690261), and Open Project of Jiangsu Provincial Key Laboratory of Coastal Wetland Bioresources and Environmental Protection and Jiangsu Key Laboratory for Bioresources of Saline Soil.

References

Atlas RM, Bartha R (1972) Degradation and mineralization of petroleum in sea water: limitation by nitrogen and phosphorous. Biotechnol Bioeng 14(3):309–318

Banat IM, Makkar RS, Cameotra SS (2000) Potential commercial applications of microbial surfactants. Appl Microbiol Biotechnol 53(5):495–508

Bordoloi NK, Konwar BK (2008) Microbial surfactant-enhanced mineral oil recovery under laboratory conditions. Colloids Surf B Biointerfaces 63(1):73–82. https://doi.org/10.1016/j.colsurfb.2007.11.006

Dubey K, Juwarkar A (2001) Distillery and curd whey wastes as viable alternative sources for biosurfactant production. World J Microbiol Biotechnol 17(1):61–69

Hatha A, Edward G, Rahman K (2007) Microbial biosurfactants – review. J Atmos Sci 3(2):1–17

Henkel M, Müller MM, Kügler JH, Lovaglio RB, Contiero J, Syldatk C, Hausmann R (2012) Rhamnolipids as biosurfactants from renewable resources: concepts for next-generation rhamnolipid production. Process Biochem 47(8):1207–1219

Huang X, Chen X, Liu J, Lu L (2009) Recent progress on rhamnolipid produced from fermentation of waste edible oils. Microbiology 36(11):1738–1743 (in Chinese)

Jain RM, Mody K, Joshi N, Mishra A, Jha B (2013) Effect of unconventional carbon sources on biosurfactant production and its application in bioremediation. Int J Biol Macromol 62 (11):52–58

Jamal P, Mir S, Alam MZ, Wan NW (2014) Isolation and selection of new biosurfactant producing bacteria from degraded palm kernel cake under liquid state fermentation[J]. J Oleo Sci 63(8):795–804

Li Y, Wu D (1993) A study of major seawater chemical constituents in part of Bohai sea. J Qingdao Univ 6(4):41–48

Li G, Huang W, Lerner DN, Zhang X (2000) Enrichment of degrading microbes and bioremediation of petrochemical contaminants in polluted soil. Water Res 34(15):3845–3853

Liang X, Yao B, Sha R, Zhang H, Meng Q (2009) Production of rhamnolipids by *Pseudomonas aeruginosa* under high salt concentration conditions. Chin Sci Pap Online 4(6):418–422 (in Chinese)

Miao L, Zhang X, Zhou C, Zhang F, Chai X (2013) Study on production of biosurfactants by seawater fermentation. J Qilu Univ Technol (Nat Sci Ed)(4):26–30 (in Chinese)

Mukherjee S, Das P, Sen R (2006) Towards commercial production of microbial surfactants. Trends Biotechnol 24(11):509–515. https://doi.org/10.1016/j.tibtech.2006.09.005

Neto DC, Meira JA, Tiburtius E, Zamora PP, Bugay C, Mitchell DA, Krieger N (2009) Production of rhamnolipids in solid-state cultivation: characterization, downstream processing and application in the cleaning of contaminated soils. Biotechnol J 4(4):748–755

Nitschke M, Costa SG, Haddad R, G Gonçalves LA, Eberlin MN, Contiero J (2005) Oil wastes as unconventional substrates for rhamnolipid biosurfactant production by *Pseudomonas aeruginosa* LBI. Biotechnol Prog 21(5):1562–1566

Noah KS, Bruhn DF, Bala GA (2005) Surfactin production from potato process effluent by Bacillus subtilis in a chemostat. Appl Biochem Biotechnol 121(124):465–473. doi:ABAB:122:1-3:0465 [pii]

Orban E, Di Lena G, Nevigato T, Casini I, Caproni R, Santaroni G, Giulini G (2007) Nutritional and commercial quality of the striped venus clam, *Chamelea gallina*, from the Adriatic sea. Food Chem 101(3):1063–1070

Pirog TP, Shevchuk TA, Shuliakova MA (2012) Glycerol metabolism in surfactants producers Acinetobacter calcaaceticus IMV B-7241 and Rhodococcus erythropolis IMV Ac-5017. Mikrobiol Z 74(4):29–36

Ramani K, Jain SC, Mandal AB, Sekaran G (2012) Microbial induced lipoprotein biosurfactant from slaughterhouse lipid waste and its application to the removal of metal ions from aqueous solution. Colloids Surf B Biointerfaces 97:254–263. https://doi.org/10.1016/j.colsurfb.2012.03.022

Raza ZA, Khan MS, Khalid ZM, Rehman A (2006) Production kinetics and tensioactive characteristics of biosurfactant from a *Pseudomonas aeruginosa* mutant grown on waste frying oils. Biotechnol Lett 28(20):1623–1631. https://doi.org/10.1007/s10529-006-9134-3

Sarachat T, Pornsunthorntawee O, Chavadej S, Rujiravanit R (2010) Purification and concentration of a rhamnolipid biosurfactant produced by *Pseudomonas aeruginosa* SP4 using foam fractionation. Bioresour Technol 101(1):324–330

Shen Z, Yang HY, Yan XT, Nan FY, Guo YP, Xie HZ (2011) The application and development trends of biosurfactants in petroleum industry. Appl Chem Ind 40(10):1842–1846

Silva MAM, Silva AF, Rufino RD, Luna JM, Santos VA, Sarubbo LA (2017) Production of biosurfactants by pseudomonas species for application in the petroleum industry. Water Environ Res 89(2):117–126

Sodagari M, Invally K, Ju L-K (2017) Maximize rhamnolipid production with low foaming and high yield. Enzym Microb Technol

We Q, Liu L, Zhan R, Wei X, Zang J (2010) Distribution features of the chemical parameters in the southern Yellow sea in summer. Period Ocean Univ China 40(1):82–88 (in Chinese)

Wu H, Wang W, Han S (2007) Research advance on rhamnolipid biosurfactants. Microbiology 34 (1):148–152 (in Chinese)

Xiong J, Song W, Ye J (2007) Research advance on decrease the cost of biodiesel production. Chem Ind Eng Prog 26(6):774–776 (in Chinese)

Yao J, Min E (2010) Hazardous effects and resource utilization of waste oil. New Energy 30(5):1–6 (in Chinese)

Zhang X (2013) Enhancement of hydrocarbon degrading and biosurfactant production of *Pseudomonas aeruginosa* strain Z41 by joint employment of low energy ion beam and ultraviolet irradiations. Res J Biotechnol 8(12):31–36

Zhang X (2014) A pilot study on screening and simplified fermentation of a biosurfactant producing strain with seawater. Res J Biotechnol 9(3):74–79

Zhang X, Lu D (2013) Response surface analysis of rhamnolipid production by *Pseudomonas aeruginosa* strain Z41 with two response values. Afr J Microbiol Res 7(22):2757–2763

Zhang X, Xiang T (2010a) Application of physical factors in genetic improvement of biosurfactant producing strains. J Huzhou Vocat Technol Coll 8(1):1–5 (in Chinese)

Zhang X, Xiang T (2010b) Genetic modification of MEOR bacterium bacillus licheniformis H strain by low energy ion beam irradiation. Open Biotechnol J 4:14–17

Zhang X, Xu D, Yang G, Zhang H, Li J, Shim H (2012a) Isolation and characterization of rhamnolipid producing *Pseudomonas aeruginosa* strains from waste edible oils. Afr J Microbiol Res 6(7):1466–1471

Zhang X, Xu D, Zhu C, Lundaa T, Scherr KE (2012b) Isolation and identification of biosurfactant producing and crude oil degrading *Pseudomonas Aeruginosa* strains. Chem Eng J 209:138–146

Zhao X, Jiang B (2004) Brief review on MEOR technology. Pet Sci 1(4):17–23

9 Bottlenecks in Commercialization and Future Prospects of Beneficial Halotolerant Microorganisms for Saline Soils

Bushra Tabassum, Adeyinka Olawale Samuel, Muhammad Umar Bhatti, Neelam Fatima, Naila Shahid, and Idrees Ahmad Nasir

Abstract

Salinity stress is a major abiotic stress that limits crop productivity. One way to alleviate the effect of salinity stress on plants is to use salt-tolerant microorganisms. Mechanisms employed by these salt-tolerant microbes not only drive the plant to grow well in saline environment but also boost its growth by producing various hormones, solubilizing phosphate, and fixing nitrogen. Inconsistent field efficacy of salt-tolerant microbial formulations, versatility for various crops, and economic aspects to develop such product by industry are the bottlenecks in commercialization of these bioproducts. There is no commercial formulation of salt-tolerant microbes currently available. For sustained agriculture, formulations based on compatible halotolerant microbes must be developed and marketed. This chapter focuses on the current status and prospects of commercialization of microbial products based on halotolerant microbes for increased crop production in saline soils.

Keywords

Halotolerant microorganisms · Saline soil-based agriculture · Bioproducts · PGPR · Biofertilizer

9.1 Introduction

Salt-dependent and salt-tolerant microbes with structural and metabolic adaptations to thrive in various salt concentrations cut across Archaea, Bacteria, and Eukarya domain. They can be categorized as extreme halophiles, moderate halophiles, and halotolerant based on salt concentration requirement. Those that survive within

B. Tabassum (✉) · A. O. Samuel · M. U. Bhatti · N. Fatima · N. Shahid · I. A. Nasir
National Centre of Excellence in Molecular Biology, University of the Punjab, Lahore, Pakistan

M. Kumar et al. (eds.), *Saline Soil-based Agriculture by Halotolerant Microorganisms*, https://doi.org/10.1007/978-981-13-8335-9_9

2.5–5.2 M salt concentration are referred to as extreme halophiles, while those that fall within 0.5–2.5 M are moderate halophiles. The degree of salt accommodation is similar in both except little change in some structural function. For instance, N-glycosidically linked saccharides were found to differ significantly between *Halobacterium halobium* (extreme halophiles) and *Halobacterium volcanii* (moderate halophile) (Mengele and Sumper 1992). The stability of extreme halophile *Halobacterium halobium* and *Halobacterium salinarum* in hypersaline environment was attributed to the presence of di-acidic phospholipid, archaetidylglycerol methylphosphate (PGP-Me) which prevent the leakage of cell content (Tenchov et al. 2006). Several studies have focused on the adaptation of halophile microbes to salt environment and the mechanisms employed (Paul et al. 2008; Argandona et al. 2010; Becker et al. 2014). Halophilic and halotolerant microorganisms employed two basic osmoadaptation mechanisms to thrive in high salt concentration environment which allow them to manage ionic strength and water stress. In Salt-in-cytoplasm osmoadaptation mechanism, the salt concentration in the cytoplasm is raised to a similar concentration as in the salty environment, while during the organic osmolyte mechanism, the cell wall is maintained by accumulation of uncharged, highly water-soluble, organic solutes like sugars, polyols, and amino acids in response to an osmotic stress. These nonionic, highly water-soluble compounds do not disturb the metabolism, even at high salt concentrations. The effort in isolation and characterization of different halophilics and halotolerant microorganism has led to their beneficial mining toward the enhancement of salinity tolerance in crop plants.

9.2 Salinity Upsets Vegetative and Reproductive Development in Crop Plants

A soil with electrical conductivity (EC) of the saturation extract (EC_e) higher than 4 dS m^{-1} (approximately 40 mM NaCl) in the root zone and with exchangeable sodium of about 15% is referred to as saline soil. Soil salinity deteriorates soil quality and consequently reduces its agricultural potentials. The world continues to lose hectares of farm soil daily to salt-induced degradation. Salinity upsets the vegetative and reproductive development in crops. It inhibits plant growth either by osmotic or ionic effects (Läuchli and Epstein 1990). The osmotic effect phase constrains water uptake because in the saline soil solution, the osmotic pressure lowers its potential energy. The ionic phase effects arise when specific ions buildup over a period of time in the plant lead to ion toxicity or ion imbalance (Munns and Tester 2008). High salinity reduces leaf expansion, stomata closure, photosynthesis, and biomass loss due to water deficit caused by osmotic imbalance (James et al. 2011; Rahnama et al. 2010). Most salt-tolerant plants have developed osmotic adjustment mechanism that reduces the cellular osmotic potential by net solute accumulation, which permits them to survive in salt-affected soil. Other adaptive mechanisms include cell wall modification, reactive oxygen species (ROS) detoxification, transport proteins, K^+ and NO_3^-homeostasis, vacuolar compartmentation, compatible solutes, etc.

Vacuole sequestration of Na^+ detoxifies Na^+ with the help of NHXs, while the SOS signaling pathway exports Na^+ out of the cell. Studies have shown that the NHX-type proteins are also important for compartmentalization of K^+ into vacuoles and for cellular pH homeostasis. The overexpression of AtNHX1 in *Arabidopsis* was attributed to Na^+ compartmentation in the vacuoles and has been proven to improved salt tolerance *Arabidopsis* (Apse et al. 1999), tomato, and canola (Zhang and Blumwald 2001; Zhang et al. 2001). Similarly ROS-scavenging gene overexpressed in transgenic plant resulted in low cellular damage and improved shoot and root growth in salt-affected soil (Roy et al. 2014).

9.3 Managing Salinity in Agricultural Land

Factors that contribute to the increasing salinity include weathering of primary minerals and higher evapotranspiration rates, the increasing irrigation, excessive fertilization, and processes of desertification (Ramadoss et al. 2013). Basically, salt-affected soils are categorized as saline, sodic, and saline-alkali based on salt accumulation. The salt-affected soils hamper crop yields and subsequently result in huge economic damages. Saline-sodic fields effect on crops in the Indus basin in Pakistan result to overall average loss of 32% in wheat and about 48% loss in rice (Murtaza 2013). Khodarahmpour et al. (2012) reported 77.4% reduction in germination percentage, 32.4% germination rate, and 95% seed vigor at 240 mM level of salinity in maize.

Several methods have been employed for salt-affected soil mitigation toward sustainable soil management and improved food production. However, attention is given recently to microbial colonizing rhizosphere, bacteria, fungi, actinomycetes, protozoa, and algae. But bacteria are the most abundant microbial present in the rhizosphere. Specific halotolerant microbes are able to colonize the roots of plants and enhance crop performance and plant growth in salt-affected soil. Rigorous research has been carried out to identify the halophile strains that are suitable for plant growth-promoting activities. Basically the following characteristics, nitrogen fixation, phosphate solubilization, ACC (1-aminocyclopropane-1-carboxylate) deaminase activity, and production of siderophores and phytohormones, are usually investigated to confer the plant growth-promoting ability of specific halophiles.

9.3.1 Biological Nitrogen Fixation

Biological nitrogen fixation (BNF) is the reduction of atmospheric nitrogen (N=N) to ammonia in the presence of nitrogenase complex that contains of two metalloprotein components of the molybdenum-iron (MoFe) protein and iron (Fe) protein (Hu and Ribbe 2015). Although BNF is a general activity in microorganism, they are case of difference in mechanism mode. The N_2-fixing system of non-heterocystous cyanobacteria contrasts from that of heterocystous cyanobacteria significantly (Bergman et al. 2013). Nitrogenase activity of cyanobacterium *Trichodesmium* was limited under aerobic conditions (Staal et al. 2007). Unlike other organisms BNF mechanism in halophile or halotolerant requires Na^+. Attempts have been made to establish whether Na^+ deficiency directly or indirectly renders nitrogenase into inactive conformation and whether this can be reversed by preincubation under acetylene (Apte and Thomas 1984). The loss of supply of ATP in the absence of Na^+ indicated that Na^+ plays an important role in maintaining supply of ATP to support various energy-linked functions, which have also been observed in *Pseudomonas stutzeri* (Kodama and Taniguchi 1976, 1977). Several nitrogen fixation halophile bacteria that survive above 5% NACl have been isolated and characterized (Zahran et al. 1995; Kang et al. 2015).

9.3.2 Phosphorus Solubilization

Phosphorus (P) is a very important element for plant growth, and a very little amount is available for plant growth because it usually presents in insoluble form. Recent research has shown that phosphate-solubilizing halotolerant plant growth-promoting rhizobacteria (PGPR) enhance P availability to plants. Various mechanisms like chelation, ion exchange, enzymolysis, and acidification by secreting low molecular weight organic acids have been used to explain how microbes solubilize insoluble phosphates and increase the available P for optimum plant growth (Sharma et al. 2013; Etesami 2018). Srinivasan et al. (2012) isolated several phosphate-solubilizing bacteria (PSB) and phosphate-solubilizing fungi (PSF) from soil-affected soil. They reported that *Aerococcus* sp. strain $PSBCRG_1$-1 showed the maximum P-solubilization (12.12%) which was significantly superior over all other isolates irrespective of NaCl concentrations, while *Aspergillus* sp. strain PSFNRH-2 (20.81%) recorded the maximum P_i release irrespective of the NaCl concentrations and was significantly superior over all other PSF.

9.3.3 ACC Deaminase Activity

Several microbes found in plant rhizosphere demonstrated the ability of ACC deaminase production by the convection of the ethylene precursor (ACC) to α-ketobutyrate and ammonium, reducing the ethylene levels in host plants and making more nitrogen available for plant utilization. *Enterobacter cloacae* and *Pseudomonas sp.* were found to produce and secrete large amounts of ACC deaminase at about 10% salinity level NaCl (Trung et al. 2016). Similarly the role of ACC deaminase from *Ps. fluorescens* strain TDK1 in alleviating saline stress was evaluated in groundnut, and significant growth was reported in groundnut inoculated with *P. fluorescens* strain TDK1 unlike other *Pseudomonas* strains without ACC deaminase activity (Saravanakumar and Samiyappan 2007). Recently the molecular interaction of ACC deaminase-producing endophytic *Streptomyces* sp. GMKU 336 and its ACC deaminase-deficient mutant was investigated. These bacterial strains were inoculated into *Oryza sativa* L. cv. KDML105 under salt stress (150 mM NaCl) conditions. It was found that *Streptomyces* sp. GMKU

336 significantly increased plant growth, chlorophyll, proline, K^+, Ca^+, and water contents but decreased ethylene, ROS, Na^+, and Na^+/K^+ ratio when compared to plants not inoculated and those inoculated with the ACC deaminase-deficient mutant. Furthermore, genes involved in the ethylene pathway, *ACO1* and *EREBP1*, were significantly downregulated (Jaemsaeng et al. 2018).

9.3.4 Production of Siderophores

Siderophores are low molecular weight iron-chelating compounds produced by microorganisms (bacteria and fungi) to combat low iron stress and regulate its uptake at various concentrations. They are categorized as catecholates, hydroxamates, and hydroxycarboxylates based on their functional groups. Low iron triggers a cascade of gene activation, allowing the cell to survive due to the synthesis of important proteins involved in siderophore synthesis and transport and release of chelates. Siderophores mobilize Fe and are effective at solubilizing Fe for plant uptake and do not leach or decompose as readily as synthetic chelates (Siebner-Freibach et al. 2004).

9.3.5 Phytohormones

Plant hormones or phytohormones are naturally organic substances which impact physiological processes of plants at low concentrations. They affect differentiation and development of plant growth via the regulation of diverse processes (Tabassum et al. 2017). Plant hormones are categorized as auxins, gibberellins (GA), cytokinins (CK), abscisic acid (ABA), and ethylene (Davies 2004; Jha et al. 2015). The increase in auxin (IAA) which caused a reduction of the uptake of toxic ions by plants improved plant growth in high saline conditions (Chakraborty et al. 2011). Two halotolerant strains of *P. extremorientalis* and *P. chlororaphis* inoculated with *Phaseolus vulgaris* under high salinity conditions demonstrated the impact of plant hormones like IAA, ABA, and GA_3 on amelioration of salinity stress in crop plant growth (Egamberdieva 2011). Similarly Ul-Hassan and Bano (2014) reported an increase in the level of IAA and ABA in leaves of wheat grown in saline soil upon inoculation with isolates of *Pseudomonas* sp. and *Bacillus cereus*. A reduction in IAA concentration in leaves of soybean was reported after 48 h of inoculation while uninoculated plants showed an increase in IAA content with salt stress (Asim et al. 2013).

9.4 Application of Halotolerant PGPR in Agriculture to Help Plant Survive in Saline Environments

Halotolerant PGPRs are known to involve in array of agricultural-related applications as detailed in Table 9.1 (Etesami and Beattie 2018), some of which are outlined below:

Table 9.1 Summarized halotolerant strains and their particular role in plant growth promotion

Halotolerant/halophilic bacteria	Mode of action	NaCl concentration	Experimental plant	Effects on plant	References
Serratia sp. SL-12	–	150–200 mM	Wheat	Significant increase in plant growth, as measured by parameters such as shoot/root length, fresh/dry weight, and photosynthetic pigment accumulation	Singh and Jha (2016)
Pseudomonas fluorescens 002	Indole-3-acetic acid (IAA) secretion and ACC-deaminase activity	150 mM	Maize	Increase in primary, lateral, and seminal root lengths and numbers, as well as root dry mass	Zerrouk et al. (2016)
Klebsiella sp. SBP-8	–	150,175, and 200 mM	Maize	Increased in plant biomass and chlorophyll content. The bacterial strain caused Na^+ exclusion (65%) and increased uptake of K^+ (84.21%) in the host plant	Singh and Jha (2015)
Hartmannibacter diazotrophicus E19T	ACC-deaminase production	1%, 2%, and 3%	Barley	Significant increase in root (308%) and shoot (189%) dry weight	Suarez et al. (2015)
Bacillus amyloliquefaciens NBRISN13 (SN13)	IAA and ACC deaminase production, phosphate solubilization, and proline accumulation	200 mM	Rice	Upgradation of *NADPMe2*, *EREBP*, *SOSI*, *BADH* and *SERK1* and increase of osmoprotectant	Nautiyal et al. (2013)
Bacillus amyloliquefaciens SQR9	Sequestering Na^+ into vacuoles, expelling Na^+ from roots, accumulation	100 mM	Maize	Upregulation of *RBCS*, *RBCL*, H^+-*PPase*, *HKT1*, *NHX1*, *NHX2*, and *NHX3* and significant promotion of the growth of maize	Chen et al. (2016)

	of total solute sugar, and enhancing antioxidant content			seedlings and increase of the chlorophyll Upregulation of the expression of genes related to salt tolerance and downregulation of the expression of genes related to ABA in plants	
Achromobacter piechaudii	IAA and ACC-deaminase activity	172 mM	Tomatoes	Increase in the fresh and dry weights of tomato seedlings grown	Mayak et al. (2004)
P. fluorescens YsS6 and *P. migulae* 8R6	IAA and ACC-deaminase activity	165 and 185 mM	Tomatoes	Increase of fresh and dry biomass, chlorophyll contents, and number of flowers and buds than the other treatments	Ali et al. (2014)
Azospirillum brasilense	Nitrogen fixation	40, 80, or 120 mM	White clover	Increase in plant growth under both nonsaline and saline environments	Khalid et al. (2017)
Consortium containing strains *Bacillus alcalophilus* EA1, *B. thuringiensis* EA3 and *Gracilibacillus saliphilus*	Phosphate solubilization and siderophores production	0, 510, and 1030 mM	*Arthrocnemum macrostachyum*	Mitigated the effects of high salinity on plant growth and physiological performance	Navarro-Torre et al (2017)
Micrococcus yunnanensis, *Planococcus rifietoensis*, and *Variovorax paradoxus*	ACC deaminase	50–125 mM	Sugar beet	Enhanced salt stress tolerance remarkably in sugar beet, resulting in greater seed germination and plant biomass, and higher photosynthetic capacity	Zhou et al. (2017)
Dietzia natronolimnaea STR1	Modulation of ABA-signaling, SOS pathway, ion transporters, and antioxidant machinery	50–150 mM	Wheat	Enhanced expression of *TaST*, a salt stress-induced gene	Bharti et al. (2016)

(continued)

Table 9.1 (continued)

Halotolerant/halophilic bacteria	Mode of action	NaCl concentration	Experimental plant	Effects on plant	References
Klebsiella sp. strain	Modulating expression profile of *rbcL* and *WRKY1* genes	100 mM	Oat	Increase in physiological parameter such as shoot length, root length, shoot dry weight, root dry weight, and relative water content (RWC) in inoculated plants	Sapre et al. (2018)
P. fluorescens	ACC-deaminase activity, siderophore, and IAA production	5%	Cucumber	Alleviated the negative impacts of salinity on plant growth	Nadeem et al. (2016)
Enterobacter sp. UPMR18	Antioxidant enzyme activities and upregulation of ROS pathway genes, and ACC-deaminase activity		Okra	Increase in germination percentage, growth parameters, and chlorophyll content	Habib et al. (2016)
Klebsiella, Pseudomonas, Agrobacterium, and *Ochrobactrum*	IAA and phosphate solubilization	1–20%	Peanut	A significant increase in total nitrogen (N) content (up to 76%)	Sharma et al. (2016)
Brachybacterium saurashtrense JG-06, *Brevibacterium casei* JG-08, and *Haererohalobacter* JG-11	Higher K^+/Na^+ ratio and higher IAA concentration, Ca^{2+}, phosphorus, and nitrogen content	100 mM	*Arachis hypogaea*	Increase in plant length, shoot length, root length, shoot dry weight, root dry weight, and total biomass	Shukla et al. (2012)
Bacillus licheniformis A2	IAA and phosphate solubilization	50 mM	Peanut	Increase in fresh biomass, total length, and root length by 28%, 24%, and 17%, respectively	Goswami et al. (2014)
Bacillus endophyticus, B. tequilensis, Planococcus rifietoensis, Variovorax paradoxus, and *Arthrobacter agilis*	IAA and phosphate solubilization	50–500 mM	Glasswort	Increase in plant growth under saline stress conditions	Zhao et al. (2016)

Bacillus sp. and *Arthrobacter pascens*	Phosphate solubilization and siderophore production	100 mmol/L	Maize	Increase in accumulation of osmolytes, including sugar and proline, and antioxidant enzymes activity	Ullah and Bano (2015)
Enterobacter asburiae, Moraxella pluranimalium, and *Pseudomonas stutzeri*	ACC-deaminase activity	0.25 M	*Solanum lycopersicum*	Increase in seedlings fresh and dry biomass	Raheem and Ali (2015)
Bacillus sp., *Z. halotolerans*, *S. succinus*, *B. gibsonii*, *O. oncorhynchi*, *Halomonas* sp., and *Thalassobacillus* sp.	IAA production, nitrogen fixation, and ACC-deaminase activity	200 mM	*Triticum aestivum*	Increase in the root and shoot length and total fresh weight of the plant	Orhan (2016)
Enterobacter sp. and *Bacillus* sp.	ACC-deaminase activity, exopolysaccharide secretion, and IAA production	0 or 400 mM	*Chenopodium quinoa*	Improved plant water relations and decreased Na^+ uptake, which consequently reduced osmotic and ionic stress	Yang et al. (2016)

(i) Upregulate the activity of key enzymes, thereby activating plant antioxidant defense machinery (Jha and Subramanian 2013; Islam et al. 2016; Qin et al. 2016); (ii) fix atmospheric nitrogen by improvements in plant nutrition (Dodd and Perez-Alfocea 2012; Etesami and Beattie 2017; Etesami 2018); (iii) maintain a high K^+/Na^+ ratio, thereby enhancing the affectivity of inoculated plants (Giri et al. 2007; Zuccarini and Okurowska 2008; Shukla et al. 2012; Islam et al. 2016; Etesami 2018); (iv) augment soil structure by promoting soil aggregation that is achieved by exopolysaccharides (EPS)-producing-halotolerant PGPRs (Watanabe et al. 2003; Nunkaew et al. 2015); (v) help in altering root architecture and morphology, hydraulic conductance, and hormonal status (Arora et al. 2006, 2012); and (vi) enhance plant biomass and defense against drought stress, which are possible due to emission of stress-related volatile compounds (Timmusk et al. 2014).

9.5 Stages of Commercialization

At present, no halotolerant formulation is available commercially. For commercialization of halotolerant microbial product, several crucial steps are required and are prerequisite starting from field survey, selection of microbes, their compatibility with each other, geographical conditions in that particular area, performance study of prepared formulation in laboratory conditions, and efficacy testing in the field. Briefly, stages for commercialization of a halotolerant consortium are underlined as:

9.5.1 Screening and Selection of Microbe

Availability and isolation of suitable microbes are trivial for commercialization of a successful biocontrol/biostimulator product. This selection procedure ranges from a specific environment and early trials to differentiate a microbe's biocontrol/biostimulator capability. Although no screening method can be singled out, a plausible selection strategy could be followed based upon the pathosystem (plant-pathogen-environment) of interest (Fravel 2005). This may include microbes that colonize phyllosphere (Enya et al. 2007, Yoshida et al. 2012)/rhizosphere or the ones that protect/promote a harvested crop (Janisiewicz and Korsten 2002).

Evaluation of suitable consortium of microbes is vital for selecting a potential microbe. It is worth noting that even in the presence of highly efficient and robust screening protocols, just less than 1% of candidate bacteria make efficacious products (Bailey and Falk 2011). Thus, it is the need of the hour to come up hefty microbial consortia to identify specific biocontrol/ biostimulator applicants. This can be achieved via both target-directed and broad sampling techniques.

A simple model of microbial control/stimulator may consist of construction of microbe databases. Such database would contain information on the isolates, which could be aligned with required biocontrol/biostimulator compatibility, pathogen targets, host plants, mode of action, and environmental niche, to name a few (Glare et al. 2012). Every resulting subset can then be exposed to a chain of standard bioassays to establish biocontrol/biostimulator ability and on-field competence. Importance of field setting as a marker of approval and application cannot be neglected. It is common practice to establish biocontrol/biostimulator agents using in vitro screening; however, a cautious approach should be taken while correlating in vitro inhibition/promotion and field performance (Burr et al. 1996; Milus and Rothrock 1997; Fravel 2005). There may be diverse screening strategies and approaches, but they all should converge to form a microbial agent that is effective, eco-friendly, and low on cost (Köhl et al. 2011; Ravensberg 2011).

9.5.2 Mass Production

To commercially produce such microbial agents, certain factors should be kept in mind. First of all, a large (approx. 200 km^2), centralized location with better road accessibility and unhindered electric supply is recommended. Secondly, one must ensure uninterrupted supply of all the raw materials which may consist of mother cultures, carrier materials (explained later), pet bottles, cardboard, and growth materials. Thirdly, to set up a production facility of 250,000 l/year, almost 1 acre of land is required. This land may be used to set up laboratory, installing tube well, office, parking, and other facilities. The whole site must have a barbed wire fence with suitable entry gates. Initially, 2000 m^2 of area can be covered for regular operations.

A separate and closed biofertilizing unit must be constructed. This production unit should be equipped with specific channels for microbial lab, production (fermentation) area, packaging area, storage area, and marketing way. Latest and efficient laboratory equipment, such as fermenters, culture medium tank, fermenter assembly, autoclaves, boiler, broth dispensers for sterilization, demineralizing plant, air compressor, etc., should be installed. All such equipment should cater the overall production capacity. Steps for mass production of halotolerant microbial suspension are shown in Fig. 9.1.

9.5.3 Preparation of Mother or Starter Cultures

Starter cultures of selected strains are obtained after determining their functioning in greenhouse and at field levels. The pure culture of efficient strain of plant growth-promoting organism is grown on respective agar medium and preserved in the laboratory. This mother cultures are further reproduced in larger flasks. After sterilization, each flask containing suitable broth is inoculated with the mother culture in 1:5 proportions aseptically and grown in ideal conditions. This broth culture with population of 10^9–10^{10} cells per ml should not be stored more than 24 h or stored at 4 °C. Fermenters are used for large-scale production of microbial products.

For the production of liquid bioproduct, the broth from the fermenters directly goes to the automatic filling machine and gets packed in, properly labeled, pet bottles

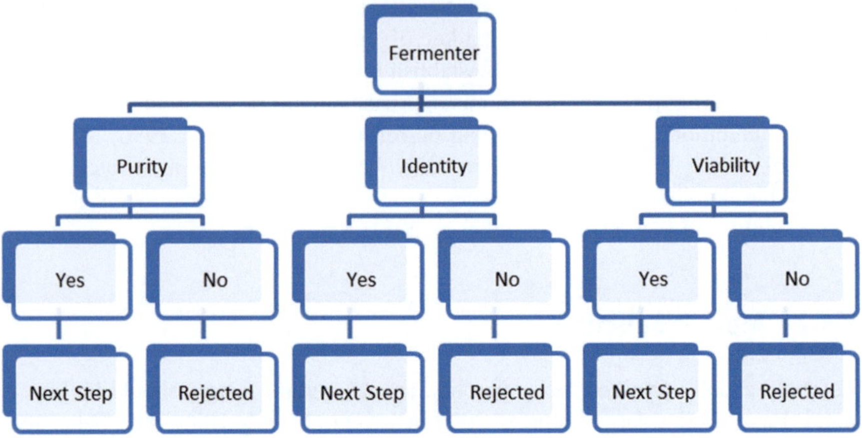

Fig. 9.1 Steps for mass production of halotolerant microbial suspension

(varying volumes) as per the demand. The microbial count of the inoculants has to be checked at the time of manufacturing. The viable cell count in the inoculants should be maintained, for proper quality standards. The inoculants shall be stored by the manufacture in a cool place away from direct heat. There are some quality control standards for some microbial species, but there is no systemic certification. Therefore, it is completely a voluntary monitoring system of quality checkup. It is highly recommended that every unit should come up with proper arrangements and measures for this purpose.

9.5.4 Carrier Material

Inoculations of biocontrol agents/biostimulator in seed or soil require carrier materials. These materials are usually ground into powdered form with particle size of 10–40 μm. A suitable carrier material is the one which (i) is inert and nontoxic to inoculant bacterial strain, (ii) has fair moisture absorbance ability, (iii) can be processed easily and does not contain lump-forming materials, (iv) can be sterilized effortlessly by either autoclaving or gamma irradiation, (v) is available in sizeable amounts, (vi) is cost-effective, (vii) has higher seed adhesion capability, and (viii) can buffer the overall pH values. Above all, it must not be toxic to plants. Peat, along with lignite, bentonite, and charcoal, is commonly used as carrier material for seed inoculation. A lot of countries use rhizobial inoculant, based on peat; information regarding its properties and effects is readily available.

For soil inoculation, carrier material in powdered or meshed form (0.5–1.5 mm) is generally used. Such granular forms of peat, perlite, charcoal, or soil aggregates are appropriate for soil inoculation. Carrier selection with respect to persistence of inoculant bacteria depends on many different factors which include the following: (i) once the seeds are coated with bacteria, they're left to dry so that bacteria could survive drying conditions. After drying, the coated seeds are then sown; (ii) inoculant bacteria must also survive during the storage phase; and (iii) inoculant bacteria should be well equipped to compete with natural soil microbiota for nutrients and habitat. It must also save itself from grazing protozoans. This survival trait requires microporous structure, which is found in soil aggregates and charcoal. Therefore, such materials are automatic choices as carriers for soil inoculant. After the carrier molecule is selected and before its usage, it undergoes a rigorous sterilization. This ensures that higher number of bacteria remains attached with carrier for longer duration of storage. Gamma irradiation is considered best suited for carrier sterilization. Such technique does not alter physical or chemical properties of the material. Another, most accessible and commonly used technique is autoclaving. However, sometimes carriers undergo changes in their properties and risk of the production of toxic substances for certain bacterial strains.

Furthermore, studies have shown that aggregate-based inoculants that were mixed with soil depict extremely enhanced survival as compared with those mixed with soil without any particular carrier material. Also, such plants exhibited significantly higher growth patterns in comparison with inoculants based on peat. Therefore, it is highly recommended that soil aggregates might be appropriate carrier materials for preparing cheap and effective, especially rhizobial, inoculants (El-Fattah et al. 2013). Steps for selection of a suitable carrier material are shown in Fig. 9.2.

For solid formulations, there is a requirement of lignite/bentonite/charcoal/peat of desiredquality in powder form (70–100 mesh). However, the solid formulation technology is gradually becoming obsolete because of the issues relating to the quality and stability of the product. The solid-state formulation is sensitive to

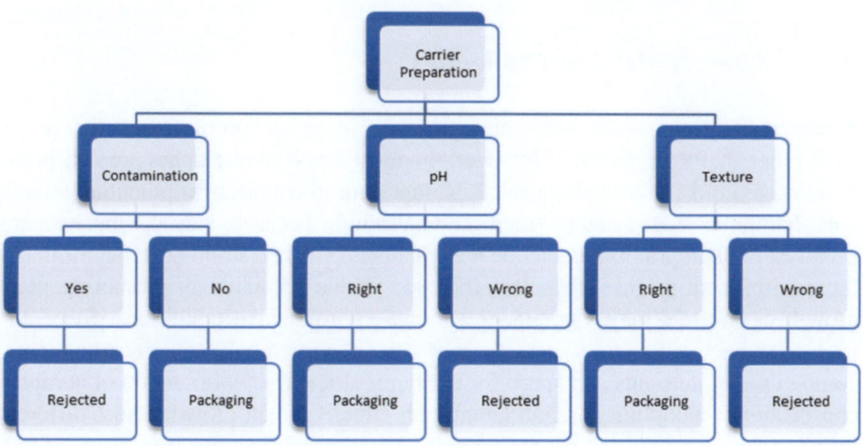

Fig. 9.2 Steps for selection of a suitable carrier material and its desired characteristics for mass production of halotolerant microbial suspension

temperature, and the microbial count comes below the threshold limit. In comparison to the liquid formulation, it is a much stable technology, and the self-life of the product can be maintained very well up to 12 months.

9.5.5 Laboratory Testing

The selected bacteria are first evaluated in pots in a control environment with subsequent application in farmer fields. Extensive trials are done before the product is to be marketed.

Although PGPR have the potential scope in commercialization, there are some safety issues associated with environment and humans. The commercial companies should therefore prefer the microbial consortia with zero impact on human/environment health. The government should also regulate the screening procedure and laws for selecting halotolerant strains for commercial purpose. The safety concerns need to be clarified before the large-scale acceptance, registration, and adoption of PGPR for pest and disease management.

9.6 Bottlenecks in Commercialization

As attention is recently drawn to the potentials of biofertilizer, farmers are not aware of the opportunities and potentials of halotolerant PGPR to ameliorate soil salinity. Even when they are aware, the acceptance level remains low because the technology is still in infant stages and certain challenges must be overcome.

9.6.1 Characterization Challenges

A reasonable progress has been achieved in the identification of distinct halophiles from other PGPR organisms. However, the most reported researches are still in the basic stage unlike other halophiles. It is important to characterize potentially available halophile and evaluate their compatibility, because various mechanisms involved in plant promotion may be host plant-specific and strain-specific. Similarly the environment might affect the efficiency of the halophile in promoting plant growth as impacts may vary under various environmental conditions. Since the root rhizosphere contains several microbes, the identified halophile must be a good competitor for nutrients and space for optimum efficiency. Formulation of complex consortium of halophile or halotolerant with effective plant growth across different saline region is paramount for the success of commercialization.

9.6.2 Production Challenges

The lack in awareness could be as result of poor or low production of beneficial halophiles. Although the same processes are involved in general production of PGPR, the production on a large scale remains very challenging. Although the setup cost of the producing plant is less expensive unlike chemical fertilizer, little manufactories are recently available for its production. For the optimum utilization of halophile microbes, it is penitent that the government should make an enabling environment by promoting biofertilizer. Also establishing a sustainable policy, which will boost the large-scale production of identified potential halophiles, will increase crop production output in saline soil. More resource should be allocated to improve on the level of awareness of the beneficial halophiles.

It must be noted that a successful microbial product is a result of multiple factors (Ravensberg 2011). A study conducted by Gelernter and Lomer (2000) suggests an outline for evaluating successful biocontrol/biostimulator products, which enlist the following key properties of effective products:

(i) Efficacy of the product in laboratory, greenhouse, and field is a vital step to achieve reproducible biological agent (Nicot et al. 2011; Whipps 2001). Along with efficacy, the impact of mutation, viruses, and phase variation should also be taken into account especially during industrial scale manufacturing (Takors 2012). The effectiveness of the microbes in the laboratory and in the greenhouse is not necessarily the same when applied on the field level. The efficacy of data of the microbes in the lab lacks their interaction with complex plant and soil environment, and thus, on-field efficacy of the product also needs to be improved. Along with the on-field effectiveness issue, there are other efficacy challenges involved, too, which include the high-scale production of biological agents/biostimulator and their extensive distribution in the field. Also, there is a question mark on the genetic stability of the certain strains of bacteria for halotolerance because it is very different to go from laboratory level to industry level.

(ii) Microbial strains should also effectively colonize the plants; their association with roots is also an important step (Barea 2015; Compant et al. 2010). The production of halotolerant bacteria, to be used as inoculum to improve plant growth in saline soils, needs great care from the laboratory to the farm. It also requires a proper study and assessment about the market it is going to hit. The commercialization of every bioproduct to be used by the farmers needs to pass several stages before getting approved. This includes the selection of a certain strain of microorganism to be available for commercialization. Also, that strain should have the activity that a particular crop requires. Similarly, the weather conditions of certain area also affect the activity of the bioproduct. The change in weather conditions results in increased temperature which not only harms the plant but also influences the activity of microbes by resulting in insufficient biological activity.

(iii) Plants use the microbes according to their needs, growth, and development and the environment that they are growing in, along with the way they respond to the biotic and abiotic stresses and the availability of the nutrients in the soil. So, the effectiveness of any microbial strain inhabiting the plant would be affected by all these factors. All these factors help the selection of specific microbial strain for colonization for an adequate support to the plant in meeting its functional needs. The microbial strains used as halotolerant in increasing plant resistance to salinity in saline soils are required to have consortia capable of adapting varying degree of salt concentration.

(iv) Practicality is another essential factor in the success of an inoculant. Compatibility of the product with production practices of farmers should also be streamlined. The specificity related to crop of the microbial agents is not dependent on their colonization in root. So, some strains can be, for a certain crop, less specific, and for another crop, they can be much more specific with the same concentration. Once the bacterial strains are colonized in the roots, they produce certain metabolites which help the plant cope with the issue they are dealing with by interacting with the plant. Thus, a combination of microbes prepared for a certain crop as in sugar cane and is showing a great activity, it would not be effective when it is applied on another crop, for example, cotton. The reason could be the incapability of the strains with the crop; also, the growth pattern of a certain crop can also hinder the activity and survival of the biological agents.

(v) Products with extended shelf life and well protected against harsh conditions are more sought after. Such formulations tend to enhance microbial life on roots and leaves (Xavier et al. 2004; Warrior et al. 2002; Leggett et al. 2015; Ravensberg 2011). For the faster commercialization of biological products, their formulations need to be of broad spectrum, consistent in their action, and with an increased shelf life. There are also some very serious factors which need to be addressed for the development of a better biological agent that include inoculant formulation, bulk level multiplication, selection of the suitable strain, selection of the carrier, and marketing and packaging.

The proper formulation of microbial inoculant plays a critical role in the stability of the microbial agent, its protection at the target site, increasing its efficacy, and helping the delivery. One of the most important factors in the race of decreasing the promotion of the inoculants is the development and formulation of these biological products.

(vi) Persistence of microbial product in the field is a major issue (Chutia et al. 2007). Because these products have more specific targets, their shelf and active life are usually less than chemical fertilizer/pesticides (van Lenteren 2012). Thus, the efficacy of a product presents a compromise between immediate short-lived impact and persistence in the environment (Barea 2015).

(vii) Developing such biological control agents/biostimulator often require higher costs, which hinder the success rate of these products (Dalpé and Monreal 2004). Therefore, production costs should align with investment returns and economic aspect of such microbial products (Nicot et al. 2011).

Conclusively, effective utilization of halotolerant microbes for management of saline soils will demand a rational choice of organism as well as technical improvements in upscaling and formulation techniques. Genetic engineering of plants expressing traits from halotolerant microbes may help crop plants survive and withstand in saline environments. Therefore, a single bacterial strain with multiple characteristics will facilitate plant to grow in saline environments and simultaneously promote plant growth. It has become apparent that halotolerant strains employ several mechanisms to promote plant growth, although studies should be focused on the relative contribution of each mechanism responsible for effective plant growth promotion. However, carefully controlled field trials of crop plants inoculated with halotolerant inoculants are mandatory for maximum commercial exploitation of these strains. In conclusion, the success of microbial inoculant-producing industries will depend on measures including advanced business management, product marketing, extensive education, and research.

References

Ali S, Charles TC, Glick BR (2014) Amelioration of high salinity stress damage by plant growth-promoting bacterial endophytes that contain ACC deaminase. Plant Physiol Biochem 80:160–167

Apse MP, Aharon GS, Snedden WA, Blumwald E (1999) Salt tolerance conferred by overexpression of a vacuolar Na^{+}/H^{+} antiport in Arabidopsis. Science 285:1256–1258

Apte SK, Thomas J (1984) Effect of sodium on nitrogen fixation in Anabaena torulosaand *Plectonema boryanum*. J Gen Microbiol 130:1161–1168

Argandona M, Nieto JJ, Iglesias-Guerra F, Calderon MI, Garcia-Estepa R, Vargas C (2010) Interplay between iron homeostasis and the osmotic stress response in the halophilic bacterium Chromohalobacter salexigens. Appl Environ Microbiol 76:3575–3589. https://doi.org/10.1128/aem.03136-09

Arora N, Singhal V, Maheshwari D (2006) Salinity-induced accumulation of poly-β-hydroxybutyrate in rhizobia indicating its role in cell protection. World J Microbiol Biotechnol 22:603–606

Arora NK, Tewari S, Singh S, Lal N, Maheshwari DK (2012) PGPR for protection of plant health under saline conditions. In: Bacteria in agrobiology: stress management. Springer, Heidelberg, pp 239–258

Asim M, Aslam M, Bano A, Munir M, Majeed A, Abbas SH (2013) Role of phytohormones in root nodulation and yield of soybean under salt stress. Am J Res Commun 1:191–208

Bailey KL, Falk S (2011) Turning research on microbial bioherbicides into commercial products–a Phoma story. Pest Technol 5:73–79

Barea J (2015) Future challenges and perspectives for applying microbial biotechnology in sustainable agriculture based on a better understanding of plant-microbiome interactions. J Soil Sci Plant Nutr 15:261–282

Becker EA et al (2014) Phylogenetically driven sequencing of extremely halophilic archaea reveals strategies for static and dynamic osmo-response. PLoS Genet 10:e1004784. https://doi.org/10.1371/journal.pgen.1004784

Bergman B, Sandh G, Lin S, Larsson J, Carpenter EJ (2013) Trichodesmium–a widespread marine cyanobacterium with unusual nitrogen fixation properties. FEMS Microbiol Rev 37:286–302

Bharti N, Pandey SS, Barnawal D, Patel VK, Kalra A (2016) Plant growth promoting rhizobacteria Dietzia natronolimnaea modulates the expression of stress responsive genes providing protection of wheat from salinity stress. Sci Rep 6:34768. https://doi.org/10.1038/srep34768

Burr T, Matteson M, Smith C, Corral-Garcia M, Huang T-C (1996) Effectiveness of bacteria and yeasts from apple orchards as biological control agents of apple scab. Biol Control 6:151–157
Chakraborty U, Roy S, Chakraborty AP, Dey P, Chakraborty B (2011) Plant growth promotion and amelioration of salinity stress in crop plants by a salt-tolerant bacterium. Recent Res Sci Technol 3:11
Chen L, Liu Y, Wu G, Veronican Njeri K, Shen Q, Zhang N, Zhang R (2016) Induced maize salt tolerance by rhizosphere inoculation of *Bacillus amyloliquefaciens* SQR9. Physiol Plant 158 (1):34–44
Chutia M, Mahanta J, Bhattacheryya N, Bhuyan M, Boruah P, Sarma T (2007) Microbial herbicides for weed management: prospects, progress and constraints. Plant Pathol J 6:210–218
Compant S, Clément C, Sessitsch A (2010) Plant growth-promoting bacteria in the rhizo-and endosphere of plants: their role, colonization, mechanisms involved and prospects for utilization. Soil Biol Biochem 42:669–678
Dalpé Y, Monreal M (2004) Arbuscular mycorrhiza inoculum to support sustainable cropping systems. Online. In, 2004
Davies PJ (2004) The plant hormones: their nature, occurrence, and functions. In: Davis PJ (ed) Plant hormones biosynthesis, signal transduction, action, vol 1. Kluwer Academic/ Springer, Dordrecht, pp 1–15
Dodd IC, Pérez-Alfocea F (2012) Microbial amelioration of crop salinity stress. J Exp Bot 63 (9):3415–3428
Egamberdieva D (2011) Survival of Pseudomonas extremorientalis TSAU20 and P. chlororaphis TSAU13 in the rhizosphere of common bean (*Phaseolus vulgaris*) under saline conditions. Plant Soil Environ 57:122–127
El-Fattah DAA, Eweda WE, Zayed MS, Hassanein MK (2013) Effect of carrier materials, sterilization method, and storage temperature on survival and biological activities of *Azotobacter chroococcum* inoculant. Ann Agric Sci 58:111–118
Enya J, Shinohara H, Yoshida S, Tsukiboshi T, Negishi H, Suyama K, Tsushima S (2007) Culturable leaf-associated bacteria on tomato plants and their potential as biological control agents. Microb Ecol 53:524–536
Etesami H (2018) Can interaction between silicon and plant growth promoting rhizobacteria benefit in alleviating abiotic and biotic stresses in crop plants? Agric Ecosyst Environ 253:98–112. https://doi.org/10.1016/j.agee.2017.11.007
Etesami H, Beattie GA (2017) Plant-microbe interactions in adaptation of agricultural crops to abiotic stress conditions. In: Probiotics and plant health. Springer, pp 163–200
Etesami H, Beattie GA (2018) Mining halophytes for plant growth-promoting halotolerant bacteria to enhance the salinity tolerance of non-halophytic crops. Front Microbiol 9:148
Fravel D (2005) Commercialization and implementation of biocontrol. Annu Rev Phytopathol 43:337–359
Gelernter WD, Lomer CJ (2000) Success in biological control of above-ground insects by pathogens. In: Gurr G, Wratten S (eds) Biological control: measures of success. Springer, Dordrecht, pp 297–322
Giri B, Kapoor R, Mukerji KG (2007) Improved tolerance of *Acacia nilotica* to salt stress by arbuscular mycorrhiza, Glomus fasciculatum may be partly related to elevated K/Na ratios in root and shoot tissues. Microb Ecol 54(4):753–760
Glare T et al (2012) Have biopesticides come of age? Trends Biotechnol 30:250–258
Goswami D, Dhandhukia P, Patel P, Thakker JN (2014) Screening of PGPR from saline desert of Kutch: growth promotion in Arachis hypogea by Bacillus licheniformis A2. Microbiol Res 169:66–75
Habib SH, Kausar H, Saud HM (2016) Plant growth-promoting rhizobacteria enhance salinity stress tolerance in okra through ROS-scavenging enzymes. Biomed Res Int 2016:6284547
Hu Y, Ribbe MW (2015) Nitrogenase and homologs. J Biol Inorg Chem 20(2):435–445

Islam F, Yasmeen T, Arif MS, Ali S, Ali B, Hameed S, Zhou W (2016) Plant growth promoting bacteria confer salt tolerance in *Vigna radiata* by up-regulating antioxidant defense and biological soil fertility. Plant Growth Regul 80:23–36

Jaemsaeng R, Jantasuriyarat C, Thamchaipenet A (2018) Molecular interaction of 1-aminocyclopropane-1-carboxylate deaminase (ACCD)-producing endophytic Streptomyces sp. GMKU 336 towards salt-stress resistance of *Oryza sativa* L. cv. KDML105. Sci Rep 8:1950

James RA, Blake C, Byrt CS, Munns R (2011) Major genes for Na+ exclusion, Nax1 and Nax2 (wheat HKT1;4 and HKT1;5), decrease Na+ accumulation in bread wheat leaves under saline and waterlogged conditions. J Exp Bot 62:2939–2947. https://doi.org/10.1093/jxb/err003

Janisiewicz WJ, Korsten L (2002) Biological control of postharvest diseases of fruits. Annu Rev Phytopathol 40:411–441

Jha Y, Subramanian R (2013) Paddy plants inoculated with PGPR show better growth physiology and nutrient content under saline condition. Chilean J Agric Res 73:213–219

Jha B, Singh VK, Weiss A, Hartmann A, Schmid M (2015) Zhihengliuella somnathii sp. nov., a halotolerant actinobacterium from the rhizosphere of a halophyte Salicornia brachiata. Int J Syst Evol Microbiol 65:3137–3142

Kang SR, Srinivasan S, Lee S-S (2015) Vibrio oceanisediminis sp. nov., a nitrogen-fixing bacterium isolated from an artificial oil-spill marine sediment. Int J Syst Evol Microbiol 65:3552–3557

Khalid M, Bilal M, Hassani D, Iqbal HMN, Wang H, Huang D (2017) Mitigation of salt stress in white clover (*Trifolium repens*) by Azospirillum brasilense and its inoculation effect. Bot Stud 58(1):5

Khodarahmpour Z, Ifar M, Motamedi M (2012) Effects of NaCl salinity on maize (*Zea mays* L.) at germination and early seedling stage. Afr J Biotechnol 11:298–304

Kodama T, Taniguchi S (1976) Sodium-dependent growth and respiration of a nonhalophilic bacterium, Pseudomonas stutzeri. J Gen Microbiol 96(1):17–24

Kodama T, Taniguchi S (1977) Sodium-controlled coupling of respiration to energy-linked functions in *Pseudomonas stutzeri*. Microbiology 98(2):503–510

Köhl J, Postma J, Nicot P, Ruocco M, Blum B (2011) Stepwise screening of microorganisms for commercial use in biological control of plant-pathogenic fungi and bacteria. Biol Control 57:1–12

Läuchli A, Epstein E (1990) Plant responses to saline and sodic conditions. Agric Salinity Assess Manag 71:113–137

Leggett M, Newlands N, Greenshields D, West L, Inman S, Koivunen M (2015) Maize yield response to a phosphorus-solubilizing microbial inoculant in field trials. J Agric Sci 153:1464–1478

Mayak S, Tirosh T, Glick BR (2004) Plant growth-promoting bacteria confer resistance in tomato plants to salt stress. Plant Physiol Biochem: PPB 42(6):565–572. https://doi.org/10.1016/j.plaphy.2004.05.009

Mengele R, Sumper M (1992) Drastic differences in glycosylation of related S-layer glycoproteins from moderate and extreme halophiles. J Biol Chem 267:8182–8185

Milus E, Rothrock C (1997) Efficacy of bacterial seed treatments for controlling Pythium root rot of winter wheat. Plant Dis 81:180–184

Munns R, Tester M (2008) Mechanisms of salinity tolerance. Annu Rev Plant Biol 59:651–681. https://doi.org/10.1146/annurev.arplant.59.032607.092911

Murtaza G (2013) Economic aspects of growing rice and wheat crops on salt-affected soils in the Indus Basin of Pakistan (unpublished data). Institute of Soil and Environmental Sciences, University of Agriculture, Faisalabad, Pakistan

Nadeem SM, Ahmad M, Naveed M, Imran M, Zahir ZA, Crowley DE (2016) Relationship between in vitro characterization and comparative efficacy of plant growth-promoting rhizobacteria for improving cucumber salt tolerance. Arch Microbiol 198:379–387. https://doi.org/10.1007/s00203-016-1197-5

Nautiyal CS, Srivastava S, Chauhan PS, Seem K, Mishra A, Sopory SK (2013) Plant growth-promoting bacteria *Bacillus amyloliquefaciens* NBRISN13 modulates gene expression profile of leaf and rhizosphere community in rice during salt stress. Plant Physiol Biochem 66:1–9

Navarro-Torre S et al (2017) Assessing the role of endophytic bacteria in the halophyte Arthrocnemum macrostachyum salt tolerance. Plant Biol (Stuttg) 19:249–256. https://doi.org/10.1111/plb.12521

Nicot P, Blum B, Köhl J, Ruocco M (2011) Perspectives for future research-and-development projects on biological control of plant pests and diseases. Class Augmentative Biol Control Against Dis Pests: Crit Status Anal Rev Factors Influencing Success 2011:68–70

Nunkaew T, Kantachote D, Nitoda T, Kanzaki H, Ritchie RJ (2015) Characterization of exopolymeric substances from selected Rhodopseudomonas palustris strains and their ability to adsorb sodium ions. Carbohydr Polym 115:334–341

Orhan F (2016) Alleviation of salt stress by halotolerant and halophilic plant growth-promoting bacteria in wheat (*Triticum aestivum*). Braz J Microbiol 47:621–627

Paul S, Bag SK, Das S, Harvill ET, Dutta C (2008) Molecular signature of hypersaline adaptation: insights from genome and proteome composition of halophilic prokaryotes. Genome Biol 9: R70. https://doi.org/10.1186/gb-2008-9-4-r70

Qin Y, Druzhinina IS, Pan X, Yuan Z (2016) Microbially mediated plant salt tolerance and microbiome-based solutions for saline agriculture. Biotechnol Adv 34:1245–1259

Raheem A, Ali B (2015) Halotolerant rhizobacteria: beneficial plant metabolites and growth enhancement of *Triticum aestivum* L. in salt-amended soils. Arch Agron Soil Sci 61:1691–1705

Rahnama A, James RA, Poustini K, Munns R (2010) Stomatal conductance as a screen for osmotic stress tolerance in durum wheat growing in saline soil. Funct Plant Biol 37:255–263

Ramadoss D, Lakkineni VK, Bose P, Ali S, Annapurna K (2013) Mitigation of salt stress in wheat seedlings by halotolerant bacteria isolated from saline habitats. Springerplus 2:6. https://doi.org/10.1186/2193-1801-2-6

Ravensberg WJ (2011) A roadmap to the successful development and commercialization of microbial pest control products for control of arthropods, vol 10. Springer Science & Business Media, Dordrecht

Roy SJ, Negrao S, Tester M (2014) Salt resistant crop plants. Curr Opin Biotechnol 26:115–124

Sapre S, Gontia-Mishra I, Tiwari S (2018) Klebsiella sp. confers enhanced tolerance to salinity and plant growth promotion in oat seedlings (*Avena sativa*). Microbiol Res 206:25–32. https://doi.org/10.1016/j.micres.2017.09.009

Saravanakumar D, Samiyappan R (2007) ACC deaminase from *Pseudomonas fluorescens* mediated saline resistance in groundnut (Arachis hypogea) plants. J Appl Microbiol 102:1283–1292

Sharma SB, Sayyed RZ, Trivedi MH, Gobi TA (2013) Phosphate solubilizing microbes: sustainable approach for managing phosphorus deficiency in agricultural soils. Springerplus 2:587. https://doi.org/10.1186/2193-1801-2-587

Sharma S, Kulkarni J, Jha B (2016) Halotolerant rhizobacteria promote growth and enhance salinity tolerance in peanut. Front Microbiol 7:1600. https://doi.org/10.3389/fmicb.2016.01600

Shukla PS, Agarwal PK, Jha B (2012) Improved salinity tolerance of Arachis hypogaea (L.) by the interaction of halotolerant plant-growth-promoting rhizobacteria. J Plant Growth Regul 31:195–206

Siebner-Freibach H, Hadar Y, Chen Y (2004) Interaction of iron chelating agents with clay minerals. Soil Sci Soc Am J 68:470–480

Singh RP, Jha PN (2015) Plant growth promoting potential of ACC deaminase rhizospheric bacteria isolated from *Aerva javanica*: a plant adapted to saline environments. Int J Curr Microbiol App Sci 4(7):142–152

Singh RP, Jha PN (2016) A halotolerant bacterium Bacillus licheniformis HSW-16 augments induced systemic tolerance to salt stress in wheat plant (*Triticum aestivum*). Front Plant Sci 7:1890

Srinivasan R, Yandigeri MS, Kashyap S, Alagawadi AR (2012) Effect of salt on survival and P-solubilization potential of phosphate solubilizing microorganisms from salt affected soils. Saudi J Biol Sci 19:427–434

Staal M, Rabouille S, Stal LJ (2007) On the role of oxygen for nitrogen fixation in the marine cyanobacterium Trichodesmium sp. Environ Microbiol 9:727–736

Suarez C, Cardinale M, Ratering S, Steffens D, Jung S, Montoya AMZ, Geissler-Plaum R, Schnell S (2015) Plant growth-promoting effects of Hartmannibacter diazotrophicus on summer barley

(*Hordeum vulgare* L.) under salt stress. Appl Soil Ecol 95:23–30. https://doi.org/10.1016/j.apsoil.2015.04.017
Tabassum T, Farooq M, Ahmad R, Zohaib A, Wahid A (2017) Seed priming and transgenerational drought memory improves tolerance against salt stress in bread wheat. Plant Physiol Biochem 118:362–369
Takors R (2012) Scale-up of microbial processes: impacts, tools and open questions. J Biotechnol 160:3–9
Tenchov B, Vescio EM, Sprott GD, ZeidelJo ML, Mathai JC (2006) Salt tolerance of Archaeal extremely Halophilic lipid membranes. J Biol Chem 281:10016–10023. https://doi.org/10.1074/jbc.M600369200
Timmusk S et al (2014) Drought-tolerance of wheat improved by rhizosphere bacteria from harsh environments: enhanced biomass production and reduced emissions of stress volatiles. PLoS One 9:e96086. https://doi.org/10.1371/journal.pone.0096086
Trung N, Hieu H, Thuan N (2016) Screening of strong 1-aminocyclopropane-1-carboxylate deaminase producing bacteria for improving the salinity tolerance of cowpea. Appli Micro Open Access 2:2
Ul-Hassan T, Bano A (2014) Role of plant growth promoting rhizobacteria and L-tryptophan on improvement of growth, nutrient availability and yield of wheat (*Triticum aestivum*) under salt stress. Int J Appl Agric Res 4:30–39
Ullah S, Bano A (2015) Isolation of plant-growth-promoting rhizobacteria from rhizospheric soil of halophytes and their impact on maize (*Zea mays* L.) under induced soil salinity. Can J Microbiol 61:307–313
van Lenteren JC (2012) The state of commercial augmentative biological control: plenty of natural enemies, but a frustrating lack of uptake. BioControl 57:1–20. https://doi.org/10.1007/s10526-011-9395-1
Warrior P, Konduru K, Vasudevan P (2002) Formulation of biological control agents for pest and disease management. In: Biological control of crop diseases. Marcel Dekker, New York, pp 421–442
Watanabe M, Kawahara K, Sasaki K, Noparatnaraporn N (2003) Biosorption of cadmium ions using a photosynthetic bacterium, Rhodobacter sphaeroides S and a marine photosynthetic bacterium, Rhodovulum sp. and their biosorption kinetics. J Biosci Bioeng 95:374–378. https://doi.org/10.1016/S1389-1723(03)80070-1
Whipps JM (2001) Microbial interactions and biocontrol in the rhizosphere. J Exp Bot 52:487–511
Xavier IJ, Holloway G, Leggett M (2004) Development of rhizobial inoculant formulations. Crop Manag 3. https://doi.org/10.1094/cm-2004-0301-06-rv
Yang H, Hu J, Long X, Liu Z, Rengel Z (2016) Salinity altered root distribution and increased diversity of bacterial communities in the rhizosphere soil of Jerusalem artichoke. Sci Rep 6:20687
Yoshida S, Ohba A, Liang YM, Koitabashi M, Tsushima S (2012) Specificity of Pseudomonas isolates on healthy and Fusarium head blight-infected spikelets of wheat heads. Microb Ecol 64:214–225. https://doi.org/10.1007/s00248-012-0009-y
Zahran H, Ahmad M, Afkar E (1995) Isolation and characterization of nitrogen-fixing moderate halophilic bacteria from saline soils of Egypt. J Basic Microbiol 35:269–275
Zerrouk IZ, Benchabane M, Khelifi L, Yokawa K, Ludwig-Müller J, Baluska F (2016) A Pseudomonas strain isolated from date-palm rhizospheres improves root growth and promotes root formation in maize exposed to salt and aluminum stress. J Plant Physiol 191:111–119
Zhang HX, Blumwald E (2001) Transgenic salt-tolerant tomato plants accumulate salt in foliage but not in fruit. Nat Biotechnol 19:765–768. https://doi.org/10.1038/90824
Zhang HX, Hodson JN, Williams JP, Blumwald E (2001) Engineering salt-tolerant Brassica plants: characterization of yield and seed oil quality in transgenic plants with increased vacuolar sodium accumulation. Proc Natl Acad Sci U S A 98:12832–12836. https://doi.org/10.1073/pnas.231476498

Zhao S, Zhou N, Zhao Z-Y, Zhang K, Wu G-H, Tian C-Y (2016) Isolation of endophytic plant growth-promoting bacteria associated with the halophyte Salicornia europaea and evaluation of their promoting activity under salt stress. Curr Microbiol 73:574–581

Zhou N, Zhao S, Tian CY (2017) Effect of halotolerant rhizobacteria isolated from halophytes on the growth of sugar beet (*Beta vulgaris* L.) under salt stress. FEMS Microbiol Lett 364. https://doi.org/10.1093/femsle/fnx091

Zuccarini P, Okurowska P (2008) Effects of mycorrhizal colonization and fertilization on growth and photosynthesis of sweet basil under salt stress. J Plant Nutr 31(3):497–513

Role of Halotolerant Microbes in Plant Growth Promotion Under Salt Stress Conditions

10

Zahir Ahmad Zahir, Sajid Mahmood Nadeem, Muhammad Yahya Khan, Rana Binyamin, and Muhammad Rashid Waqas

Abstract

The salt-tolerant microorganisms also referred to as halotolerant including bacteria and fungi have the ability to promote growth of plant in salty environment. Presence of certain particular traits like exopolysaccharide production ability, 1-aminocyclopropane-1-carboxylate (ACC) deaminase, production of osmolytes, enhancing nutrient acquisition and activity of antioxidant enzymes as well as maintaining of K^{+}:Na^{+} ratio make them suitable candidates for promoting plant growth under saline environment. Moreover, the capability of these microbes to fix nitrogen, produce siderophores, solubilize insoluble nutrients, and provision protection against harmful pathogens further accelerates the importance of beneficial microbes in agriculture system. To make use of these environment friendly species as biofertilizers in agricultural system is increasing nowaday to facilitate the plant growth under adverse conditions. The halotolerant bacteria and fungi could be a cost-effective approach to overcome the problem of salinity stress. These halotolerant microbes can be used as biopesticides and biofertilizers and could decrease our reliance on synthetic agrochemicals. These halotolerant microbes are also the most suitable candidates for bioremediation of contaminated environment. This review chapter highlights the significant role of halotolerant microbes for improving crop growth under saline conditions and bioremediation of contaminated environment. The mechanisms used by halotolerant microbes to tolerate salinity as well as promote plant growth under salinity stress have been discussed with selected examples. Also the role of these microbes in environmental sciences has been reviewed. The areas that need further research and future perspectives of this technology have also been discussed in detail.

Z. A. Zahir (✉)
Institute of Soil & Environmental Sciences, University of Agriculture, Faisalabad, Pakistan
e-mail: zazahir@uaf.edu.pk

S. M. Nadeem · M. Y. Khan · R. Binyamin · M. R. Waqas
University of Agriculture, Faisalabad Sub-Campus Burewala-Vehari, Vehari, Pakistan

M. Kumar et al. (eds.), *Saline Soil-based Agriculture by Halotolerant Microorganisms*, https://doi.org/10.1007/978-981-13-8335-9_10

Keywords
Halotolerant · Bacteria · Mycorrhizae · Stress · Plant · Growth · Environment

10.1 Halotolerant Microbes: An Introduction

In soil, plants face various environmental stresses including salinity, drought, chilling, nutrient deficiency, pathogen, heavy metal, etc. Under stress environment, plant faces a number of abnormalities in their growth due to hormonal and nutritional imbalances, toxicity of a particular ion, and difficulty in the uptake of water due to osmotic stress. All these factors result in poor plant growth. Due to low rainfall and high temperature in arid and semiarid areas, salinity is considered as a major environmental stress. Due to shortage of good-quality canal water, the farmers have to use poor-quality underground water for fulfilling the demand of irrigation. The quality of underground water in most of the areas is not up to the mark, and the concentration of salts in soils increases due to the recycling of poor-quality drainage water for irrigation (Shakirova et al. 2003).

To survive in saline environment, plants develop a number of physiological and biochemical mechanisms including synthesis of osmolytes, ion compartmentalization, activation of antioxidant enzyme, hormone modulation, ion exclusion, and blocking the entry of sodium ion (Na^+) into the cell (Tuteja 2007; Gupta and Huang 2014). In addition to mechanisms used by the plant itself to tolerate salinity, the microbial population occurring in the soil also plays a key role to induce/enhance salinity tolerance in plants. Such microbial populations owing to their number of growth-promoting traits and particular mechanisms cause a significant positive impact on plant growth in harsh environment. Among these microbial populations, certain bacterial and mycorrhizal fungal strains have the capability to survive in saline environment. Such microbes are termed as halotolerant and halophilic microbes. Halotolerance is the ability of living organisms to maintain their growth under salinity, and these organisms can grow under saline conditions, whereas halophiles live in highly saline environments and require saline conditions for their proper growth. These microbes have evolved particular characteristics that make them suitable candidate for their use in stress agriculture as well as their use on commercial basis and in biotechnological processes. These microbes also play significant role in cleaning contaminated environment through their effective role in bioremediation processes.

Various halotolerant microbes including bacteria and fungi have the ability to survive in saline conditions at variable concentration of salts. For example, *Staphylococcus epidermidis*, *Bacillus cereus*, and *Halomonas salina* are halotolerant bacteria that have the ability to survive up to 20% salt concentration (Behera et al. 2012, 2014). Some of the well-known strains of halotolerant bacteria include *Bacillus aquimaris* VITP4, *Bacillus clausii* I-52, *Bacillus subtilis* RSKK96, and *Bacillus licheniformis* Shahed-07 (Joo and Chang 2005; Rasooli et al. 2008; Shivan and

Jayaraman 2009; Akcan and Uyar 2011). Mandal (2014) reported that among 27 strains of *Rhizobium trifolii*, 5 strains were resistant to 3% salinity. Akhter et al. (2012) while studying the salt tolerance mechanism of *Azotobacter* spp. reported that out of 15 isolates, 5 were able to tolerate 6% NaCl and 2 strains maintained their growth at 10% NaCl concentration. Prior to this, Tippannavar et al. (1989) also reported the salinity tolerance ability of *Azotobacter*. Halotolerant proteases have the ability to perform well at low as well as at high concentration of salts. The capability of halotolerant bacteria to tolerate salinity variable concentration of salts, i.e., 3–30% NaCl, makes them a suitable candidate for their application in stress agriculture (Ventosa et al. 1998). Recently, Banik et al. (2018) isolated halotolerant *Bacillus* and *Halobacillus* sp. from saline environment. They reported that these bacteria also promoted groundnut growth in saline as well as metal-contaminated conditions.

Similar to halotolerant bacteria, fungal species like *Aureobasidum pullulans* and *Wallemia ichthyophaga* are considered as the most suitable organisms for halotolerance studies (Gunde-Cimerman and Zalar 2014). Sengupta and Chaudhuri (1990) and Hildebrandt et al. (2001) reported the presence of mycorrhizal fungus in salt-marsh plants. Landwehr et al. (2002) also reported about the occurrence of mycorrhizal spores in alkaline soils. Such salt-tolerant fungi may be useful for improving crop growth under saline conditions. Manga et al. (2017) while studying the impact of eight mycorrhizal fungi on the growth of *Acacia seyal* under salt stress demonstrated that *Rhizophagus intraradices* enhanced the growth of *A. seyal* at 680 mM NaCl by improving mineral uptake. They further reported that presence of spores was independent of type of soil and salts; however, degree of colonization varied among different mycorrhizal species.

Among different halotolerant fungi, *Glomus* spp. is the most commonly occurring mycorrhizal fungus (Allen and Cunningham 1983; Wang et al. 2004). Aliasgharzadeh et al. (2001) also reported that *Glomus* spp. were present in salt-affected soils of Tabriz plains. They found the occurrence of *G. etunicatum*, *G. versiform*, and *G. intraradices* in these soils having high EC_e, i.e., 16 dS m^{-1}. The presence of *Glomus* sp. in the saline conditions has been reported by Wang et al. (2004). The occurrence of these species in salty environment enables them to facilitate plant growth by protecting them from negative impact of salinity. The work of Porras-Soriano et al. (2009) also supported this concept, who tested the plant growth-promoting potential of three *Glomus* spp. (*G. intraradices*, *G. mosseae*, and *G. claroideum*). They found that among these three species, *G. mosseae* was the most efficient strain in terms of enhancing salinity tolerance of olive trees.

The salinity tolerance ability of same fungi may be varied with respect to their stage of growth. As it has been seen in case of *Glomus* sp. that these strains colonized the plant root in the presence of 300 mM NaCl, however, the spores of this fungus did not grow at this salinity level (Juniper and Abbott 2006). They demonstrated that it might be due to energy difference between different growth stages as well as the critical limit of water and energy that is required to initiate their germination.

The above discussion shows that a number of bacteria and mycorrhizal fungi that have the capability to survive in salty environment. These species not only maintain their growth in saline condition but are also helpful to lighten the impact of stress conditions on plant growth and development. The mechanisms used by these microbes to tolerate saline conditions and promote the plant growth by reducing the salinity impact are discussed in the next section.

10.2 Mechanisms of Salinity Tolerance in Halotolerant Microbes

In order to promote growth of plant under stress, the microbes must have the capacity to sustain their growth in such harsh environment. There are a number of mechanisms used by halotolerant microbes (bacteria and mycorrhizal fungus) to tolerate saline environment and promote plant growth. The adaptation of these microbes to saline conditions and then improving plant growth are related to their particular traits and some specified mechanisms which provide them competitive advantage over other soil microbial population.

Kunte (2012) while studying the mechanisms of osmoadoptation demonstrated that halophiles use two basic mechanisms to cope with salinity. These mechanisms include salt-in-cytoplasm and the organic-osmolyte mechanism. According to their view, these mechanisms enable the halophilic microbes to flourish in saline environment even in the presence of salt solutions with varying levels of salt concentration. In salt-in-cytoplasm mechanism, the cytoplasm of halophilic organism remains free from sodium (Na^+), and they accumulate potassium ion (K^+) in their cell. The accumulation of potassium results in exposure of cytoplasm to an increased ionic strength. In case of organic-osmolyte mechanism, the bacteria accumulate organic solutes like sugars, polyols, and amino acids. These nonionic solutes don't disturb the metabolism even at high concentration. According to them, compared to halophiles, microbes which are using salt-in-cytoplasm mechanism, the organisms using organic-osmolyte mechanism are more flexible because they can grow at high and low salt concentration. However, organic-osmolyte mechanism requires more ATP compared to salt-in-mechanism.

To survive in stress environment, the halophilic bacteria accumulate compatible solutes like glycine, betaine, proline, ectoine, and glutamate, whereas accumulation of glycerol is well known in most of the fungi (Blomberg and Adler 1992; Burg et al. 2007). Many bacteria and a number of eukaryotic organisms accumulate compatible solutes to maintain intracellular Na^+ concentrations below the toxic levels. In halotolerant bacteria, in order to maintain osmotic balance due to low intracellular salt concentration, they accumulate various organic solutes in cytoplasm. These solutes keep away the cells from plasmolysis in the presence of salty environment (Kempf and Bremer 1998). In heterotrophic eubacteria, ectoine is the most abundant osmolytes (Galinski 1995) that was discovered in halophilic *Ectothiorhodospira halochloris* (Galinski et al. 1985). In addition to glycine and betaine, trehalose also plays a key part in organisms to tolerate stress tolerance (Turan et al. 2012).

According to Mendpara et al. (2013) among six bacterial spp. isolated from agricultural soil, two species *Exiguobacterium* sp.and *Serratia* sp.showed salt tolerance up to 10% NaCl. They demonstrated that high salt tolerance ability of these isolates might be due to the synthesis of betaine, ectoine, and trehalose as reported by earlier workers (Oren 2008; Kondepudi and Chandra 2011). Recently, Patel et al. (2018) studied the mechanism used by halotolerant *Exiguobacterium profundum* PHM11to tolerate salt condition. They reported that this strain accumulates L-proline to tolerate salinity. They further reported that salinity-tolerant ability of this strain was also related to fine-tuning gene expression profiles of related metabolic pathways. They observed that maximum growth, biomass, and metabolite production occurred at 100 mM salinity and minimum at 1500 mM. *Exiguobacterium* sp. has already been reported to be present in cold and hot environments with temperature range from −12 to55 °C (Vishnivetskaya et al. 2009).

Osmotolerance strategy of *Colwellia psychrerythraea* 34H and *Psychrobacter* sp. 7E was studied by Firth et al. (2016). They used ^{14}C-choline (precursor to glycine, betaine) and observed that this compatible solute allowed bacteria to survive in variable salt conditions, i.e., short-term, long-term, or abrupt change in salinity. They demonstrated that choline contained nitrogen and when bacteria respired, the regeneration of ammonium might enhance specific biogeochemical process possibly algal productivity but particularly nitrification.

One of the mechanisms used by the bacteria to survive under salinity is the accumulation of potassium. The K^+ ion accumulation in the cytoplasm of halotolerant microbes is coupled with glutamate that is accumulated as a counterion. This accumulation of K^+ ions with glutamate has been reported in *Escherichia coli* and other bacteria as well (Oren 1999; Ventosa et al. 1998; Dinnbier et al. 1988). Similarly, tolerance of up to 11% salinity concentration by the bacterial strains has been mentioned by Amir et al. (1993). They noted the accumulation of sodium ions inside the cell, and to maintain intracellular balance of sodium ions, bacteria accumulated K^+ ion and osmolytes like proline. However, accumulation of high K^+ may disturb the normal functioning of cellular enzymes, and to solve this problem, bacteria use another strategy that is the accumulation of compatible solute (Lucht and Bremer 1994; Galinski 1995; Ventosa et al. 1998; Saum and Muller 2007).

Sandhya and Ali (2015) reported that halotolerant bacteria can survive under harsh environment by synthesizing specific metabolites. The function of PDH45 (pea DNA helicase 45) in salt tolerance of various crops is well documented by various reasercher (Sanan-Mishra et al. 2005; Sahoo et al. 2012; Augustine et al. 2015). However, its role in bacterial salt tolerance has been studied by Tajrishi et al. (2011). They documented that salinity tolerance ability of *Escherichia coli* BL21 was due to functionally active PDH45 gene. They also observed that stress response is specific one, because bacteria could not grow when NaCl was replaced with LiCl. It means this gene is specific for Na^+ and enhanced bacterial stress in salinity caused by Na^+.

The accumulation of proline, glycine, glutamate, and trehalose was also observed in salinity-tolerant strain of *Azospirillum brasilense* (Tripathi et al. 2002). They demonstrated that proline plays significant role in bacterial osmotolerance. They found that with increase in osmotic stress, the accumulation of osmolytes shifts from glutamate to proline. The ability of *Chromohalobacter salexigens* to tolerate salinity was studied by Reina-Bueno et al. (2012) with respect to the role of trehalose in this regard. They observed that the said mutant strain was not affected by salinity when it was deficient in trehalose-6-phosphate synthase gene; however, its double mutant deficient of ectoine and trehalose showed sensitivity to salinity. They reported that trehalose plays a role in bacterial salinity tolerance. In addition to help in osmoadaptation, compatible solutes also played an important role in the protection of macromolecules, tissues, cells, and organs (da Costa et al. 1998; Welsh 2000a, b).

Tiquia et al. (2007) isolated 127 halophilic bacterial strains that belonged to genera including *Staphylococcus*, *Bacillus*, *Paenabacillus*, *Halobacillus*, *Clostridium*, and *Halomonas*. They found that most of these strains were positive for gelatinase, β-galactosidase, and tryptophane deaminase enzymes. They observed that arginine dihydrolase was the primary enzyme found in these isolates, and according to their view, these enzymes were present in halotolerant bacteria that helped them to survive in salty conditions. Upadhyay et al. (2012) studied the salinity tolerance ability of the bacterial strains and reported that among about 33% isolates that were able to survive at 8% NaCl, only 19% showed plant growth promoting traits at higher NaCl concentration. They studied the mechanism of osmotolerance in bacteria and found that high concentration of proline and protein was present in salt-tolerant strains.

Similar to bacteria, certain fungal strains can also tolerate saline conditions. Similar to bacteria these strains also adopt various mechanisms that enable them to survive in salty environment. Petrovic et al. (2002) mentioned that glycerol is one of the most important compatible solutes that played an important role in salinity tolerance of *Hortaea werneckii*. According to Plemenitas et al. (2014), high osmolarity glycerol (HOG) pathway is the main pathway that is responsible for cellular stress. They reported that *Wallemia ichthyophaga* and *Hortaea werneckii* maintain low intracellular Na^+ concentrations by the production of compatible organic solutes along with glycerol.

To adopt stress condition, fungi also brought changes in membrane properties that are one of the important mechanisms to survive in harsh environment. One of the mechanisms used by fungi to survive in the saline environment is the maintenance of membrane fluidity. It has been observed that plasma membrane of salt-tolerant *H. werneckii* remained more fluid under varying levels of salinity (Turk et al. 2004, 2007). They observed that under salinity stress, the total sterol content of plasma membrane of *H. werneckii* remained unchanged and their membrane fluidity was due to modification in the structure of component phospholipids. Fungi also accumulate polyols and free amino acids as well as their derivatives to tolerate stress condition.

The above discussion shows that halotolerant microbes use a number of mechanisms to tolerate saline conditions. One of the common mechanisms used by these bacteria is the accumulation of compatible solutes. Some species also accumulate K^+ and Ca^{+2} to maintain ion balance. The presence of some specific enzymes and metabolites that protect the microbes from harmful effects of salinity has also been observed.

10.3 Mechanisms of Plant Growth Promotion and Effectiveness in Stress Agriculture

Environment of rhizosphere differs from bulk soil with respect to its physicochemical properties. In this environment, a number of interactions take place between plant and microbes. These interactions may be antagonistic and synergistic one that depends upon the kind of strains involved, and the plant species interact with this strain. The interactions may take place between plant and bacteria or plant and fungus and/or between fungus and bacteria. In synergistic interactions, which are a kind of mutulistic association, both partners help each other by providing benefits, and ultimately, improvement in plant growth-promoting traits of microbes occurred (Finlay 2007; Beattie 2007; Franche et al. 2009; Nadeem et al. 2014; Shin et al. 2016).

Soil microbes and plants interact with each other for growth and survival. Plants have developed beneficial association with different microorganisms in rhizosphere which ultimately reduces the stress outcomes (Badri et al. 2009). Halotolerant microbes including plant growth-promoting rhizobacteria (PGPR) and fungi can play significant role by protecting the plant from salinity-induced negative impact (Lugtenberg and Kamilova 2009). As far as the mechanism of plant growth promotion by these microbes is concerned, generally, the halotolerant microbes reduce the harmful effect of salinity on plant growth by improving the production of osmolytes, exopolysaccharides, siderophores, and phytohormones, enhancing antioxidant enzymes activities, solubilizing the essential nutrients, improving nutrient acquisition, water use efficiency, and root hydraulic conductivity. Some selected recent examples of the effectiveness of halotolerant bacteria and mycorrhizae on plant growth in saline environment have been presented in Tables 10.1 and 10.2, respectively.

The bacteria in the rhizosphere can improve plant growth by colonizing the plant roots and these bacteria can promote the plant growth by number of their direct and indirect mechanism (Muthukumarasamy et al. 2007). Solubilization of phosphate and potassium by the bacteria, presence of ACC-deaminase, and production of indole-3-acetic acid (IAA), exopolysaccharides, hydrogen cyanide (HCN), proline, and siderophores facilitate the plant growth promotion under saline conditions (Mohamed and Gomaa 2012; Nunkaew et al. 2014; Palaniyandi et al. 2014; Munoz et al. 2014; Kang et al. 2014a, b). Among a number of bacterial strains, *Bacillus* and *Pseudomonas* are widely used for enhancing plant stress tolerance against salinity. *Pseudomonas putida* UW4 showed significant increase in shoot fresh weight of canola compared to uninoculated plants at 150 mM NaCl (Cheng et al. 2007). Salinity-tolerant bacterial strains could be a good option to reduce salinity-induced impact on growth of plant. Due to the presence of salinity-tolerant gene, these bacteria have the ability to develop successful interaction with plant. Reduction in deleterious effect of salinity and improved plant growth of wheat have been observed when it was inoculated with *gfp*-tagged *Azospirllum lipoferum* (Bacilio et al. 2004).

Table 10.1 Inoculation of PGPR for inducing salinity tolerance in plants

Crop	Bacterial strain	Response/mechanisms	Reference
Oryza sativa	*Curtobacterium albidum* SRV4 strain	Improved proline contens, membrane stabilization index, and efficacy of photosynthetic pigments Enhanced antioxidative enzymatic activity and K^+ uptake in paddy plant correlated Reduce stress-induced ethylene by ACC deaminase activity and Na+ availability by EPS production	Vimal et al. (2018)
	Enterobacter sp. P23	Decrease in antioxidant enzymes and stress-induced ethylene Phosphate solubilization IAA production and siderophore production	Sarkar et al. (2018)
	Bacillus amyloliquefaciens NBRI-SN13 (SN13)	Improved the membrane integrity Improved accumulation of osmoprotectants Improved the expression of marker genes	Tiwari et al. (2017)
	Lysinibacillus sp. (BPC2) and *Pseudomonas aeruginosa* (PRR1 and PHL3)	Increased plant dry weight, plant fresh weight, and secondary root number Increased chlorophyll contents Phosphate solubilization IAA production and siderophore production	Kumar et al. (2017)
Brassica napus L.	*Enterobacter cloacae* HSNJ4	Increased chlorophyll contents, amount of lateral roots, shoot length, and root length Decreased MDA content Increased activities of antioxidant enzyme and proline content More endogenous IAA content Reduced ethylene emission by ACC deaminase activity	Li et al. (2017)
Citrullus lanatus, *Arabidopsis*	*Bacillus sp.* LBEndo1 and *Pseudomonas lini* KBEcto4	More growth both under normal conditions and in salinity stress	Palacio-Rodríguez et al. (2017)

(continued)

Table 10.1 (continued)

Crop	Bacterial strain	Response/mechanisms	Reference
thaliana, and *Cucumis sativus*		Expression of inductive gene construct in response to auxins and more auxin contents in roots	
		Phosphate solubilization	
		IAA production and siderophore production	
Capsicum annuum	*Pseudomonas frederiksbergensis* OS261	Reduced ethylene emission by ACC deaminase activity	Chatterjee et al. (2017)
		More activities of antioxidant enzymes	
		Reduction in concentration of H^+ ion	
Arabidopsis thaliana	*Burkholderia phytofirmans* PsJN	More proline contents	Pinedo et al. (2015)
		Upregulated the gene expression that are associated with ROS scavenging (*Ascorbate peroxidase* 2)	
		Upregulated the gene expression that are responsible for abscisic acid signaling (*Relative to Dessication*, *RD29A* and *RD29B*)	
		Upregulated the gene expression that is responsible for detoxification (*glyoxalase I 7*)	
		Downregulated the gene expression that are responsible for jasmonic acid biosynthesis (*lipoxygenase* 2)	
		Altered the expression pattern of ion homeostasis-related genes (*high-affinity* K^+ *transporter 1, Arabidopsis* K^+ *transporter 1*, *Arabidopsis salt overly sensitive 1*, and *sodium hydrogen exchanger* 2)	
Arachis hypogaea	*Genera of Pseudomonas, Klebsiella, Ochrobactrum*, and *Agrobacterium*	Increased total nitrogen and more growth	Sharma et al. (2016)
		Balancing of ion homeostasis	

(continued)

Table 10.1 (continued)

Crop	Bacterial strain	Response/mechanisms	Reference
		Reduced ROS accumulation	
		IAA production and phosphate solubilization	
		ACC deaminase activity	
Triticum aestivum L	*Stenotrophomonas maltophilia* SBP-9	Increased chlorophyll content	Singh and Jha (2017a, b)
		Decreased MDA as well as proline contents	
		More activities of antioxidant enzymes like CAT, SOD, and POD	
		Decreased concentration of Na^+ contents roots and shoots	
		Increased the K^+ uptake	
	Rhizobium tropici strain CIAT 899 and *Azospirillum brasilense strains* Ab-V5 and Ab-V6	Augmented the activities of antioxidant enzymes responsible for detoxification of ROS like CAT, SOD, and APX	Fukami et al. (2018)
		Decreased MDA and poroline contents	
		Modulated the gene expression (in general upregulated) that are responsible for activities of antioxidant enzymes like *SOD2*, *APX1*, *SOD4*, and *CAT1* in leaves while *APX2* in roots	
		Downregulated the gene expression that is related to pathogenesis like *prp2*, *PR1*, and *prp4*	
		Downregulated the expression of gene *hsp70* in leaves and roots that is related to heat shock protein	
	Klebsiella sp. SBP-8	Increased contents of total proteins, soluble sugars as well as proline	Singh et al. (2015) and Singh and Jha (2017a, b)
		Decreased the concentration of salinity-induced MDA content	
		More activities of antioxidant enzymes like POX, CAT, and SOD	

(continued)

Table 10.1 (continued)

Crop	Bacterial strain	Response/mechanisms	Reference
		Na^+ exclusion and increased uptake of K^+ in the host plant	
		Increased plant biomass and chlorophyll content	
Beta vulgaris L.	*Variovorax paradoxus, Planococcus rifietoensis,* and *Micrococcus yunnanensis*	More plant biomass and seed germination	Zhou et al. (2017)
		More photosynthesis	
		Decreased production of stress-induced ethylene	
		Decreased host ACC by ACC-deaminase activity	
Solanum lycopersicum	*Bacillus pumilus* AM11 *and Exiguobacterium* sp. AM25	Higher biomass	Ali et al. (2017)
		Increased photosynthetic rate	
		Increased pigment accumulation	
		Decreased lipid peroxidation,	
		Increased glutathione, catalase, and peroxidase activities	
Pisum sativum L.	*Planomicrobium sp.* MSSA-10	Decreased levels of ROS	Shahid et al. (2018)
		Enhanced antioxidative enzyme activities	
		Increase in chlorophyll and protein contents	
		Phosphate solubilization	
		Auxin production	
		ACC deaminase activity	

In order to minimize the adverse effect of salinity, plants accumulate ions in its vacuole, and ion balance is maintained by accumulation of counterions. For example, to neutralize the impact of Na^+ ions accumulated in the vacuole, plant sequestered organic solutes and K^+ into the cytoplasm (Hasegawa et al. 2000). Inoculation of plant with bacteria under salinity stress causes reduction in Na^+/K^+ ratios. The inoculated bacteria enhance the uptake of K^+ and decrease Na^+ concentration in the tissue. The low value of Na^+/K^+ ratios shows the capability of bacteria to enhance uptake of K^+ (Govindarajan et al. 2006).

Enhancing the accumulation of organic solutes in plant under salinity stress is also a well-adopted mechanism used by bacteria to improve plant growth. Under stress, bacteria induce the plants to produce compatible solutes. These low molecular weight compounds protect the cell from hyperosmotic stress. Accumulation of

Table 10.2 Inoculation of mycorrhizal fungi for inducing salinity tolerance in plants

Crop	Fungal strain	Response/mechanism	Reference
Sesbania sesban	*Funneliformis mosseae* (syn. *Glomus mosseae*); *Rhizophagus intraradices* (syn. *Glomus intraradices*) and *Claroideoglomus etunicatum* (syn. *Glomus etunicatum*)	Improvement in nodules number and weight	Abd-Allah et al. (2015)
		Enhanced nitrogenase activity	
		Reduced oxidative damage by scavenging ROS	
		Increase level of phytohormones (IAA, IBA and GA_3) and non enzymatic antioxidants	
Oryza sativa	*Claroideoglomus etunicatum* (isolate EEZ 163)	Extrusion of Na^+ from the cytoplasm	Porcel et al. (2016)
		Sequestration of Na^+ into vacuole and its unloading from xylem	
		Decreased root-to-shoot Na^+ distribution	
	Claroideoglomus etunicatum (isolate EEZ 163)	Enhanced photosynthetic, stomatal conductance, and transpiration	Porcel et al. (2015)
		Enhancing the quantum yield of PSII photochemistry and reducing the quantum yield of non-photochemical quenching	
Triticum aestivum L.	*Glomus mosseae*, *G. intraradices*, and *G. etunicatum*	Regulate the uptake of sodium (Na^+) and chloride (Cl^-)	Mardukhi et al. (2015)
		Increased the uptake of essential macro- and micronutrients	
Nicotiana tabacum	*Rhizophagus irregularis*	Increased respiration via cytochrome oxidase pathway (COP)	Del-Saz et al. (2017)
		Higher synthesis of ATP	
		Higher biomass production	
		Improved P accumulation and favoring both ATP production and shoot growth	
Cicer arietinum L.	*Funneliformis mosseae*	Improved growth, yield, and nutrient uptake	Garg and Bhandari (2016)
		Protection of chlorophyll pigment from damage	
		Increased the activity of RUBISCO (ribulose 1,5-bisphosphate carboxylase oxygenase)	
Osmium basilicum	*Glomus deserticola*	Improved balance between K/Na and between Ca/Na	Elhindi et al. (2017)
		Improvement in chlorophyll content, photosynthetic	

(continued)

Table 10.2 (continued)

Crop	Fungal strain	Response/mechanism	Reference
		efficiency gas exchange, and water use efficiency	
	Glomus clarum Nicol. & Schenck	Produced more biomass	Elhindi et al. (2016)
		Increased the accumulation of nutrients, reducing sugars, total soluble carbohydrates, photosynthetic pigments, proline, and protein	
		Reduced the Na^+ accumulation	
Cajanus cajan (L.) Millsp.	*Funneliformis mossseae* and *Rhizophagus irregularis*	Improved nutrient uptake	Garg and Pandey (2015)
		Significant improvement in biomass and yield	
		Improved membrane stability	
	Funneliformis mosseae	Strengthen the antioxidant system	Garg and Chandel (2015)
		Modulated the nonenzymatic components (GSH-GSSG cycling, their ratio, and total glutathione)	
Sesbania cannabina	*Funneliformis mosseae (BGC NM03D)*	Decreased H_2O_2 concentration	Kong et al. (2017)
		The induction of strigolactone accumulation	
		Regulated strigolactone-induced salt stress tolerance	
Puccinellia tenuiflora	*Funneliformis mosseae* and *Claroideoglomus etunicatum*	Increased plant biomass	Liu et al. (2018)
		Increased shoot P and K concentrations,	
		Reduced shoot Na^+ concentrations	
Arundo donax,	*Funneliformis mosseae and Rhizophagus irregularis*	Increased the proline accumulation and	Pollastri et al. (2017)
		Higher isoprene emission and higher H_2O_2	
Citrus macrophylla	*G. mosseae and G. intraradices*	Decreased intercellular CO_2 levels and oxidative stress	Navarro et al. (2015)
		Increased total chlorophyll	
		Increased the water percentage in leaves	
Leymus chinensis	*Glomus mosseae* (BGC HEB02)	Increased plant biomass,	Lin et al. (2017)
		Higher photosynthetic parameters and	
		Increased contents of photosynthetic pigments	

glycine, proline, betaine, glutamate, and trehalose has been observed in *Azospirillum* to minimize the salinity effect (Tripathi et al. 2002). Inoculation of PGPR significantly enhanced the concentration of soluble sugar and proline in the leaves and root of salinity-stressed plant (Azarmi et al. 2016). The accumulations of these compatible solutes enable the plant to maintain growth. Similarly, the role of soluble protein for maintaining osmotic adjustment, internal water potential, and cell turgidity has also been reported (Tester and Davenport 2003). Plant growth under salinity is also improved by inoculating salinity-tolerant bacterial strains that increase the water potential and decrease electrolyte leakage as well as reduce the sodium concentration (Kang et al. 2014a, b).

Syntheses of the exopolysaccharides (EPS) by the rhizospheric bacteria promote soil aggregation. High population density of exopolysaccharide-producing bacteria in the rhizosphere could be helpful for minimizing the salinity-induced impact on plant. Stabilization of soil structure and aggregation formation were also observed through the exopolysaccharides produced by the bacteria (Gouzou et al. 1993). The inoculation of exopolysaccharides producing *Pseudomonas mendocina* improves the plant growth in saline environment by reducing Na^+ availability for plant uptake (Kohler et al. 2006). Exopolysaccharides produced by the bacteria play a significant role to reduce the Na^+ availability for uptake of plant. Decrease in Na^+ uptake was also observed by Ashraf et al. (2004) due to the inoculation of bacterial strain having the ability to produce exopolysaccharides.

Bacteria produce different compounds including phytohormones such as auxins, gibberellins, cytokinin, as well as siderophores and antimicrobial compound which play a role in mitigating the negative impact of biotic and abiotic stress (de-Salamon et al. 2001; Mitter et al. 2013; Cassan et al. 2014). Better seedling growth and relatively higher auxin contents were observed when wheat seeds were inoculated with salt-tolerant bacteria and then grown in saline conditions (Afrasayab et al. 2010). Mazhar et al. (2013) also reported the beneficial effect of auxins producing cyanobacteria. The inoculation enhanced the seedling biomass by causing positive impact on root and shoot length.

Under salinity stress, high ethylene produced negatively affects the root length of plant, and inhibition in root elongation is observed (Abeles et al. 1992). Among different growth promoting traits of rhizospheric bacteria, the presence of ACC-deaminase enzyme could be one of the best traits that play a key role in minimizing the stressed induce ethylene on root growth and ultimately plant maintain its proper growth in salty environment (Nadeem et al. 2010a, b; Glick 2010). The bacteria containing this enzyme, i.e., ACC-deaminase, cause reduction in internal plant ethylene concentration by hydrolyzing its immediate precursor, i.e., 1-aminocyclopropane-1-carboxylate (ACC), and therefore promote plant growth by stimulating root elongation (Glick et al. 1995, 1999). An improvement in cucumber growth has been reported in the presence of 75 mM salt concentration due to inoculation of bacteria containing ACC-deaminase (Gamalero et al. 2010). Similarly, in addition to its positive effect on root length and seedling biomass of tomato, enhancement in germination percentage has also been observed by inoculation with salt-tolerant bacteria containing ACC-deaminase (Chookietwattana and Maneewan,

2012). Salt-tolerant *Planococcus rifietoensis* promoted the wheat growth and yield under saline conditions (Rajput et al. 2013). The strain was able to grow up to 65 g/L NaCl concentration and showed its potential to hydrolyze ACC by using it as sole source of nitrogen as well as solubilizing phosphorus from insoluble source (tricalcium phosphate). On overall basis, 37% increase in plant growth was observed under salt stress. However, in the presence of inorganic tricalcium phosphate, the growth enhancement was 63%, and more than 60% increase in growth was noted when ACC was added in the medium. This showed that growth-promoting ability of strain also depends upon other factors. A number of other studies conducted by various workers in the saline environment also mentioned the potential of ACC-deaminase bacteria in lowering the level of stress-induced ethylene (Mayak et al. 2004; Glick et al. 2007; Nadeem et al. 2010a, b, Shin et al. 2016).

Salinity causes oxidative damage to plants by producing reactive oxygen species (Zhu et al. 2007). Inoculation with halotolerant bacteria enables the plant to mitigate the impact of reactive oxygen species (ROS) by enhancing antioxidant enzyme activities. It has been noted that bacterial inoculation enhanced the salinity tolerance of okra by scavenging the reactive oxygen species and improving water use efficiency (Habib et al. 2016). A much higher concentration of antioxidant enzymes including peroxidase, catalase, and nitrate reductase has been observed under salinity stress in plant inoculated with PGPR (Sen and Chandrasekhar 2015). Prior to this, Jha and Subramanian (2013) also found an improvement in growth parameter through inoculation with PGPR under salinity. The inoculated bacteria regularized the ion concentration and enhanced the activities of antioxidant enzymes.

Similar to bacteria, arbuscular mycorrhizal (AM) association with plants is also one of the most widespread beneficial associations that protect the plants from biotic and abiotic stresses by their direct and indirect mechanisms. Inoculation of *Citrus reticulata* (red tangerine) by *Paraglomus occultum* and *Glomus mosseae* enhanced the salinity tolerance by improving root morphology, photosynthetic activity, and maintenance of ion balance (Wu et al. 2010). Similarly, increase in growth, fresh and dry weight of root, as well as nutrient and water absorption were also reported by other researcher (Ghoulam et al. 2002; Cho et al. 2006; Al-Karaki 2006).

Mechanisms used by mycorrhizal fungus to increase resistance of host plants against salinity were intensively investigated (Liu et al. 2015a, b; Wu et al. 2014; Meng et al. 2015; De Almeida et al. 2016). Mycorrhizal hyphae colonize the plant root and establish symbiotic relationship. Fungus penetrates into the cortex and forms specialized structures, i.e., arbuscules, and after colonization of the root of host plant, the mycelium of the fungus penetrates into soil in search of water and nutrients (Breuninger and Requena 2004; He and Nara 2007). Ectomycorrhizal and endomycorrhizal fungi significantly improve seedling growth and biomass of the inoculated plant under salt-stressed condition (Diouf et al. 2005). Mycorrhizae dilute the negative impact imposed by the salinity on plant by enhancing the antioxidant system to mitigate the harmful effect of reactive oxygen species and reducing oxidative damage (Rabie and Almadini 2005; Manchanda and Garg 2011; Wu et al. 2014; Ahmad et al. 2015a, b), improving nutrient acquisition especially phosphorus (Meng et al. 2015; Evelin et al. 2009), inducing acquired systemic tolerance (Hashem et al. 2016a), enhancing water uptake (Ruiz-Lozano and Azcon 2000), improving osmolyte production (Ibrahim et al. 2011; Evelin et al. 2013), reducing the uptake of harmful ions such as Cl^- and Na^+ (Al-Karaki 2006; Daei et al. 2009), maintaining ion balance by improving uptake of nutrients and stimulating selective uptake (Evelin et al. 2012), improving water use efficiency (Hajiboland et al. 2010), and altering root morphology to increase water and nutrient absorption (Aroca et al. 2013; Ahanger et al. 2014).

Mycorrhizal fungi reduced the salinity-imposed negative impact on growth of *Cucumis sativus* L (Hashem et al. 2018). They demonstrated that fungus strengthens the plant antioxidant system and therefore reduces the harmful impact of ROS by accelerating the activity of antioxidant system. Earlier studies conducted by various workers also showed that mycorrhizal association is helpful for maintaining plant growth in the salt-stressed environment by improving the antioxidant system of plant (Alguacil et al. 2003; He et al. 2007). Sarwat et al. (2016) demonstrated that salinity tolerance of *B. juncea* seedlings was due to improvement in antioxidant system, reorganization of the physiological and biochemical parameters, as well as production of secondary metabolites and phytohormones. Yang et al. (2014) also documented an enhancement in the activity of ascorbate peroxidase and catalase enzymes in mycorrhizal inoculated plants compared to non-mycorrhizal in saline conditions. Abd-Allah et al. (2015) while studying the impact of myccorhizae on *Sesbania Sesban* under saline conditions observed that an improvement in the growth and yield parameters of *Sesbania* was due to activation of antioxidant system and nonenzymatic antioxidant as well as production of phytohormones.

Superoxide dismutase (SOD) stabilizes the membrane of plant cells by regulating the ROS. Plants having symbiotic relationship with *R. irregularis* have greater SOD activity that reduced the level of lipid peroxidation and provide protection to the plant membrane against damage. Lower ROS in plants having mycorrhizal connection is due to rapid ROS-scavenging enzymes, i.e., catalase and guaiacol peroxidase (Pandey and Garg 2017). Higher isoprene emission linked with more accumulation of ROS and more H_2O_2, especially in plants colonized by *R. irregularis* (Pollastri et al. 2017). In tomato, due to mycorrhizal association, antioxidants, e.g., peroxidase, catalase, etc., are produced which cause degradation of ROS under salt stress condition (He et al. 2007).

Salt stress on plants affects various physiological processes of plant such as gas exchange, efficiency of photosynthetic system, water and nutrient absorption, and permeability of membrane (Aroca et al. 2006; Porcel et al. 2006). Mycorrhizae improve the plant physiological processes by enhancing water absorption owing to its positive role in enhancing root hydraulic conductivity (Ruiz-Lozano and Azcon 2000; Ruiz-Lozano 2003). Similarly, root colonization of *G. intraradices* dilutes the impact of salinity and protected the leaf from dehydration. Lower water saturation deficit and higher turgor potential in mycorrhizal plants also improved the water status of the plant (Sheng et al. 2008; Al-Garni 2006). Relatively high water content has been observed in mycorrhizae-inoculated plants that was due to improved root hydraulic conductivity at low water potential (Sheng et al. 2008; Jahromi et al. 2008; Kapoor et al. 2008). These changes in physiological processes improve the growth of plants by protecting them from negative impact of toxic ions as it was observed in case of Solanum lycopersicum where mycorrhizae mitigated the negative effect of salinity on plants (Balliu et al. 2015; Hashem et al. 2016a, b, c). Better root colonization of *Cucurbita pepo* plants was observed by inoculation with *Glomus intraradices* under salinity stress which improved growth, yield, and quality of fruit compared to uninoculated by enhancing water and nutrient status (Colla et al. 2008). Root colonization of *Lotus glaber* by mycorrhizal fungi also facilitated the growth of said plant (Sannazzaro et al. 2007). High value of electrical conductivity was observed in the root of mycorrhizal plants (Garg and Manchanda 2008). Maize inoculated with mycorrhizal showed higher electrolyte permeability of root plasma membrane that was due to high production of antioxidant enzyme and uptake of phosphorus (Feng et al. 2002).

Rate of photosynthesis is reduced under salinity stress. Mycorrhizal plants have more ability to fix CO_2 that's why their growth improved (Querejeta et al. 2007). Inoculation of mycorrhizae restored the chlorophyll activity in the presence of salinity that was due to the activity of specific enzymes (Hajbagheri and Enteshari 2011). High chlorophyll content has been observed under saline condition in the leaves of mycorrhizal plants (Sannazzaro et al. 2006; Sheng et al. 2008). Mycorrhizal symbiosis increased the chlorophyll contents and photosynthetic capacity of corn crop under salt stress conditions (Sheng et al. 2008). Efficiency of photosynthetic system enhanced due to regularizing the energy bifurcation and improving efficiency of photochemical and non-photochemical PSII. Borde et al. (2010) also observed the enhancement of photosynthesis rate in garlic plants under salt stress. In mycorrhizal plant, absorption of Mg was increased that improved the synthesis of chlorophyll. For better functioning of photosynthetic machinery, the mycorrhizae inhibited the Na^+ transport and therefore increased photosynthesis rate (García-Garrido and Ocampo 2002). Recently, Sharma et al. (2017) observed that inoculating the plant with mycorrhizae enhanced the growth of plant by causing positive impact on nutrient uptake, photosynthetic pigments, phosphatase, and peroxidase activity. They found that mycorrhizae also decreased the membrane damage. Elhindi et al. (2017) also demonstrated that mycorrhizae enhanced the plant growth under saline condition that was linked with gas exchange, photosynthetic efficiency, and water use efficiency.

Mycorrhizae can also minimize the negative impact of stress on plant by altering the hormone level of the plant. Abscisic acid that is an important growth hormone plays a significant role in salinity tolerance. Mycorrhizae have the ability to alter the ABA level of inoculated plant (Bothe et al. 1994; Estrada-Luna and Davies 2003) that could be helpful for maintaining growth of plant in saline conditions. Similarly, the higher cytokinin level was also observed in root and shoots in plants inoculated with mycorrhizae (Allen et al. 1980). Allen et al. (1980) also reported the improvement in growth due to improvement in the content of cytokinin.

The accumulation of carbohydrate under stress is another mechanism to maintain proper plant growth. The inoculation of *Phragmites australis* with mycorrhizae *Glomus fasciculatum* caused a significant increase in carbohydrate (Al-Garni 2006). The work of Porcel and Ruiz-Lozano (2004) also reported similar results when soybean was inoculated with *Glomus intraradices*. Evelin et al. (2009) observed high soluble sugar content in mycorrhizal plant when compared with non-mycorrhizal one.

Salinity reduces the absorption of mineral nutrients particularly phosphorus due to its precipitation with Ca^{+2} and Mg^{+2}. Due to very small diameter of the mycelium, it can penetrate into soil and extend outside the rhizosphere and therefore increase the root surface area for nutrient absorption. Such fine roots and improved root surface area are very helpful for the plant to uptake nutrients and water from the stress environment. According to an estimate, up to 80% phosphorus (P) requirements of plant are fulfilled by the activity of external hyphae of mycorrhizae (Matamoros et al. 1999). Therefore mycorrhizal plant has competitive advantage over non-myccorhizal plant and has better ability to survive in the stress conditions. Plenchette and Duponnois (2005) and Sharifi et al. (2007) also reported the improvement in plant growth by improving nutrient acquisition particularly the availability of phosphorus.

Mycorrhizal fungi are also important in maintaining ion homeostasis during salt stress condition (Estrada et al. 2013). Mycorrhizal fungus enhanced halotolerance of plants through improving plant nutrition, more K^+/Na^+ ratio in plant tissue, improved photosynthesis and water use efficiency (WUE), production of osmolytes, compartmentalization for ion storage, and protecting enzyme activity (Rabie and Almadini 2005; Al-Karaki 2006; Porcel et al. 2012). High ratio of K/Na is an indication of salinity tolerance, and inoculation with mycorrhizae is also helpful to maintain a high K/Na ratio in saline conditions (Giri et al. 2007; Zhang et al. 2011). Mycorrhizal fungus *Glomus deserticola* improved water use efficiency and balance between K/Na and between Ca/Na in sweet basil plants during saline conditions (Elhindi et al. 2017). Yang et al. (2014) also observed relatively high K/Na ratio in mycorrhizal plants as compared to non-mycorrhizal plants. Mycorrhizal fungi reduce the salinity stress by discouraging the uptakes of Na^+ and Cl^-, their transport toward plasma membrane, tonoplast, and their movement toward shoots (Al-Karaki 2006; Lee et al. 2015).

Mycorrhizal fungus has the ability to minimize the salinity-induced impact and enhance crop productivity (Evelin et al. 2009; Garg and Pandey 2015). Mycorrhizal fungi form association with legume plant *Sesbania cannabina* which improved its resistance to soil salinity (Ren et al. 2016). Fine-tuned coordination is needed at molecular level for successful establishment of plant-mycorrhizae association (Andreo-Jimenez et al. 2015). Mycorrhizal fungus increases strigolactone (SL) levels of plant and is responsible for the increased salt resistance and even restored the biomass and photophysiological damage (Kong et al. 2017). Strigolactone role as signal molecule in rhizosphere has been already proven and detected in plant root extracts (Xie et al. 2010). In lettuce plants, increase in strigolactones levels was also noticed during salt stress in mycorrhizal-inoculated *Rhizophagus irregularis* (Aroca et al. 2013). It has been observed that plants not only produce a single strigolactone but also blend of various SLs depending on a plant species (Xie et al. 2010; Ruyter-Spira et al. 2013). Different studies have proved that less H_2O_2 concentration is also a mechanism used by fungi to support crop growth against salt stress (Hajiboland et al. 2010; Garg and Bhandari 2012). One important thing to remember is that increased concentration of H_2O_2 acts as an oxidative agent, whereas a low concentration may act as a signaling molecule (Xia et al. 2009; Torres and Dangl 2005). In *Sesbania cannabina* seedlings, accumulation of H_2O_2 was noticed against salt stress after mycorrhizal fungal colonization. Strigolactone induces halotolerance by a complex set of signal transcription pathways. In this H_2O_2 is used as a common signal molecule and is produced as a result of NADPH oxidase activation during salt stress conditions in mycorrhizal plants (Kong et al. 2017). Plant roots are the part of plant which mostly experience the salinity stress and so produce more reactive oxygen species (ROS) as compared to leaves. ROS ($O_2^{\ -}$, H_2O_2) have high potential to cause peroxidation of lipids membrane present in the intercellular organelles and cell membrane (Pedranzani et al. 2016).

Mycorrhizal fungal species *Funneliformis mosseae* and *Rhizophagus irregularis* improve plant's performance and prevent damage under salt stress by increasing proline production which is one of the well-known scavengers of free radicals and acts as buffer during stresses (Porcel et al. 2012; Pollastri et al. 2017) and is known as the best compatible solutes those balancing the salt accumulated in plants and being currently used by biotechnologists for salt tolerance in plants (Szabados and Savoure 2010; Deinlein et al. 2014). Other scientists have already mentioned high proline content in mycorrhizal plants (Jindal et al. 1993; Sharifi et al. 2007). Sharifi et al. (2007) demonstrated that a high concentration of proline in the root of mycorrhizal soybean plant compared to shoot under saline conditions might be due to the reason that roots play a key role in water absorption and maintain osmotic balance between internal and external environment. Glycine betaine is another osmoprotectant produced by several plants during stress conditions including salinity stress. Betaines play important part in stabilizing the enzymes, protein complexes, and membrane integrity against salt stress. Considerably more betaine contents were recorded in mycorrhizal plants (Al-Garni 2006).

Piriformospora indica produced cytokinin and was involved in halotolerance of host plant through its cross talk with abscisic acid (Crafts and Miller 1974; Nishiyama et al. 2011). Abscisic acid is a phytohormone important for plant growth, development, and even abiotic stresses, i.e., salinity. ABA accumulation has been reported to be altered in plants having mycorrhizal association (Estrada-Luna and Davies 2003). Fungi also produce ACC-deaminase in order to degrade ethylene precursor ACC in order to rescue normal plant growth and regulate expression of hkt1leading to Na^+ homeostasis during salt stress (Contreras-Cornejo et al. 2009; Viterbo et al. 2010).

Jatropha curcas inoculated with mycorrhizae showed normal water content in leaves and improved water conductance in root of stress plant (Kumar et al. 2015). In a greenhouse experiment, two date palm cultivars (Nakhla hamra and Tijib) inoculated with five different species of mycorrhizae (*G. intraradices*, *G. aggregatum*, *G. mosseae*, *G. verriculosum*, *G. fasciculatum*) and subjected to salinity (Diatta et al. 2014). The cultivar showed variable response to salinity and mycorrhizae species. It has been observed that Nakhla hamra inoculated with *G. intraradices* showed better growth under salinity stress, while Tijib grew better when inoculated with *G. fasciculatum*. Salt stress ability of cultivar was due to accumulation of compatible solute, i.e., proline. The concentration of proline depends upon the species of mycorrhizae and the intensity of the salinity.

Recently, Manga et al. (2017) also found variable response regarding the performance of mycorrhizae under saline environment. While testing the efficacy of eight mycorrhizal fungi to enhance salinity tolerance of *Acacia seyal*, they observed that efficiency of mycorrhizae was variable in improving mineral nutrition and it depends upon the type of mycorrhizae associated with seedlings as well as salinity level. They suggested that interaction between plant and mycorrhizae can be modulated by the type and diversity of mycorrhizae as well as intensity of abiotic stress. It has observed that inoculation of mycorrhizae not only promoted the plant yield but quality of fruit also improved (Kaya et al. 2009).

It has also been observed that all fungi do not have the similar ability to reduce the salinity-induced impact on plant. In some cases mycorrhizae don't save the plant from harmful effect of salinity due to the reason that mycorrhizae were not able to maintain its proper growth and its activity affected under salt stress conditions (Juniper and Abbott 2006; Sheng et al. 2008). There is variability among fungal species to grow in stress environment. Nazareth et al. (2012), while studying the metal resistance of halotolerant fungi isolated from mangroves and salterns of Goa, India, reported that species belonging to *Penicillium* showed highest levels of halotolerance as well as resistance to heavy metals compared to *Aspergillus*. Species belonging to genera *Paecilomyces* showed a lower halotolerance. They demonstrated that mechanism of tolerance in fungi was phylogenetically related.

10.4 Synergistic Application of Halotolerant Microbes

In spite of the use of PGPR since long, their survival in the rhizosphere by competing with native microflora as well as their abilities for effective colonization of the rhizosphere remain unexplored and need to be further investigated for making this approach successful (Bashan 1998). Moreover, higher concentration of salts in soil could further cause problems for microbes to maintain their activity and survival (Bremer and Kramer 2000). Performance of single microbial strains/agents often remains inconsistent in field evaluation by reason of less population counts of introduced single microbial strains and their competition with aboriginal microbial population (Felici et al. 2008). Additionally, different environmental factors (including both biotic and abiotic) and rapid reduction of introduced microbial population counts also lead to inoculum failure. Soil microbiostasis causes rapid decrease of introduced microbial strains (Johannes et al. 1997). Effectiveness and stability of single microbial strain used for inoculation could also be one of the major problems of inoculums failure. Hence, more efforts are needed to increase the survivability of introduced microbes in the rhizosphere especially under unfavorable environmental conditions.

The use of beneficial microbes in their two or more combinations increases their reliability because of better established microbial community in the rhizosphere, multiple synergistic mechanisms, and improved survivability in various kinds of environmental conditions (Larkin and Fravel 1998). Multi-strain inoculation of microbes has pronounced effect for improving plant growth as compared to individual inoculations of microbial strains. This additive stimulatory effect of multi-strain microbial inoculation is due to synergy between two or more beneficial bacteria and/or beneficial fungi/mycorrhizae which possess various plant growth-promoting mechanisms/activities like production of phytohormones, enhancing availability of nutrients (Raja et al. 2006), nitrogenase activity (Alam et al. 2001), and colonization of endosphere by endophytic bacteria (Govindarajan et al. 2007). Mixed inoculation of microbes not only improved growth of plant in normal condition, but they also induced systemic tolerance under stressed conditions (Ruiz-Sanchez et al. 2011). The impact of co-inoculation of bacteria and mycorrhizal fungus on growth of plant in saline conditions has been mentioned in Table 10.3.

Multiple ranges of metabolic activities of microbial consortium are mainly responsible for synergistic effects of microbes to improve plant growth as compared to their soil inoculation (Adesemoye and Kloepper 2009). Combined additive effect of multi-strain microbes of consortium on plant growth could be enhanced through biosynthesis of phytohormones, N_2 fixation, and increased release of sugars and amino acids from roots (Raja et al. 2006; Ruiz-Sanchez et al. 2011). More grain weight and grains per spike in wheat were observed in response to combined inoculation of arbuscular mycorrhizae and Azotobacter as compared to their sole inoculations as well as un-inoculated control. This improvement in grain yield of wheat could be linked to better efficiency of microbial inoculants to explore additional soil volume (Bahrani et al. 2010). Combined application of *Mesohizobium* and *Bacillus* enhanced the number of tillers, spike length, and grain yield of wheat due to better uptake of phosphorus. Akhtar et al. (2013) documented that co-inoculation of *Rhizobia* and *Bacillus* enhanced the phosphorus availability. Further, it has been observed that co-aggregation of *Azospirillum brasilense* and *Methylobacterium* sp. results in improved tomato growth as compared to their individual inoculation of these strains (Joe et al. 2014). Induced salinity tolerance in lettuce has been reported by co-inoculation of *Serratia* sp. and *Rhizobium* sp. with improved activities of ascorbate peroxidase and glutathione reductase. Besides mineral contents and photosynthetic rate were also more by combined inoculation as compared to use of single strains (Han and Lee 2005).

Afrasayab et al. (2000) also observed that multi-strain inoculation of microbes caused more improvement in growth of wheat under saline condition which might be attributed to more auxins, soluble protein contents and K^+ contents, as well as decreased concentration of Na^+ contents. Bulut (2013) reported that combined use of different microbial strains results in the improvement of wheat growth almost equal to that was observed in response of chemical fertilizers. This increase in growth of wheat due to multi-strain bacterial inoculation was due to more uptake of nitrogen as compared to individual use of these bacterial strains. Bacterial consortium of *Pseudomonas flourescens*, *Azospirillum lipoferum*, and *Bacillus megaterium* enhanced the efficiency of these bacterial strains for effective

Table 10.3 Co-inoculation of PGPR and mycorrhizal fungi for inducing salinity tolerance in plants

Crop	Fungal strain	PGPB strain	Response/ mechanism	Reference
Cajanus cajan	*Funneliformis mosseae* and *Rhizophagus irregularis*	*Sinorhizobium fredii strain AR-4*	Higher biomass accumulation, nodulation, N_2 fixation, and N and P uptake	Garg and Pandey (2016)
			Higher AM colonization	
			Higher trehalose in nodules due to decreased trehalase and increased trehalose-6-P synthase, trehalose-6-phosphatase activities	
Zea mays L.	*Glomus etunicatum*	*Methylobacterium oryzae CBMB20*	Increased dry biomass, AMF root colonization, and nutrient accumulation	Lee et al. (2015)
			Reduced proline content and Na uptake	
	Gigaspora margarita S–23 and *Claroideoglomus lamellosum S–11*	*Pseudomonas koreensis* S2CB35	Improved plant dry weight, nutrient content of shoot and root tissues	Selvakumar et al. (2018)
			Reduced the accumulation of proline in shoots and Na^+ in roots	
			Exhibited high K^+/ Na^+ ratios in roots	
			Altered the expression of *ZmAKT2*, *ZmSOS1*, and *ZmSKOR* genes	
Acacia gerrardii Benth	*Claroideoglomus etunicatum*; *Rhizophagus intraradices*; and *Funneliformis mosseae*	*Bacillus subtilis* (BERA 71)	Enhanced AMF root colonization	Hashem et al. (2016b)
			Increased in total lipids, phenols, and fiber content	
			Increased content of osmoprotectants such as glycine, betaine, and proline	

(continued)

Table 10.3 (continued)

Crop	Fungal strain	PGPB strain	Response/ mechanism	Reference
			Reduced lipid peroxidation	
Mentha arvensis L.	*Glomus aggregatum, Glomus mosseae, Glomus fasciculatum*, and *Glomus intraradices*	*Halomonas desiderata* (STR8) and *Exiguobacterium oxidotolerans* (STR36)	Highest fresh weights	Bharti et al. (2016a, b)
			Higher oil yield	
			Better AMF colonization	
			Increased nitrogen accretion	
Avena sativa	*Glomus intraradices*	*Acinetobacter* sp.	Decreased the MDA and free proline contents in plant	Xun et al. (2015)
			Augmented the activities of the antioxidant enzyme (SOD, POD, and CAT)	
			Increased the activities of soil enzyme (i.e., urease, sucrase, and dehydrogenase)	
Vicia faba	*Acaulospora laevis, Glomus geosporum, Glomus mosseae, and Scutellospora armeniaca*	*Rhizobium leguminosarum* biovar *viciae* STDF-Egypt 19 (R)	Increased in number and mass of nodules, nitrogenase activity, leghaemoglobin content of nodule	Abd-Alla et al. (2014)
			Improved mycorrhizal colonization	
			Increased dry mass of root and shoot	
Phaseolus vulgaris L.	*Glomus mosseae* (Nicol. & Gerd.)	*Pseudomonas fluorescence*	Higher concentrations of shoots K^+ and lower concentrations of shoots Na^+	Younesi and Moradi (2014)
			Increased shoots proline concentration	
			Increased the antioxidant enzyme activities (POX and CAT)	

(continued)

Table 10.3 (continued)

Crop	Fungal strain	PGPB strain	Response/ mechanism	Reference
Ocimum basilicum	*Glomus irradicans*	*Bacillus subtilis*	Positive effects on growth, oil %, oil yield, and nutrient uptake	Abdel-Rahman et al. (2011)
	Glomus intraradices (Gi)	*Dietzia natronolimnaea* STR1	A positive effect on the indigenous microbial community structure	Bharti et al. (2016a, b)
			Increased the essential oil content, oil yield and herb yield	

colonization with improved rice growth by improved biosynthesis of amino acid, reducing sugars and phytohormones, as well as improved capacity to fix atmospheric nitrogen (Raja et al. 2006).

Most of the legumes are sensitive to salinity, and symbiotic association of mycorrhizae with legumes in salinity causes a significant effect on plant and enhances the ability of inoculated plant to fixation (Garg and Chandel 2011). A study conducted to assess the effectiveness of co-inoculation of mycorrhizae and rhizobium to enhance the salinity tolerance of alfalfa showed that combined application of two strains increased the yield of alfalfa and a significant increase in number and weight of nodule was observed (Zhu et al. 2016). The co-inoculated plants showed more proline content in shoot. The synergistic relationship between mycorrhizae and bacteria caused an increase in nitrogen and phosphorus content that enhanced plant nutrition (Giri et al. 2002).

Koc et al. (2016) also studied the role of combined application of bacteria and mycorrhizae in salinity stress tolerance of strawberry. Increasing salt concentration reduced all plant growth parameters. The combined application of mycorrhizae and bacteria ameliorated the salinity-induced negative impact on strawberry. They observed that co-inoculation protected the plant from salinity damage by increasing proline and anthocyanin levels. They observed that effect was more pronounced at higher salinity level. In an earlier study conducted by Rabie et al. (2005), more effectiveness of mycorrhizae was observed on lettuce at high salinity level. In many cases, co-inoculation proved useful for improving crop growth; however, in addition to co-inoculation of two microbial strains, combined application of mycorrhizae with polyamine also mitigated the negative effect of salinity. Abdel-Fattah et al. (2013) evaluated the effect of *Glomus mosseae* and spermine on growth and physiological aspects of wheat (*Triticum aestivum*) by conducting a pot experiment in saline conditions. It has been observed that dual application of mycorrhizae and spermine caused positive effect on growth of wheat. The wheat salt tolerance was associated with better root colonization, enhancement of proline, and protein and chlorophyll content.

Nutritional imbalance under stress condition affects the growth of plant negatively. High Na^+ content in saline conditions inhibits the uptake of essential nutrients like K^+ and Ca^{+2}. Inoculation of plant with mycorrhizae and bacteria proved useful to enhance the uptake of nutrient and maintain nutrient balance (Gamalero et al. 2010). Najafi et al. (2012) also reported that co-inoculation of PGPR-mycorrhizae enhanced the water and nutrient absorption. Due to reduction of nutrient availability particularly P, the application of mycorrhizae could be helpful for improving nutrition. Inoculation with mycorrhizae enhanced the production of antioxidant enzymes, nitrogen fixation, nodule formation, and ultimately growth and yield of legumes (Garg and Manchanda 2008). Similarly, mycorrhizae improved the availability and uptake of nitrogen (Govindarajulu et al. 2005), and according to their view, it might be due to the reasons that microbial inoculation brought changes in nitrogen metabolism by changing the activity of enzymes associated with it. In addition to nitrogen and phosphorus, application of mycorrhizae also enhanced the availability and uptake of essential plant nutrients including Ca, Mg, K, Cu, Mn, Zn, and Fe (Al-Karaki et al. 2001; Mardukhi et al. 2011).

Co-inoculation of mycorrhizae with bacteria improves nodulation of subterranean clover and ultimately enhanced the growth (Meyer and Linderman 1986). Mycorrhizae can alleviate salinity-imposed harmful effects on the nitrogen fixation process and cause significant increased nodules (Giri and Mukerji 2004; Manchanda 2008). Improved nitrogen fixation under salinity stress was due to improvement in nitrogenase activity and content of leghemoglobin. Soliman et al. (2012) while studying the effect of mycorrhizal fungus with *Rhizobium* observed that co-inoculation protected the *A. saligna* plant from salinity-induced negative impact by decreasing the uptake of Na^+ and accumulation of proline.

The interaction between endophytic bacteria and mycorrhizae has been studied by Hashem et al. (2016b, c) to assess the effectiveness of co-inoculation on the growth of *Acacia gerrardii* under salt stress. An increase in root and shoot weight and number of nodules and leghemoglobin was recorded due to co-inoculation. Also, inoculated and uninoculated plant showed a significant difference regarding the activities of nitrogenase and nitrite reductase. Inoculation protected the salt-stressed *A. gerrardii* by enhancing the phosphatase activities and uptake of nitrogen, phosphorus, potassium, magnesium, and calcium while reducing the concentration of sodium and chloride. Similarly, dual inoculation of mycorrhizae and phosphate-solubilizing fungus (PSF) caused significant impact on phosphate content, phosphatase activity, and uptake of essential nutrients under NaCl stress (Xueming et al. 2014). They also observed better root colonization of inoculated plant than uninoculated one. Phosphate-solubilizing fungus can solubilize the precipitated P along with K^+ and Ca^{+2} and therefore enhance the availability of phosphorus in saline conditions (Zhang et al. 2013).

Mycorrhizosphere that is a zone surrounding the plant roots and fungal hyphae serve as a place of fungal and bacterial interaction (Johansson et al. 2004). The mutual interactions that occurred among microbial population in this zone not only affect the plant growth positively but also promoted the survival of each other (Artursson et al. 2006; Yusran et al., 2009). In this interaction, bacteria by producing certain compound enhance the cell permeability, stimulate the growth of hyphae, and facilitate the fungus to penetrate the root penetration (Jeffries et al. 2003). In most of the cases, these interactions are two way and both partners facilitate each other. Bacterial interactions with fungi facilitate in the development of mycosymbionts and increase the ability of fungi to colonize the plant roots effectively (Hildebrandt et al. 2002; Jaderlund et al. 2008). For example, co-inoculation of *Paenibacillus brasilensis* with mycorrhizal fungus *G. mosseae* enhanced fungal root colonization of clover (Artursson 2005) and mycorrhizae enhance the phosphorus-solubilizing and nitrogen-fixing ability of the bacteria (Linderman 1992).

The association between mycorrhizae and plants is strongly affected by environmental conditions including salinity (Gupta et al. 2000; Cavagnaro et al. 2015), and efficiency of mycorrhizae under such harsh condition can be further increased by the dual inoculation of mycorrhizae with associative bacteria. Salt-tolerant *Bradyrhizobium* caused significant impact on the *Vigna radiata* L in saline conditions when it was co-inoculated with *Glomus mosseae* (Singh et al. 2011). They observed that application of *Bradyrhizobium* enhanced shoot and root length, number of nodules, total plant biomass, as well as population and colonization of arbuscular mycorrhizal fungal (AMF) compared to sole application of myccorhizae. However, marked increase in growth and yield parameters of *Vigna radiata* has been observed by the dual application of these two strains. This indicates that a synergistic association of arbuscular mycorrhizal fungi and rhizobia could be very effective to increase the efficiency of plant for nitrogen fixation that ultimately causes positive effect growth and yield of legumes.

From the above discussion, it can be concluded that co-inoculation of bacteria and mycorrhizae are useful for promoting the growth of plant under salinity stress. In such synergistic interactions, both partners help each other for expressing their growth-promoting traits. Dual inoculation seems to be more effective under stress conditions because under harsh environment, single strain sometimes do not show its potential due to unfavorable environment.

10.5 Roles of Halotolerant Microbes in Environmental Science

Environmental contamination is increasing nowaday due to anthropogenic activities, progressive growth of population, industrialization, and urbanization (Tabak et al. 2005; Ahemad et al. 2009). The industries which manufacture pesticide, pharmaceuticals, herbicides, oil, and gas recovery processes produce wastewaters with salinity fluctuating from time to time. Conventional microbiological treatment processes don't function at these salinity variations because of inadaptability of microbes to varying salt concentration. Due to involvement of halophilic bacteria in biogeochemical process, these also act as a biological mediator (Capone 2002). Das et al. (2011) studied the distribution of halophilic microbes with increasing depth of Sundarban Mangrove forest. These microbes are involved in the process of biotransformation biomineralization of soil minerals. According to their observation, the top soil microbes were more tolerant to salinity fluctuation than the microbes present in the middle and bottom soil layer. They also observed that relatively more salinity tolerance ability of anaerobic microbes than top aerobic microbes present in the bottom segment of soil.

The report of Lowe et al. (1993) indicates that anaerobic bacteria have better ability to cope with extreme environmental conditions like pH, temperature, and salinity. Anaerobic microbes having better energy-conserving mechanisms than aerobic bacteria that make them suitable candidate to adapt harsh environmental conditions effectively. According to their view, fluctuation in water and salinity level occurring due to global warming will disturb the activities of aerobic bacteria more than anaerobic ones.

Microorganisms are ideally suited for cleaning contaminated environment because certain microbial species containing particular enzymes enable to survive in contaminated environment by using certain compounds as a food source. Halotolerant microbes have the ability to tolerate adverse environment, and by virtue of their particular enzymes, these can be used as bioremediation agents. Qin et al. (2012) studied bioremediation of petroleum hydrocarbons in saline-alkaline conditions. They observed that salinity had great impact on bioremediation. At initial stage, moderate salinity enhanced the biodegradation; however, gradual increase in salinity decreased the rate of degradation. The addition of mixed culture of two fungal strains (*Trichosporon* sp. and *Penicillium citrinum*) isolated from saline-alkaline conditions improved the biodegradation efficiency and dehydrogenase activity. There was a positive correlation between degradation rate and activity of dehydrogenase enzyme. They suggested that addition of salt-tolerant inocula is effective for enhancing the bioremediation of hydrocarbon in saline environment.

Bioremediation potential of salt-resistant *Bacillus* strains was studied by Abdughafurovich et al. (2010) under aerobic condition. The study employed the use of tritium-labeled polychlorinated biphenyls (PCBs) congeners and tracing the tritium label in cultures with high salt content. The results showed that most of the *Bacillus* strains were able to absorb a part of the radioactivity. The results of gas chromatography showed the transformation of PCBs. They demonstrated that such strains could be very suitable for bioremediation of contaminated salt-affected soils. Response of sea water microbial inocula with and without biosurfactants and micronutrients was evaluated to enhance the bioremediation of hydrocarbon-contaminated mangrove swamp (Okoro 2010). A significant increase in bioremediation was observed where microbial inoculum was applied with biosurfactants and micronutrients.

The work of Arakaki et al. (2013) indicated that salinity inhibits the growth of seven out of ten fungus strains and three strains *Polyporus* sp, *Dacryopinax elegans*, and *Datronia stereoides* produced high biomass in saline medium. When glycerol and mannitol were added in the medium, it enhanced the activity of *D. elegans* regarding the biomass production and ligninases. It degraded diuron in the presence of NaCl as well as polyols. Similarly, among 26 halotolerant isolates, 2 aerobic *Bacillus* spp. showed resistance and potential for Se (IV) reduction. Both strains completely reduced the Se (IV) to selenite within 40 h. Nanosized spherical particles of selenium were found in and around the bacterial cell which were further confirmed by confocal micrograph study. They demonstrated that these microbial strains could be an important source for bioremediation of Se.

Due to rapid increase in textile industry, a number of synthetic dyes are present in the effluent, and the removal of these dyes from the wastewater is a challenge to protect the soil from contamination. These dyes not only degrade the soil quality but also cause negative impact on underground water quality. The use of bacteria to remove these dyes from the wastewater is one of the promising technologies. Salinity caused significant impact on bioremediation processes, and the process of degradation is affected due to negative impact of salinity on microbial population participating in degradation process. Therefore due to the presence of number of salts in such water, not all the bacterial strains could be effective for bioremediation of such dyes. The use of halotolerant bacteria could be very effective in such conditions due to eco-friendly and an inexpensive strategy (Parshetti et al. 2006). A number of bacterial strains have been isolated which can degrade azo dyes. Some of these strains include *Exiguobacterium*, *Klebsiella pneumoniae*, *Acinetobacter*, and *Gracilibacillus* (Salah et al. 2007; Tan et al. 2009; Bibi et al. 2012; Agarwal and Singh 2012).

Isolating microorganism from the contaminated environment and then using these microbes in the same environment for bioremediation purpose is better strategy to clean the environment. The isolation of bacterial strains from saline conditions and then their application to remove azo dyes have been evaluated by Lalnunhlimi and Krishnaswamy (2016). A consortium of bacterium at various concentrations (100–300 mg/L) of azo dyes was applied to decolorize Direct Blue 151 (DB151) and Direct Red 31 (DR 31). Carbon and nitrogen sources are given to the bacterium as food, and growth was with required pH and temperature. They observed that bacterium consortium was very effective for decolorization of azo dyes and the effect was more pronounced when sucrose and yeast extract was given as carbon and nitrogen source to bacterium. Biodegradation ability of halotolerant bacteria was evaluated by Nicholson and Fathepure (2004) under aerobic conditions. For this purpose, culture of halophilic bacteria dominated by *Marinobacter* spp. was used to assess its potential to degrade benzene, ethylbenzene, and toluene and xylene compound. It has been observed that this bacterial culture completely degraded benzene, ethylbenzene, toluene, and xylenes. The culture also mineralized the benzene that was evident from conversion of (^{14}C) benzene to ($^{14}CO_2$). They reported that activity of culture was enhanced markedly by the addition of yeast extract as a growth-promoting nutrient. Similarly, the ability of a halophilic archaea to degrade aromatic hydrocarbons and a wide range of n-alkanes in the presence of salinity have been documented by Oren et al. (1992) and Bertrand et al. (1990).

Feng et al. (2014) isolated salt-tolerant strain *Pichia Kudriavzevii*, and the effect of salinity, initial concentration of azo dye acid red B, temperature, and pH on decolorization performance of this strain was evaluated. On the basis of results, they concluded that degradation efficiency of this strain was more than 94% and optimal condition for its performance were salinity, 10%; initial concentration of acid red B solutions,100 mg/L; temperature, 33 °C; and pH, 5.0. Asad et al. (2007) observed that among 27 strains of halotolerant bacteria, 3 showed remarkable ability to decolorize the azo dyes at various concentrations of NaCl and a wide range of pH and temperature. These strains have potential to decolorize a pure as well as mixture of dyes. They suggested that the decolorization was due to degradation that was the result of reduction of azo bonds.

Polycyclic aromatic hydrocarbons cause harmful impact in aquatic environment due to their carcinogenic nature in marine organisms (Perugini et al. 2007). Due to their hydrophobic nature, these compounds can be adsorbed on a particulate matter, and marine sediments act as sinks for such compounds (Yu et al. 2005; Osuji and Ezebuiro 2006). Biodegradation of polycyclic aromatic hydrocarbon such as naphthalene, fluorene, anthracene, and phenanthrene by a halotolerant bacterial consortium consisting of *Ochrobactrum* sp., *Enterobacter cloacae*, and *Stenotrophomonas maltophilia* was studied by Arulazhagan et al. (2010). Various concentrations of PAHs, i.e., 5, 10, 20, 50, and 100 ppm and a high level of NaCl (30 g/L), were used to assess the degradation potential of the consortium. The study showed that consortium degrades the PAHs within 4 days, and addition of yeast extract further enhanced the degradation.

Sowmya et al. (2014) isolated 35 halophilic bacterial strains and categorized these strains into moderately and extremely halotolerant. The isolated halophilic bacterial strains were tested against lead and cadmium. The strains showed more tolerance against lead than cadmium. And this tolerance was dependent on salt concentration of the medium. Removal efficiency of selected isolates regarding heavy metals showed that a maximum reduction of cadmium was less than lead, i.e., 37 and 99%, respectively. They demonstrated that variable response of strains were due to the toxicity of metals to strains and variable characteristics of two strains. A positive response was observed when a fungus was used to remediate petroleum-contaminated saline –alkaline soil (Qin et al. 2012). Zhang et al. (2016) isolated salt-tolerant bacteria (*Corynebacterium variabile*). The isolated bacterium with biochar as a carrier material was applied to see the bioremediation potential of said strain. The bacteria caused significant effect on degradation of n-alkanes and polycyclic aromatic hydrocarbons. Similarly, bioremediation potential of halotolerant fungus *Embellisia phragmospora* was studied by Jenab et al. (2016) against crude oil and pyrene. The strain was able to degrade 77, 83.4, and 77.4% of petroleum at salinity of 0, 2.5, and 5%, respectively, during 7 days, whereas pyrene removal was 82.4, 88.3, and 95.2% at these salinity levels.

The above-discussed review showed that in addition to improving plant growth under salinity stress conditions, the halotolerant microbes are also successfully used to remediate the contaminated environment. These microbes have a wide range of characteristics that enable them to decontaminate the variety of soil contaminants. Owing to their environment-friendly traits and less cost, these microbes could be a suitable inoculum for their use on commercial basis.

10.6 Concluding Remarks and Future Prospectus

The above-discussed review showed that plants and microbes both are affected by the harmful effect of salinity. Most of the studies reported in the literature indicate that microbes have developed special mechanisms that enable them to survive and grow in saline environment. Although the plants itself have defense mechanisms against salinity, the interactions of plants with microbes further accelerate the survival of plants in harsh environment. Owing to their number of growth promotion mechanisms, these halotolerant microbes play a significant role in plant tolerance against salinity. Lowering of stressed-induced ethylene by ACC-deaminase, decreasing the Na availability by exopolysaccharides, enhancing the activity antioxidant system, and producing osmolytes are the key traits which dilute the negative effect of salinity, and plant showed better growth in saline conditions. Symbiotic association of plant with these microbial strains enables the plant to maintain growth by enhancing its capacity to produce compatible solute, activity of antioxidant enzyme, nutrient acquisition, and water use efficiency. The review mentioned in this chapter showed the significant positive impact of microbial inoculation on plant, and it has been noted that inoculated plants were more resistant to salinity. It has also observed that halotolerant microbes are equally important to degrade harmful compounds present in contaminated environment. These strains have competitive advantage over non-halotolerant microbes due to their ability to cope with harsh environmental conditions and presence of some particular traits that enable them to accelerate the process of bioremediation effectively.

It has also observed from the literature that most of the work conducted under axenic conditions with control environmental conditions. However, less work has been done to evaluate the efficacy of these microbes in natural environment. For the application of halotolerant bacteria as biofertilizers and bioremediation agents, several aspects like its survival, optimum environmental conditions, and ability to compete with indigenous population are still needed to be explored. Similarly, instead of using single strain, the effectiveness of multi-strain inoculum is needed to be evaluated in laboratory as well as in field. Also before using multi-strain inoculum, their compatibility with each other needs to be tested so that maximum benefits can be obtained. The application of halotolerant microbes in the field of biotechnology is another area that requires special attention so that maximum benefits can be obtained from this naturally occurring population.

References

Abd-Alla MH, El-Enany AW, Nafady NA, Khalaf DM, Morsy FM (2014) Synergistic interaction of *Rhizobium leguminosarum* bv. *viciae* and arbuscular mycorrhizal fungi as a plant growth promoting biofertilizers for faba bean (*Vicia faba* L.) in alkaline soil. Microbiol Res 169:49–58

Abd-Allah EF, Hashem A, Alqarawi AA, Bahkali AH, Alwhibi MS (2015) Enhancing growth performance and systemic acquired resi,stance of medicinal plant *Sesbania sesban* (L.) Merr using arbuscular mycorrhizal fungi under salt stress. Saudi J Biol Sci 22:274–283

Abdel-Fattah GM, Ibrahim AH, Al-Amri SM, Shoker AE (2013) Synergistic effect of arbuscular mycorrhizal fungi and spermine on amelioration of salinity stress of wheat (*Triticum aestivum* L. cv. gimiza 9). Aust J Crop Sci 7:1525

Abdel-Rahman SS, Abdel-Kader AA, Khalil SE (2011) Response of three sweet basil cultivars to inoculation with *Bacillus subtilis* and arbuscular mycorrhizal fungi under salt stress conditions. Nat Sci 9:93–111

Abdughafurovich RB, Andreevich KA, Adrian L, Tashpulatovich YK (2010) Biodegradation of tritium labeled polychlorinated biphenyls (PCBs) by local salt tolerant mesophylic *Bacillus* strains. J Environ Protect 1:420–425

Abeles FB, Morgan PW, Saltveit ME Jr (1992) Ethylene in plant biology. Academic, London

Adesemoye AO, Kloepper JW (2009) Plant microbes interactions in enhanced fertilizer-use efficiency. Appl Microbiol Biotechnol 85:1–12

Afrasayab S, Faisal M, Hasnain S (2000) Synergistic growth stimulatory effect of mixed culture bacterial inoculation on the early growth of *Triticum aestivum* var, Inqalab 91 under NaCl salinity. Pak J Biol Sci 3:1016–1023

Afrasayab S, Faisal M, Hasnain S (2010) Comparative study of wild and transformed salt tolerant bacterial strains on *Triticum aestivum* growth under salt stress. Braz J Microbiol 41:946–955

Agarwal T, Singh R (2012) Bioremedial potentials of a moderately halophilic soil bacterium. J Pharm Biomed Sci 19:1–6

Ahanger MA, Hashem A, Abd-Allah EF, Ahmad P (2014) Arbuscular mycorrhiza in crop improvement under environmental stress. In: Ahmad P, Rasool S (eds) Emerging technologies and management of crop stress tolerance, vol 2. Academic, Amsterdam, pp 69–95

Ahemad M, Khan MS, Zaidi A, Wani PA (2009) Remediation of herbicides contaminated soil using microbes. In: Khan MS, Zaidi A, Musarrat J (eds) Microbes in sustainable agriculture. Nova Science Publishers, New York, pp 261–284

Ahmad P, Hashem A, Abd-Allah EF, Alqarawi AA, John R, Egamberdieva D, Gucel S (2015a) Role of *Trichoderma harzianum* in mitigating NaCl stress in Indian mustard (*Brassica juncea* L) through antioxidative defense system. Front Plant Sci 6:1–15

Ahmad P, Sarwat M, Bhat NA, Wani MR, Kazi AG, Tranl S (2015b) Alleviation of cadmium toxicity in *Brassica juncea* L. (Czern. &Coss.) by calcium application involves various physiological and biochemical strategies. PLoS One 10(1):e0114571

Akcan N, Uyar F (2011) Production of extracellular alkaline protease from Bacillus subtilis RSKK96 with solid state fermentation. Eurasia J Biosci 5(64):72

Akhtar N, ArshadI SMA, QureshiMA SJ, Ali L (2013) Co-inoculation with *Rhizobium* and *Bacillus* sp. to improve the phosphorus availability and yield of wheat (*Triticum aestivum* L.). J Anim Plant Sci 23:190–197

Akhter MS, Hossain SJ, Amir-Hossain SK, Datta RK (2012) Isolation and characterization of salinity tolerant *Azotobacter* sp. Greener J Biol Sci 2:42–51

Alam MS, Cui Z, Yamagishi T, Ishii R (2001) Grain yield and related physiological characteristics of rice plants (*Oryza sativa* L.) inoculated with free-living rhizobacteria. Plant Prod Sci 4:126–130

Al-Garni SMS (2006) Increased heavy metal tolerance of cowpea plants by dual inoculation of an arbuscular mycorrhizal fungi and nitrogen-fixer *Rhizobium* bacterium. Afr J Biotechnol 5:133–142

Alguacil MM, Hernandez JA, Caravaca F, Portillo B, Roldan A (2003) Antioxidant enzyme activities in shoots from three mycorrhizal shrub species afforested in a degraded semi-arid soil. Physiol Plant 118:562–570

Ali A, Shahzad R, Khan AL, Halo BA, Al-Yahyai R, Al-Harrasi A, Al-Rawahi A, Lee IJ (2017) Endophytic bacterial diversity of *Avicennia marina* helps to confer resistance against salinity stress in *Solanum lycopersicum*. J Plant Interact 12:312–322

Aliasgharzadeh N, Rastin SN, Towfighi H, Alizadeh A (2001) Occurrence of arbuscular mycorrhizal fungi in saline soils of the Tabriz Plain of Iran in relation to some physical and chemical properties of soil. Mycorrhiza 11:119–122

Al-Karaki GN (2006) Nursery inoculation of tomato with arbuscular mycorrhizal fungi and subsequent performance under irrigation with saline water. Sci Hortic 109:1–7

Al-Karaki GN, Hammad R, Rusan M (2001) Response of two tomato cultivars differing in salt tolerance to inoculation with mycorrhizal fungi under salt stress. Mycorrhiza 11:43–47

Allen A, Cunningham GL (1983) Effects of vesicular–arbuscular mycorrhizae on *Distichlis spicata* under three salinity levels. New Phytol 93:227–236

Allen MF, Moore JTS, Christensen M (1980) Phytohormone changes in *Bouteloua gracilis* infected by vesicular–arbuscular mycorrhizae: I. Cytokinin increases in the host plant. Can J Bot 58:371–374

Amir R, Ahmed N, Talat M (1993) Salt-tolerance in mangrove soil bacteria. Pak J Mar Sci 2:129–135

Andreo-Jimenez B, Ruyter-Spira C, Bouwmeester H, Lopez-Raez JA (2015) Ecological relevance of strigolactones in nutrient uptake and other abiotic stresses, and in plant-microbe interactions below-ground. Plant Soil 394:1–19

Arakaki RL, Monteiro DA, Boscolo M, Dasilva R, Gomes E (2013) Halotolerance, ligninase production and herbicide degradation ability of basidiomycetes strains. Braz J Microbiol 44:1207–1214

Aroca R, Ferrante A, Vernieri P, Chrispeels MJ (2006) Drought, abscisic acid and transpiration rate effects on the regulation of PIP aquaporin gene expression and abundance in *Phaseolus vulgaris* plants. Ann Bot 98:1301–1310

Aroca R, Ruiz-Lozano JM, Zamarreno AM, Paz JA, Garcia-Mina JM, Pozo MJ, Lopez-Raez JA (2013) Arbuscular mycorrhizal symbiosis influences strigolactone production under salinity and alleviates salt stress in lettuce plants. J Plant Physiol 170:47–55

Artursson V (2005) Bacterial–fungal interactions highlighted using microbiomics: potential application for plant growth enhancement (dissertation). Swedish University of Agricultural Sciences

Artursson V, Finlay RD, Jansson JK (2006) Interactions between arbuscular mycorrhizal fungi and bacteria and their potential for stimulating plant growth. Environ Microbiol 8:1–10

Arulazhagan P, Vasudevan N, Yeom IT (2010) Biodegradation of polycyclic aromatic hydrocarbon by a halotolerant bacterial consortium isolated from marine environment. Int J Environ Sci Technol 7:639–652

Asad S, Amoozegar MA, Pourbabaee A, Sarbolouki MN, Dastgheib SMM (2007) Decolorization of textile azo dyes by newly isolated halophilic and halotolerant bacteria. Bioresour Technol 98:2082–2088

Ashraf M, Hasnain S, Berge O, Mahmood T (2004) Inoculating wheat seedlings with exopolysaccharide-producing bacteria restricts sodium uptake and stimulates plant growth under salt stress. Biol Fertil Soils 40(3)

Augustine SM, Narayan JA, Syamaladevi DP, Appunu C, Chakravarthi M, Ravichandran V, Tuteja N, Subramonian N (2015) Introduction of pea DNA helicase 45 into sugarcane (*Saccharum* spp. Hybrid) enhances cell membrane thermostability and upregulation of stress-responsive genes leads to abiotic stress tolerance. Mol Biotechnol 57:475–488

Azarmi F, Mozafari V, Abbaszadeh Dahaji P, Hamidpour M (2016) Biochemical, physiological and antioxidant enzymatic activity responses of pistachio seedlings treated with plant growth promoting rhizobacteria and Zn to salinity stress. Acta Physiol Plant 38(1)

Bacilio M, Rodriguez H, Moreno M, Hernandez JP, Bashan Y (2004) Mitigation of salt stress in wheat seedlings by a *gfp*-tagged *Azospirillum lipoferum*. Biol Fertil Soils 40:188–193

Badri DV, Weir TL, van der Lelie D, Vivanco JM (2009) Rhizosphere chemical dialogues: plant–microbe interactions. Curr Opin Biotechnol 20:642–650

Bahrani A, Pourreza J, Joo MH (2010) Response of winter wheat to co-inoculation with *Azotobacter* and arbescular mycorrhizal fungi (AMF) under different sources of nitrogen fertilizer. J Agric Environ Sci 8:95–103

Balliu A, Glenda S, Boris R (2015) AMF inoculation enhances growth and improves the nutrient uptake rates of transplanted, salt-stressed tomato seedlings. Sustainability 7:15967–15981

Banik A, Pandya P, Patel B, Rathod C, Dangar M (2018) Characterization of halotolerant, pigmented, plant growth promoting bacteria of groundnut rhizosphere and its in-vitro evaluation of plant-microbe protocooperation to withstand salinity and metal stress. Sci Total Environ 630:231–242

Bashan Y (1998) Inoculants of plant growth promoting bacteria for use in agriculture. Biotechnol Adv 16:729–770

Beattie GA (2007) Plant associated bacteria: survey, molecular phylogeny, genomics and recent advances. In: Ganamanickam SS (ed) Plant associated bacteria. Springer, Dordrecht

Behera BK, Das P, Patra A, Sharma AP (2012) Transcriptome analysis of *Enterobacter aerogenes* KCTC 2190 in response to elevated salt stress In: Abstracts of 99th the Indian Science Congress, held at Bhubaneswar during 3–7th January

Behera BK, Das P, Maharana J, Paria P, Mandal S, Meena DK, Sharma AP, Jayarajan R, Dixit V, Verma A, Vellarikkal SK, Scaria V, Sivasubbu S, Rao AR, Mohapatra T (2014) Draft genome sequence of the extremely halophilic bacterium *Halomonas salina* strain CIFRI1, isolated from the East Coast of India. Genome Announc: e01321-14

Bertrand JC, Almallah M, Acquaviva M, Mille G (1990) Biodegradation of hydrocarbons by an extremely halophilic archaebacterium. Lett Appl Microbiol 11:260–263

Bharti N, Barnawal D, Shukla S, Tewari SK, Katiyar RS, Kalra A (2016a) Integrated application of *Exiguobacterium oxidotolerans*, *Glomus fasciculatum*, and vermicompost improves growth, yield and quality of *Mentha arvensis* in salt-stressed soils. Ind Crop Prod 83:717–728

Bharti N, Barnawal D, Wasnik K, Tewari SK, Kalra A (2016b) Co-inoculation of Dietzia natronolimnaea and *Glomus intraradices* with vermicompost positively influences Ocimum basilicum growth and resident microbial community structure in salt affected low fertility soils. Appl Soil Ecol 100:211–225

Bibi R, Arshad M, Asghar HN (2012) Optimization of factors for accelerated biodegradation of reactive Black-5 Azo dye. Int J Agric Biol 14:353–359

Blomberg A, Adler L (1992) Physiology of osmotolerance in fungi. Adv Microb Physiol 33:145–212

Borde M, Dudhane M, Jite PK (2010) AM fungi influences the photosynthetic activity, growth and antioxidant enzymes in *Allium sativum* L. under salinity condition. Not Sci Biol 2:64–71

Bothe H, Klingner A, Kaldorf M, Schmitz O, Esch H, Hundeshagen B, Kernebeck H (1994) Biochemical approaches to the study of plant-fungal interactions in Arbuscular mycorrhizas. Experientia 50:919–925

Bremer E, Kramer R (2000) Coping with osmotic challenges: osmoregulation through accumulation and release of compatible solutes in bacteria. In: Storz G, Hengge-Areonis R (eds) Bacterial stress response. ASM Press, Washington, DC, pp 79–97

Breuninger M, Requena N (2004) Recognition events in AM symbiosis: analysis of fungal gene expression at the early appressorium stage. Fungal Genet Biol 41:794–804

Bulut S (2013) Evaluation of efficiency parameters of phosphorous-solubilizing and N-fixing bacteria inoculations in wheat (*Triticum aestivum* L.). Turk J Agric For 37:734–743

Burg MB, Ferraris JD, Dmitreva NI (2007) Cellular response to hyperosmotic stresses. Physiol Rev 87:1441–1474

Capone DG (2002) Microbial nitrogen cycle. In: Aurstic CJ, Crawford RL (eds) Manual of environmental microbiology. ASM Press, Washington, DC, p 439

Cassán F, Vanderleyden J, Spaepen S (2014) Physiological and agronomical aspects of phytohormone production by model plant-growth-promoting rhizobacteria (PGPR) belonging to the genus *Azospirillum*. J Plant Growth Regul 33:440–459

Cavagnaro TR, Bender SF, Asghari HR, van der Heijden MG (2015) The role of arbuscular mycorrhizas in reducing soil nutrient loss. Trends Plant Sci 20(5):283–290

Chatterjee P, Samaddar S, Anandham R, Kang Y, Kim K, Selvakumar G, Sa T (2017) Beneficial soil bacterium *Pseudomonas frederiksbergensis* OS261 augments salt tolerance and promotes red pepper plant growth. Front Plant Sci 8:1–9

Cheng Z, Park E, Glick BR (2007) 1-Aminocyclopropane-1-carboxylate deaminase from *Pseudomonas putida* UW4 facilitates the growth of canola in the presence of salt. Can J Microbiol 53:912–918

Cho K, Toler H, Lee J, Ownley B, Stutz JC, Moore JL, Auge RM (2006) Mycorrhizal symbiosis and response of sorghum plants to combined drought and salinity stresses. J Plant Physiol 163:517–528

Chookietwattana K, Maneewan K (2012) Selection of efficient salt-tolerant bacteria containing ACC deaminase for promotion of tomato growth under salinity stress. Soil Environ 31:30–36

Colla G, Rouphael Y, Cardarelli M, Tullio M, Rivera CM, Rea E (2008) Alleviation of salt stress by arbuscular mycorrhizal in zucchini plants grown at lowand high phosphorus concentration. Biol Fertil Soil 44:501–509

Contreras-Cornejo HA, Macias-Rodriguez L, Cortes-Penagos C, Lopez-Bucio J (2009) *Trichoderma virens*, a plant beneficial fungus, enhances biomass production and promotes lateral root growth through an auxin-dependent mechanism in Arabidopsis. Plant Physiol 149:1579–1592

Crafts CB, Miller CO (1974) Detection and identification of cytokinins produced by mycorrhizal fungi. Plant Physiol 54:586–588

Da Costa MS, Santos H, Galinski EA (1998) An overview of the role and diversity of compatible solutes in bacteria and Archaea. In: Biotechnology of extremophiles. Springer, Berlin, pp 117–153

Daei G, Ardekani M, Rejali F, Teimuri S, Miransari M (2009) Alleviation of salinity stress on wheat yield, yield components, and nutrient uptake using arbuscular mycorrhizal fungi under field conditions. J Plant Physiol 166:217–225

Das S, De M, Ray R, Ganguly D, Kumar Jana T, De TK (2011) Salt tolerant culturable microbes accessible in the soil of the Sundarban mangrove forest, India. Open J Ecol 1:35

De Almeida AMM, Gomes VFF, Mendes PF, de Lacerda CF, Freitas ED (2016) Influence of salinity on the development of the banana colonized by arbuscular mycorrhizal fungi. Rev Cienc Agron 47:421–428

Deinlein U, Stephan AB, Horie T, Luo W, Xu G, Schroeder JI (2014) Plant salt-tolerance mechanisms. Trends Plant Sci 19:371–379

Del-Saz NF, Romero-Munar A, Alonso D, Aroca R, Baraza E, Flexas J, Ribas-Carbo M (2017) Respiratory ATP cost and benefit of arbuscular mycorrhizal symbiosis with Nicotiana tabacum at different growth stages and under salinity. J Plant Physiol 218:243–248

de-Salamon IEG, Hynes RK, Nelson LM (2001) Cytokin in production by plant growth promoting rhizobacteria and selected mutants. Can J Microbiol 47:404–411

Diatta ILD, Kane A, Agbangba CE, Sagna M, Diouf D, Aberlenc-Bertossi F, Duval Y, Borgel A, Sane D (2014) Inoculation with arbuscular mycorrhizal fungi improves seedlings growth of two sahelian date palm cultivars (*Phoenix dactylifera* L., cv. Nakhla hamra and cv. Tijib) under salinity stresses. Adv Biosci Biotechnol 5:64–72

Dinnbier U, Limpinsel E, Schmid R, Bakker EP (1988) Transient accumulation of potassium glutamate and its replacement by trehalose during adaptation of growing cells of *Escherichia coli* K-12 to elevated sodium chloride concentrations. Arch Microbiol 150:348–357

Diouf D, Duponnois R, Ba AT, Neyra M, Lesueur D (2005) Symbiosis of *Acacia auriculiformis* and *Acacia mangium* with mycorrhizal fungi and *Bradyrhizobium* spp. improves salt tolerance in greenhouse conditions. Funct Plant Biol 32:1143–1152

Elhindi K, Sharaf El Din A, Abdel-Salam E, Elgorban A (2016) Amelioration of salinity stress in different basil (*Ocimum basilicum* L.) varieties by vesicular-arbuscular mycorrhizal fungi. Acta Agriculturae Scandinavica. Sect B-Soil Plant Sci 66:583–592

Elhindi KM, El-Din AS, Elgorban AM (2017) The impact of arbuscular mycorrhizal fungi in mitigating salt-induced adverse effects in sweet basil (*Ocimum basilicum* L.). Saudi J Biol Sci (1):170–179

Estrada B, Aroca R, Maathuis FJ, Barea JM, Ruiz-Lozano JM (2013) Arbuscular mycorrhizal fungi native from a Mediterranean saline area enhance maize tolerance to salinity through improved ion homeostasis. Plant Cell Environ 36:1771–1782

Estrada-Luna AA, Davies FT (2003) Arbuscular mycorrhizal fungi influence water relations, gas exchange, abscisic acid and growth of micropropagated chile ancho pepper (*Capsicum annuum*) plantlets during acclimatization and postacclimatization. J Plant Physiol 160:1073–1083

Evelin H, Kapoor R, Giri B (2009) Arbuscular mycorrhizal fungi in alleviation of salt stress: a review. Ann Bot 104(7):1263–1280

Evelin H, Giri B, Kapoor R (2012) Contribution of *Glomus intraradices* inoculation to nutrient acquisition and mitigation of ionic imbalance in NaCl-stressed *Trigonella foenum-graecum*. Mycorrhiza 22:203–217

Evelin H, Giri B, Kapoor R (2013) Ultrastructural evidence for AMF mediated salt stress mitigation in *Trigonella foenum-graecum*. Mycorrhiza 23:71–86

Felici C, Vettori L, Giraldi E, Forino LMC, Toffanin A, Tagliasacchi AM, Nuti M (2008) Single and co-inoculation of *Bacillus subtilis* and *Azospirillum brasilense* on *Lycopersicum esculentum*: effects on plant growth and rhizosphere microbial community. Appl Soil Ecol 40:260–270

Feng G, Zhang F, Li X, Tian C, Tang C, Rengel Z (2002) Improved tolerance of maize plants to salt stress by arbuscular mycorrhiza is related to higher accumulation of soluble sugars in roots. Mycorrhiza 12(4):185–190

Feng C, Fang-yan C, Yu-bin T (2014) Isolation, identification of a halotolerant acid red B degrading strain and its decolorization performance. Apcbee Procedia 9:131–139

Finlay RD (2007) Ecological aspects ofmycorrhizal symbiosis with special emphasis on the functional diversity of interactions involving the extraradical mycelium. J Exp Bot 59:1115–1126

Firth E, Carpenter SD, Sørensen HL, Collins RE, Deming JW (2016) Bacterial use of choline to tolerate salinity shifts in sea-ice brines. Elementa Sci Anthropocene 4:1–15

Franche C, Lindstrom K, Elmerich C (2009) Nitrogen-fixing bacteria associated with leguminous and non-leguminous plants. Plant Soil 321:35–59

Fukami J, de la Osa C, Ollero FJ, Megías M, Hungria M (2018) Co-inoculation of maize with *Azospirillum brasilense* and *Rhizobium tropici* as a strategy to mitigate salinity stress. Funct Plant Biol 45:328–339

Galinski EA (1995) Osmoadaptation in bacteria. Adv Microb Physiol 37:273–228

Galinski EA, Pfeiffer HP, Truper HG (1985) 1, 4, 5, 6-Tetrahydro- 2-methyl-4-pyrimidinecarboxylic acid. A novel cyclic amino acid from halophilic phototrophic bacteria of the genus *Ectothiorhodospira*. Eur J Biochem 149(135):139

Gamalero E, Berta G, Massa N, Glick BR, Lingua G (2010) Interactions between *Pseudomonas putida* UW4 and *Gigaspora rosea* BEG9 and their consequences for the growth of cucumber under salt-stress conditions. J Appl Microbiol 108:236–245

García-Garrido JM, Ocampo JA (2002) Regulation of the plant defence response in arbuscular mycorrhizal symbiosis. J Exp Bot 53(373):1377–1386

Garg N, Bhandari P (2012) Influence of cadmium stress and arbuscular mycorrhizal fungi on nodule senescence in *Cajanus cajan* (L.) Millsp. Int J Phytoremediation 14:62–74

Garg N, Bhandari P (2016) Silicon nutrition and mycorrhizal inoculations improve growth, nutrient status, K^+/Na^+ ratio and yield of *Cicer arietinum* L. genotypes under salinity stress. Plant Growth Regul 78:371–387

Garg N, Chandel S (2011) Effect of mycorrhizal inoculation on growth, nitrogen fixation, and nutrient uptake in *Cicer arietinum* (L.) under salt stress. Turk J Agric For 35:205–214

Garg N, Chandel S (2015) Role of arbuscular mycorrhiza in arresting reactive oxygen species (ROS) and strengthening antioxidant defense in *Cajanus cajan* (L.) Millsp. nodules under salinity (NaCl) and cadmium (Cd) stress. Plant Growth Regul 75:521–534

Garg N, Manchanda G (2008) Effect of arbuscular mycorrhizal inoculation on salt-induced nodule senescence in *Cajanus cajan* (Pigeonpea). J Plant Growth Regul 27:115–124

Garg N, Pandey R (2015) Effectiveness of native and exotic arbuscular mycorrhizal fungi on nutrient uptake and ion homeostasis in saltstressed *Cajanus cajan* L. (Millsp.) genotypes. Mycorrhiza 25:165–180

Garg N, Pandey R (2016) High effectiveness of exotic arbuscular mycorrhizal fungi is reflected in improved rhizobial symbiosis and trehalose turnover in *Cajanus cajan* genotypes grown under salinity stress. Fungal Ecol 21:57–67

Ghoulam CA, Foursy A, Fares K (2002) Effect of salt stress on growth, inorganic ions and proline accumulation in relation to osmotic adjustment in five sugar beet cultivars. Environ Exp Bot 47:39–50

Giri B, Mukerji KG (2004) Mycorrhizal inoculant alleviates salt stress in *Sesbania aegyptiaca* and *Sesbania grandiflora* under field conditions: evidence for reduced sodium and improved magnesium uptake. Mycorrhiza 14:307–312

Giri B, Kapoor R, Mukerji KG (2002) VA mycorrhizal techniques/VAM technology in establishment of plants under salinity stress conditions. In: Mukerji KG, Manoharachary C, Chamola BP (eds) Techniques in mycorrhizal studies. Kluwer, Dordrecht, pp 313–327

Giri B, Kapoor R, Mukerji KG (2007) Improved tolerance of *Acacia nilotica* to salt stress by arbuscular mycorrhiza, Glomus fasciculatum may be partly related to elevated K/Na ratios in root and shoot tissues. Microb Ecol 54(4):753–760

Glick BR (2010) Using soil bacteria to facilitate phytoremediation. Biotechnol Adv 28:367–374

Glick BR, Karaturovic DM, Newell PC (1995) A novel procedure for rapid isolation of plant growth promoting *Pseudomonas*. Can J Microbiol 41:533–536

Glick BR, Patten CL, Holgin G, Penrose DM (1999) Biochemical and genetic mechanisms used by plant growth promoting bacteria. Imperial College Press, London, p 267

Glick BR, Todorovic B, Czarny J, Cheng Z, Duan J, McConkey B (2007) Promotion of plant growth by bacterial ACC deaminase. Crit Rev Plant Sci 26:227–242

Gouzou L, Burtin G, Philippy R, Bartoli F, Heulin T (1993) Effect of inoculation with *Bacillus polymyxa* on soil aggregation in the wheat rhizosphere: preliminary examination. Geoderma 56:479–491

Govindarajan M, Balandreau J, Muthukumarasamy R, Revathi G, Lakshminarasimhan C (2006) Improved yield of micropropagated sugarcane following inoculation by endophytic *Burkholderia vietnamiensis*. Plant Soil 280:239–252

Govindarajan M, Kwon SW, Weon HY (2007) Isolation, molecular characterization and growth-promoting activities of endophytic sugarcane diazotroph *Klebsiella* sp. GR9. World J Microbiol Biotechnol 23:997–1006

Govindarajulu M, Pfeffer PE, Jin HR et al (2005) Nitrogen transfer in the arbuscular mycorrhizal symbiosis. Nature 435:819–823

Gunde-Cimerman N, Zalar P (2014) Extremely halotolerant and halophilic fungi inhabit brine in solar salterns around the globe. Food Technol Biotechnol 52:170–179

Gupta B, Huang B (2014) Mechanism of salinity tolerance in plants: physiological, biochemical, and molecular characterization. Int J Genomics 2014:701596

Gupta V, Satyanarayana T, Garg S (2000) General aspects of mycorrhiza. In: Mukerji KG, Chamola BP, Singh JE (eds) Mycorrhizal biology. Kluwer Academic/Plenum, New York, pp 27–44

Habib SH, Kausar H, Saud HM (2016) Plant growth-promoting rhizobacteria enhance salinity stress tolerance in okra through ROS-scavenging enzymes. Biomed Res Int 1–10

Hajbagheri S, Enteshari S (2011) Effects of mycorrhizal fungi on photosynthetic pigments, root mycorrhizal colonization and morphological characteristics of salt stressed *Ocimum basilicum* L. Iran J Plant Physiol 1:215–222

Hajiboland R, Aliasgharzadeh N, Laiegh SF, Poschenrieder C (2010) Colonization with arbuscular mycorrhizal fungi improves salinity tolerance of tomato (*Solanum lycopersicum* L.) plants. Plant Soil 331:313–327

Han HS, Lee KD (2005) Plant growth promoting rhizobacteria effect on antioxidant status, photosynthesis, mineral uptake and growth of lettuce under soil salinity. Res J Agric Biol Sci 1:210–215

Hasegawa PM, Bressan RA, Zhu JK, Bohnert HJ (2000) Plant cellular and molecular responses to high salinity. Plant Mol Biol 51:463–499

Hashem A, Abd-Allah EF, Alqarawi AA, Al Huqail AA, Egamberdieva D, Wirth S (2016a) Alleviation of cadmium stress in *Solanum lycopersicum* L. by arbuscular mycorrhizal fungi via induction of acquired systemic tolerance. Saudi J Biol Sci 23:272–281

Hashem A, Abd Allah EF, Alqarawi AA, Al-Huqail AA, Shah MA (2016b) Induction of osmoregulation and modulation of salt stress in *Acacia gerrardii* benth. by arbuscular mycorrhizal fungi and *Bacillus subtilis* (BERA 71). Biomed Res Int 2016:1–11

Hashem A, Abd Allah EF, Alqarawi AA, Al-Huqail AA, Wirth S, Egamberdieva D (2016c) The interaction between arbuscular mycorrhizal fungi and endophytic bacteria enhances plant growth of *Acacia gerrardii* under salt stress. Front Microbiol 7:1–15

Hashem A, Alqarawi AA, Radhakrishnan R, Al-Arjani A-BF, Aldehaish HA, Egamberdieva D, Abd_Allah EF (2018) Arbuscular mycorrhizal fungi regulate the oxidative system, hormones and ionic equilibrium to trigger salt stress tolerance in *Cucumis sativus* L. Saudi J Biol Sci 25 (6):1102–1114

He XH, Nara K (2007) Element biofortification: can mycorrhizas potentially offer a more effective and sustainable pathway to curb human malnutrition? Trends Plant Sci 12:331–333

He Z, He C, Zhang Z, Zou Z, Wang H (2007) Changes of antioxidative enzymes and cell membrane osmosis in tomato colonized by arbuscular mycorrhizae under NaCl stress. Colloids Surf B: Biointerfaces 59:128–133

Hildebrandt U, Janetta K, Ouziad F, Renne B, Nawrath K, Bothe H (2001) Arbuscular mycorrhizal colonization of halophytes in Central European salt marshes. Mycorrhiza 10:175–183

Hildebrandt U, Janetta K, Bothe H (2002) Towards growth of arbuscular mycorrhizal fungi independent of a plant host. Appl Environ Microbiol 68:1919–1924

Ibrahim AH, Abdel-Fattah GM, Eman FM, Abdel-Aziz MH, Shokr AE (2011) Arbuscular mycorrhizal fungi and spermine alleviate the adverse effects of salinity stress on electrolyte leakage and productivity of wheat plants. Phyton 51:261–276

Jaderlund L, Arthurson V, Granhall U, Jansson JK (2008) Specific interactions between arbuscular mycorrhizal fungi and plant growth-promoting bacteria: as revealed by different combinations. FEMS Microbiol Lett 287:174–180

Jahromi F, Aroca R, Porcel R, Ruiz-Lozano JM (2008) Influence of salinity on the in vitro development of *Glomus intraradices* and on the in vivo physiological and molecular responses of mycorrhizal lettuce plants. Microbial Ecol 55:45–53

Jeffries P, Gianinazzi S, Perotto S, Turnau K, Barea J (2003) The contribution of arbuscular mycorrhizal fungi in sustainable maintenance of plant health and soil fertility. Biol Fertil Soils 37:1–16

Jenab K, Moghimi H, Hamedi J (2016) Bioremediation of crude oil and pyrene by halotolerant Fungus *Embellisia phragmospora* isolated from saline soil. International Conference and Iranian Congress of Microbiology, August 2016, University of Tehran, Tehran, Iran

Jha Y, Subramanian RB (2013) Paddy plants inoculated with PGPR show better growth physiology and nutrientcontent under saline conditions. Chil J Agric Res 73:213–219

Jindal V, Atwal A, Sekhon BS, Singh R (1993) Effect of vesicular– arbuscular mycorrhizae on metabolism of moong plants under Nacl salinity. Plant Physiol Biochem 31:475–481

Joe MM, Saravanan VS, Islam MR, Sa T (2014) Development of alginate-based aggregate inoculants of M ethylobacterium sp. and a zospirillum brasilense tested under in vitro conditions to promote plant growth. J Appl Microbiol 116(2):408–423

Johannes AVV, Overbeek LSV, Elsas JDV (1997) Fate and activity of microorganisms introduced into soil. Microbiol Mol Biol Rev 61:121–135

Johansson JF, Paul LR, Finlay RD (2004) Microbial interactions in the mycorrhizosphere and their significance for sustainable agriculture. FEMS Microbiol Ecol 48:1–13

Joo HS, Chang CS (2005) Oxidant and SDS-stable alkaline protease from a halo-tolerant *Bacillus clausii* I-52. 68: 772–783

Juniper S, Abbott LK (2006) Soil salinity delays germination and limits growth of hyphae from propagules of arbuscular mycorrhizal fungi. Mycorrhiza 16:371–379

Kang SM, Khan AL, Waqas M, You YH, Kim JH, Kim JG, Hamayun M, Lee IJ (2014a) Plant growth-promoting rhizobacteria reduce adverse effects of salinity and osmotic stress by regulating phytohormones and antioxidants in Cucumis sativus. J Plant Interact 9:673–682

Kang SM, Radhakrishnan R, Khan AL, Kim MJ, Park JM, Kim BR, Shin DH, Lee IJ (2014b) Gibberellin secreting rhizobacterium, *Pseudomonas putida* H-2-3 modulates the hormonal and stress physiology of soybean to improve the plant growth under saline and drought conditions. Plant Physiol Biochem 84:115–124

Kapoor R, Sharma D, Bhatnagar AK (2008) Arbuscular mycorrhizae in micropropagation systems and their potential applications. Sci Horticult 116:227–239

Kaya C, Ashraf M, Sonmez O, Aydemir S, Tuna LA, Cullu AM (2009) The influence of arbuscular mycorrhizal colonization on key growth parameters and fruit yield of pepper plants grown at high salinity. Sci Horticult 121:1–6

Kempf B, Bremer E (1998) Uptake and synthesis of compatible solutes as microbial stress responses to highosmolality environments. Arch Microbiol 170(319):330

Koc A, Balci G, Erturk Y, Keles H, Bakoglu N, Ercisli S (2016) Influence of arbuscular mycorrhizae and plant growth promoting rhizobacteria on proline, membrane permeability and growth of strawberry (Fragaria x ananassa) under salt stress. J Appl Bot Food Qual 89:89–97

Kohler J, Caravaca F, Carrasco L, Roldan A (2006) Contribution of *Pseudomonas mendocina* and *Glomus intraradices* to aggregate stabilization and promotion of biological fertility in rhizosphere soil of lettuce plants under field conditions. Soil Use Manag 22:298–304

Kondepudi KK, Chandra TS (2011) Identification of osmolytes from a moderately halophilic amylolytic Bacillus sp. strain TSCVKK. Eur J Exp Biol 1:113–121

Kong CC, Ren CG, Li RZ, Xie ZH, Wang JP (2017) Hydrogen peroxide and strigolactones signaling are involved in alleviation of salt stress induced by arbuscular mycorrhizal fungus in *Sesbania cannabina* seedlings. J Plant Growth Regul 36:734–742

Kumar A, Sharma S, Mishra S, Dames JF (2015) Arbuscular mycorrhizal inoculation improves growth and antioxidative response of *Jatropha curcas* (L.) under Na2SO4 salt stress. Plant Biosyst 149(2):260–269

Kumar K, Amaresan N, Madhuri K (2017) Alleviation of the adverse effect of salinity stress by inoculation of plant growth promoting rhizobacteria isolated from hot humid tropical climate. Ecol Eng 102:361–366

Kunte HJ (2012) Osmoregulation in halophilic bacteria. Extremophiles II:263–277

Lalnunhlimi S, Krishnaswamy V (2016) Decolorization of azo dyes (direct blue 151 and direct red 31) by moderately alkaliphilic bacterial consortium. Braz J Microbiol 47:39–46

Landwehr M, Wilde P, Tóth T, Biró B, Hildebrandt U, Nawrath K, Bothe H (2002) The arbuscular mycorrhizal fungus Glomus geosporum in European saline, sodic and gypsum soils. Mycorrhiza 12(4):199–211

Larkin RP, Fravel DR (1998) Efficacy of various fungal and bacterial biocontrol organisms for control of fusarium wilt of tomato. Plant Dis 82:1022–1028

Lee Y, Krishnamoorthy R, Selvakumar G, Kim K, Sa T (2015) Alleviation of salt stress in maize plant by co-inoculation of arbuscular mycorrhizal fungi and *Methylobacterium oryzae* CBMB20. J Korean Soc Appl Biol Chem 58:533–540

Li H, Lei P, Pang X, Li S, Xu H, Xu Z, Feng X (2017) Enhanced tolerance to salt stress in canola (*Brassica napus* L.) seedlings inoculated with the halotolerant Enterobacter cloacae HSNJ4. Appl Soil Ecol 119:26–34

Lin J, Wang Y, Sun S, Mu C, Yan X (2017) Effects of arbuscular mycorrhizal fungi on the growth, photosynthesis and photosynthetic pigments of *Leymus chinensis* seedlings under salt-alkali stress and nitrogen deposition. Sci Total Environ 576:234–241

Linderman RG (1992) Vesicular–arbuscular mycorrhizae and soil microbial interactions. In: Bethlenfalvay GJ, Linderman RG (eds) Mycorrhizae in sustainable agriculture. ASA, Madison, pp 1–26

Liu J, He H, Vitali M, Visentin I, Charnikhova T, Haider I, Schubert A, Ruyter-Spira C, Bouwmeester HJ, Lovisolo C, Cardinale F (2015a) Osmotic stress represses strigolactone biosynthesis in *Lotus japonicus* roots: exploring the interaction between strigolactones and ABA under abiotic stress. Planta 241:1435–1451

Liu QL, Tang JC, Bai ZH, Hecker M, Giesy JP (2015b) Distribution of petroleum degrading genes and factor analysis of petroleum contaminated soil from the Dagang Oil field China. Sci Rep 5:11068. https://doi.org/10.1038/srep11068

Liu C, Dai Z, Cui M, Lu W, Sun H (2018) Arbuscular mycorrhizal fungi alleviate boron toxicity in *Puccinellia tenuiflora* under the combined stresses of salt and drought. Environ Pollut 240:557–565

Lowe SE, Jain MK, Zeikus JG (1993) Biology, ecology, and biotechnological applications of anaerobic bacteria adapted to environmental stresses in temperature, pH, salinity, or substrates. Microbiol Biotechnol Rev 57:451–509

Lucht JM, Bremer E (1994) Adaptation of Escherichia coli to high osmolarity environments: osmoregulation of the high-affinity glycine betaine transport system ProU. FEMS Microbiol Rev 14:3–20

Lugtenberg B, Kamilova F (2009) Plant growth promoting rhizobacteria. Annu Rev Microbiol 63:541–556

Manchanda G, Garg N (2011) Alleviation of salt-induced ionic, osmotic and oxidative stresses in *Cajanus Cajan* nodules by AM inoculation. Plant Biosyst 145:88–97

Mandal HK (2014) Isolation of salt tolerant strains of *Rhizobium Trifolii*. Int J Agric Food Sci Technol 5:325–332

Manga A, Diop A, Diop TA (2017) Functional diversity of mycorrhizal fungi has differential effects on salinity tolerance of *Acacia seyal* (Del.) seedlings. Open J Soil Sci 7:315–332

Mardukhi B, Rejali F, Daei G, Ardakani MR, Malakouti MJ, Miransari M (2011) Arbuscular mycorrhizas enhance nutrient uptake in different wheat genotypes at high salinity levels under field and greenhouse conditions. C R Biol 334:564–571

Mardukhi B, Rejali F, Daei G, Ardakani MR, Malakouti MJ, Miransari M (2015) Mineral uptake of mycorrhizal wheat (*Triticum aestivum* L.) under salinity stress. Commun Sci Plant Anal 46:343–357

Matamoros MA, Baird LM, Escuredo PR, Dalton DA, Minchin FR, Iturbe-Ormaetxe I, Rubio MC, Moran JF, Gordon AJ, Becana M (1999) Stress-induced legume root nodule senescence: physiological, biochemical and structural alterations. Plant Physiol 121:97–111

Mayak S, Tirosh T, Glick BR (2004) Plant growth-promoting bacteria confer resistance in tomato plants to salt stress. Plant Physiol Biochem 42:565–572

Mazhar S, Cohen JD, Hasnain S (2013) Auxin producing non-heterocystous cyanobacteria and their impact on the growth and endogenous auxin homeostasis of wheat. J Basic Microbiol 53:996–1003

Mendpara J, Parekh V, Vaghela S, Makasana A, Kunjadia PD, Sanghvi G, Vaishnav D, Dave GS (2013) Isolation and characterization of high salt tolerant bacteria from agricultural soil. Eur J Exp Biol 3:351–358

Meng L, Zhang A, Wang F, Han X, Wang D, Li S (2015) Arbuscular mycorrhizal fungi and rhizobium facilitate nitrogen uptake and transfer in soybean/maize intercropping system. Front Plant Sci 6

Meyer JR, Linderman RG (1986) Response of subterranean clover to dual inoculationwith vesicular–arbuscular mycorrhizal fungi and a plant growth-promoting bacterium, *Pseudomonas putida*. Soil Biol Biochem 18:185–190

Mitter B, Brader G, Afzal M, Compant S, Naveed M, Trognitz F, Sessitsch A (2013) Advances in elucidating beneficial interactions between plants, soil and bacteria. Adv Agron 121:381–445

Mohamed HI, Gomaa EZ (2012) Effect of plant growth promoting *Bacillus subtilis* and *Pseudomonas fluorescens* on growth and pigment composition of radish plants (*Raphanus sativus*) under NaCl stress. Photosynthetica 50:263–272

Munoz N, Soria-Diaz ME, Manyani M, Sanchez-Matamoros RC, Serrano AG, Megias M, Lascano R (2014) Structure and biological activities of lipochitooligosaccharide nodulation signals produced by *Bradyrhizobium japonicum* USDA 138 under saline and osmotic stress. Biol Fertil Soils 50:207–215

Muthukumarasamy R, Kang UG, Park KD, Jeon WT, Park CY, Cho YS, Kwon SW, Song J, Roh DH, Revathi G (2007) Enumeration, isolation and identification of diazotrophs from Korean wetland rice varieties grown with long-term application of N and compost and their short-term inoculation effect on rice plants. J Appl Microbiol 102:981–991

Nadeem SM, Zahair ZA, Naveed M, Asghar HN, Asghar M (2010a) Rhizobacteria capable of producing ACC-deaminase may mitigate salt stress in wheat. J Soil Sci Soc Am 74:533–542

Nadeem SM, Zahir ZA, Naveed M, Ashraf M (2010b) Microbial ACC-deaminase: prospects and applications for inducing salt tolerance in plants. Crit Rev Plant Sci 29:360–393

Nadeem SM, Ahmad M, Zahir ZA, Javaid A, Ashraf M (2014) The role of mycorrhizae and plant growth promoting rhizobacteria (PGPR) in improving crop productivity under stressful environments. Biotechnol Adv 32:429–448

Najafi A, Ardakani MR, Rejali F, Sajedi N (2012) Response of winter barley to co-inoculation with *Azotobacter* and Mycorrhiza fungi influenced by plant growth promoting rhizobacteria. Ann Biol Res 3:4002–4006

Navarro JM, Morte A, Rodríguez-Morán M, Pérez-Tornero O (2015) Physiological response of Citrus macrophylla inoculated with Arbuscular mycorrhizal fungi under salt stress. Acta Hortic 1065:1351–1358

Nazareth SW, Gaitonde S, Marbaniang T (2012) Metal resistance of halotolerant fungi from mangroves and salterns of Goa, India

Nicholson CA, Fathepure BZ (2004) Biodegradation of benzene by halophilic and halotolerant bacteria under aerobic conditions. Appl Environ Microbiol 70:1222–1225

Nishiyama R, Watanabe Y, Fujita Y, Le DT, Kojima M, Werner T, Vankova R, Yamaguchi-Shinozaki K, Shinozaki K, Kakimoto T, Sakakibara H, Schmülling T, Tran LP (2011) Analysis of cytokinin mutants and regulation of cytokinin metabolic genes reveals important regulatory roles of cytokinins in drought, salt and abscisic acid responses, and abscisic acid biosynthesis. Plant Cell 23:2169–2183

Nunkaew T, Kantachote D, Kanzaki H, Nitoda T, Ritchie RJ (2014) Effects of 5-aminolevulinic acid (ALA)-containing supernatants from selected *Rhodopseudomonas palustris* strains on rice growth under NaCl stress, with mediating effects on chlorophyll, photosynthetic electron transport and antioxidative enzymes. Electron J Biotechnol 17:4–4

Okoro CC (2010) Enhanced bioremediation of hydrocarbon contaminated mangrove swamp in the Nigerian oil-rich Niger-Delta using sea water microbial inocula amended with crude biosurfactants and micronutrients. Nature Sci 8:195–206

Oren A (1999) Bioenergetic aspects of halophilism. Microbiol Mole Biol Rev 63:334–348

Oren A (2008) Microbial life at high salt concentrations: phylogenetic and metabolic diversity. Saline Syst 15:4

Oren A, Gurevich P, Azachi M, Henis Y (1992) Microbial degradation of pollutants at high salt concentrations. Biodegradation 3:387–398

Osuji LC, Ezebuiro PE (2006) Hydrocarbon contamination of a typical mangrove floor in Niger Delta, Nigeria. Int J Environ Sci Tech 3:313–332

Palacio-Rodríguez R, Coria-Arellano JL, López-Bucio J, Sánchez-Salas J, Muro-Pérez G, Castañeda-Gaytán G, Sáenz-Mata J (2017) Halophilic rhizobacteria from *Distichlis spicata* promote growth and improve salt tolerance in heterologous plant hosts. Symbiosis 73:179–189

Palaniyandi SA, Damodharan K, Yang SH, Suh JW (2014) Streptomyces sp. strain PGPA39 alleviates salt stress and promotes growth of 'Micro-Tom' tomato plants. J Appl Microbiol 117:766e773

Pandey R, Garg N (2017) High effectiveness of *Rhizophagus irregularis* is linked to superior modulation of antioxidant defence mechanisms in *Cajanus cajan* (L.) Millsp. genotypes grown under salinity stress. Mycorrhiza 27:669–682

Parshetti G, Kalme S, Saratale G, Govindwar S (2006) Biodegradation of malachite green by Kocuria rosea MTCC 1532. Acta Chim Slov 53(4):492–498

Patel VK, Srivastava R, Sharma A, Srivastava AK, Singh S, Srivastava AK, Kashyap PL, Chakdar H, Pandiyan K, Kalra A, Saxena AK (2018) Halotolerant *Exiguobacterium profundum* PHM11 tolerate salinity by accumulating L-proline and fine-tuning gene expression profiles of related metabolic pathways. Front Microbiol 9:423. https://doi.org/10.3389/fmicb.2018.00423

Pedranzani H, Rodríguez-Rivera M, Gutiérrez M, Porcel R, Hause B, Ruiz-Lozano JM (2016) Arbuscular mycorrhizal symbiosis regulates physiology and performance of *Digitaria eriantha* plants subjected to abiotic stresses by modulating antioxidant and jasmonate levels. Mycorrhiza 26:141–152

Perugini M, Visciano P, Giammarino A, Manera M, Di Nardo W, Amorena M (2007) Polycyclic aromatic hydrocarbons in marine organisms from the Adriatic Sea, Italy. Chemosphere 66:1904–1910

Petrovic U, Gunde-Cimerman N, Plemenitaš A (2002) Cellular responses to environmental salinity in the halophilic black yeast *Hortaea werneckii*. Mol Microbiol 45:665–672

Pinedo I, Ledger T, Greve M, Poupin MJ (2015) *Burkholderia phytofirmans* PsJN induces long-term metabolic and transcriptional changes involved in *Arabidopsis thaliana* salt tolerance. Front Plant Sci 6(466):466

Plemenitas A, Lenassi M, Konte T, Kejzar A, Zajc J, Gostincar C, Gunde-Cimerman N (2014) Adaptation to high salt concentrations in halotolerant/halophilic fungi: a molecular perspective. Front Microbiol 5:199

Plenchette C, Duponnois R (2005) Growth response of the saltbush *Atriplex nummularia* L. to inoculation with the arbuscular mycorrhizal fungus *Glomus intraradices*. J Arid Environ 61:535–540

Pollastri S, Savvides A, Pesando M, Lumini E, Volpe MG, Ozudogru EA, Faccio A, De Cunzo F, Michelozzi M, Lambardi M, Fotopoulos V (2017) Impact of two arbuscular mycorrhizal fungi on *Arundo donax L.* response to salt stress. Planta 9:1–3

Porcel R, Ruiz-Lozano JM (2004) Arbuscular mycorrhizal influence on leaf water potential, solute accumulation, and oxidative stress in soybean plants subjected to drought stress. J Exp Bot 55:1743–1750

Porcel R, Aroca R, Azcón R, Ruiz-Lozano JM (2006) PIP aquaporin gene expression in arbuscular mycorrhizal *Glycine max* and *Lactuca sativa* plants in relation to drought stress tolerance. Plant Mol Biol 60:389–404

Porcel R, Aroca R, Ruíz-Lozano JM (2012) Salinity stress alleviation using arbuscular mycorrhizal fungi. A review. Agron Sustain Dev 32:181–200

Porcel R, Redondo-Gomez S, Mateos-Naranjo E, Aroca R, Garcia R, Ruiz-Lozano JM (2015) Arbuscular mycorrhizal symbiosis ameliorates the optimum quantum yield of photosystem II and reduces non-photochemical quenching in rice plants subjected to salt stress. J Plant Physiol 185:75–83

Porcel R, Aroca R, Azcon R, Ruiz-Lozano JM (2016) Regulation of cation transporter genes by the arbuscular mycorrhizal symbiosis in rice plants subjected to salinity suggests improved salt tolerance due to reduced Na+ root-to-shoot distribution. Mycorrhiza 26:673–684

Porras-Soriano A, Soriano-Martín ML, Porras-Piedra A, Azcón R (2009) Arbuscular mycorrhizal fungi increased growth, nutrient uptake and tolerance to salinity in olive trees under nursery conditions. J Plant Physiol 166(13):1350–1359

Qin X, Tang JC, Li DS, Zhang QM (2012) Effect of salinity on the bioremediation of petroleum hydrocarbons in a saline-alkaline soil. Lett Appl Microbiol 55:210–217

Querejeta JI, Egerton-Warburton LM, Allen MF (2007) Hydraulic lift may buffer rhizosphere hyphae against the negative effects of severe soil drying in a California Oak savanna. Soil Biol Biochem 39:409–417

Qurashi AW, Sabri AN (2012a) Bacterial exopolysaccharide and biofilm formation stimulate chickpea growth and soil aggregation under salt stress. Braz J Microbiol 11:83–91

Qurashi AW, Sabri AN (2012b) Bacterial exopolysaccharide and biofilm formation stimulate chickpea growth and soil aggregation under salt stress. Braz J Microbiol 43:1183–1191

Rabie GH, Almadini AM (2005) Role of bioinoculants in development of salt-tolerance of *Vicia faba* plants under salinity stress. Afr J Biotechnol 4:210–222

Rabie GH, Aboul-Nasr MB, Al-Humiany A (2005) Increased salinity tolerance of cowpea plants by dual inoculation of an arbuscular mycorrhizal fungus *Glomus clarum* and a nitrogen-fixer *Azospirillum brasilense*. Microbiology 33:51–60

Raja P, Uma S, Gopal H, Govindarajan K (2006) Impact of bio inoculants consortium on rice root exudates, biological nitrogen fixation and plant growth. J Biol Sci 6:815–823

Rajput L, Imran A, Mubeen F, Hafeez FY (2013) Salt-tolerant PGPR strain *Planococcus rifietoensis* promotes the growth and yield of wheat (*Triticum aestivum* L.) cultivated in saline soil. Pak J Bot 45:1955–1962

Rasooli I, SDA A, Borna H, Barchini KA (2008) A thermostable Î±-amylase producing natural variant of Bacillus spp. isolated from soil in Iran. Am J Agric Biol Sci 3:591–596

Reina-Bueno M, Argandoña M, Salvador M, Rodriguez-Moya J, Iglesias-Guerra F, Csonka LN, Nieto JJ, Vargas C (2012) Role of trehalose in salinity and temperature tolerance in the model halophilic bacterium *Chromohalobacter salexigens*. PLoS One 7:33587

Ren CG, Bai YJ, Kong CC, Bian B, Xie ZH (2016) Synergistic interactions between salt-tolerant rhizobia and arbuscular mycorrhizal fungi on salinity tolerance of *Sesbania cannabina* plant. J Plant Growth Regul 35:1098–1107

Ruiz-Lozano JM (2003) Arbuscular mycorrhizal symbiosis and alleviation of osmotic stress: new perspectives for molecular studies. Mycorrhiza 13:309–317

Ruiz-Lozano JM, Azcon R (2000) Symbiotic efficiency and infectivity of an autochthonous arbuscular mycorrhizal *Glomus* sp. from saline soils and *Glomus deserticola* under salinity. Mycorrhiza 10:137–143

Ruiz-Sanchez A, Armada E, Munoz Y, Garcia IE, Aroca R, Ruiz-Lozano JM, Azcon R (2011) *Azospirillum* and arbuscular mycorrhizal colonization enhance rice growth and physiological traits under well-watered and drought conditions. J Plant Physiol 168:1031–1037

Ruyter-Spira C, Al-Babili S, van der Krol S, Bouwmeester H (2013) The biology of strigolactones. Trends Plant Sci 18:72–83

Sahoo RK, Gill SS, Tuteja N (2012) Pea DNA helicase 45 promotes salinity stress tolerance in IR64 rice with improved yield. Plant Signal Behav 7:1042–1046

Salah UM, Zhou JT, Qu YY, Guo JB, Wang P, Zhao L (2007) Biodecolorization of azo dye acid red B under high salinity condition. Bull Environ Contam Toxicol 79:440–444

Sanan-Mishra N, Pham XH, Sopory SK, Tuteja N (2005) Pea DNA helicase 45 overexpression in tobacco confers high salinity tolerance without affecting yield. Proc Natl Acad Sci U S A 102:509–514

Sandhya V, Ali SZ (2015) The production of exopolysaccharide by Pseudomonas putida GAP-P45 under various abiotic stress conditions and its role in soil aggregation. Microbiology 84:512–519

Sannazzaro AI, Ruiz OA, Alberto EO, Menendez AB (2006) Alleviation of salt stress in Lotus glaber by *Glomus intraradices*. Plant Soil 285:279–287

Sannazzaro AI, Echeverria M, Alberto EO, Ruiz OA, Menendez AB (2007) Modulation of polyamine balance in *Lotus glaber* by salinity and arbuscular mycorrhiza. Plant Physiol Biochem 45:39–46

Sarkar A, Ghosh PK, Pramanik K, Mitra S, Soren T, Pandey S, Mondal MH, Maiti TK (2018) A halotolerant Enterobacter sp. displaying ACC deaminase activity promotes rice seedling growth under salt stress. Res Microbiol 169:20–32

Sarwat M, Hashem A, Ahanger MA, Abd Allah EF, Alqarawi AA, Alyemeni MN, Ahmad P, Gucel S (2016) Mitigation of NaCl stress by arbuscular mycorrhizal fungi through the modulation of osmolytes, antioxidants and secondary metabolites in mustard (*Brassica juncea* L.) plants. Front Plant Sci 7:869. https://doi.org/10.3389/fpls.2016.00869

Saum SH, Müller V (2007) Salinity-dependent switching of osmolyte strategies in a moderately halophilic bacterium: glutamate induces proline biosynthesis in *Halobacillus halophilus*. J Bacteriol 189:6968–6975

Selvakumar G, Shagol CC, Kim K, Han S, Sa T (2018) Spore associated bacteria regulates maize root K^+/Na^+ ion homeostasis to promote salinity tolerance during arbuscular mycorrhizal symbiosis. BMC Plant Biol 18:109. https://doi.org/10.1186/s12870-018-1317-2

Sen S, Chandrasekhar CN (2015) Effect of PGPR on enzymatic activities of rice (*Oryza sativa* L.) under salt stress. Asian J Plant Sci Rese 5:44–48

Sengupta A, Chaudhuri S (1990) Vesicular arbuscular mycorrhiza (VAM) in pioneer salt marsh plants of the Ganges river delta in West Bengal (India). Plant Soil 122:111–113

Shahid M, Akram MS, Khan MA, Zubair M, Shah SM, Ismail M, Shabir G, Basheer S, Aslam K, Tariq M (2018) A phytobeneficial strain *Planomicrobium* sp. MSSA-10 triggered oxidative stress responsive mechanisms and regulated the growth of pea plants under induced saline environment. J Appl Microbiol 124:1566–1579

Shakirova FM, Sakhabutdinova AR, Bezrukova MV, Fatkhutdinova RA, Fatkhutdinova DR (2003) Changes in the hormonal status of wheat seedlings induced by salicylic acid and salinity. Plant Sci 164:317–322

Sharifi M, Ghorbanli M, Ebrahimzadeh H (2007) Improved growth of salinity-stressed soybean after inoculation with salt pre-treated mycorrhizal fungi. J Plant Physiol 164:1144–1151

Sharma S, Kulkarni J, Jha B (2016) Halotolerant rhizobacteria promote growth and enhance salinity tolerance in peanut. Front Microbiol. 13 7:1600

Sharma N, Aggarwal A, Yadav K (2017) Arbuscular mycorrhizal fungi enhance growth, physiological parameters and yield of salt stressed *Phaseolus mungo* (L.) Hepper. Eur J Environ Sci 7. doi: https://doi.org/10.14712/23361964.2017.1

Sheng M, Tang M, Chen H, Yang BW, Zhang FF, Huang YH (2008) Influence of arbuscular mycorrhizae on photosynthesis and water status of maize plants under salt stress. Mycorrhiza 18:287–296

Shin W, Siddikee MA, Joe MM, Benson A, Kim K, Selvakumar G, Kang Y, Jeon S, Samaddar S, Chatterjee P, Walitang D (2016) Halotolerant plant growth promoting bacteria mediated salinity stress amelioration in plants. Korean J Soil Sci Fertil 49:355–367

Shivan P, Jayaraman G (2009) Production of extracellular protease from halotolerant bacterium, *Bacillus aquimaris* strain VITP4 isolated from Kumta coast. Process Biochem 44:1088–1094

Singh RP, Jha PN (2017a) Analysis of fatty acid composition of PGPR Klebsiella sp. SBP-8 and its role in ameliorating salt stress in wheat. Symbiosis 73:213–222

Singh RP, Jha PN (2017b) The PGPR *Stenotrophomonas maltophilia* SBP-9 augments resistance against biotic and abiotic stress in wheat plants. Front Microbiol 8:1945. https://doi.org/10.3389/fmicb.2017.01945

Singh N, Samajpati N, Paul AK (2011) Dual inoculation of salt tolerant Bradyrhizobium and Glomus mosseae for improvement of *Vigna radiata* L. cultivation in saline areas of West Bengal, India. Agric Sci 2:413–423

Singh RP, Jha P, Jha PN (2015) The plant-growth-promoting bacterium Klebsiella sp. SBP-8 confers induced systemic tolerance in wheat (*Triticum aestivum*) under salt stress. J Plant Physiol 184:57–67

Soliman AS, Shanan NT, Massoud ON, Swelim DM (2012) Improving salinity tolerance of *Acacia saligna* (Labill.) plant by arbuscular mycorrhizal fungi and Rhizobium inoculation. Afr J Biotechnol 11:1259–1266

Sowmya M, Rejula MP, Rejith PG, Mohan M, Karuppiah M, Hatha AM (2014) Heavy metal tolerant halophilic bacteria from Vembanad Lake as possible source for bioremediation of lead and cadmium. J Environ Biol 35:655

Szabados L, Savouré A (2010) Proline: a multifunctional amino acid. Trends Plant Sci 15:89–97

Tabak HH, Lens P, van Hullebusch ED, Dejonghe W (2005) Developments in bioremediation of soils and sediments polluted with metals and radionuclides. Microbial processes and mechanisms affecting bioremediation of metal contamination and influencing metal toxicity and transport. Rev Environ Sci Biotechnol 4:115–156

Tajrishi MM, Vaid N, Tuteja R, Tuteja N (2011) Overexpression of a pea DNA helicase 45 in bacteria confers salinity stress tolerance. Plant Signal Behav 6:1271–1275

Tan L, Qu YY, Zhou JT, Li A, Gou M (2009) Identification and characteristics of a novel salt-tolerant *Exiguobacterium* sp. for azo dyes decolorization. Appl Biochem Biotechnol 159:728–738

Tester M, Davenport R (2003) Na^+ tolerant and Na^+ transport in higher plants. Ann Bot 91:503–527

Tippannavar CM, Venkataramana M, Reddy V, Rajashekara E (1989) Tolerance of *Azotobacter chroococcum* strains to different salt concentrations and pH levels. Farm Sys 5:24–28

Tiquia SM, Davis D, Hadid H, Kasparian S, Ismail M, Ahly S, Shim J, Singh S, Murray KS (2007) Halophilic and halotolerant bacteria from river waters and shallow groundwater along the Rouge River of southeastern Michigan. Environ Technol 28:297–307

Tiwari S, Prasad V, Chauhan PS, Lata C (2017) Bacillus amyloliquefaciens confers tolerance to various abiotic stresses and modulates plant response to phytohormones through osmoprotection and gene expression regulation in rice. Front Plant Sci. 29 8:1510. https://doi.org/10.3389/fpls.2017.01510

Torres MA, Dangl JL (2005) Functions of the respiratory burst oxidase in biotic interactions, abiotic stress and development. Curr Opin Plant Biol 8:397–403

Tripathi AK, Nagarajan T, Verma SC, Le Rudulier D (2002) Inhibition of biosynthesis and activity of nitrogenase in *Azospirillum brasilense* Sp7 under salinity stress. Curr Microbiol 44:363–367

Turan S, Cornish K, Kumar S (2012) Salinity tolerance in plants: breeding and genetic engineering. AJCS 6:1337–1348

Turk M, Mejanelle L, Sentjurc M, Grimalt JO, Gunde-Cimerman N, Plemenitas A (2004) Salt-induced changes in lipid composition and membrane fluidity of halophilic yeast-like melanized fungi. Extremophiles 8:53–61

Turk M, Abramovic Z, Plemenitas A, Gunde-Cimerman N (2007) Salt stress and plasma-membrane fluidity in selected extremophilic yeasts and yeast-like fungi. FEMS Yeast Res 7:550–557

Tuteja N (2007) Mechanisms of high salinity tolerance in plants. Methods Enzymol 428:419–438

Upadhyay SK, Maurya SK, Singh DP (2012) Salinity tolerance in free living plant growth promoting rhizobacteria. Indian J Sci Res 3:73–78

Ventosa A, Nieto JJ, Oren A (1998) Biology of moderately halophilic aerobic bacteria. Microbiol Mol Biol Rev 62:504–544

Vimal SR, Patel VK, Singh JS (2018) Plant growth promoting *Curtobacterium albidum* strain SRV4: an agriculturally important microbe to alleviate salinity stress in paddy plants. Ecol Indic. https://doi.org/10.1016/j.ecolind.2018.05.014

Vishnivetskaya TA, Sophia K, Tiedje JM (2009) The Exiguobacterium genus: biodiversity and biogeography. Extremophiles 13:541–555

Viterbo A, Landau U, Kim S, Chernin L, Chet I (2010) Characterization of ACC deaminase from the biocontrol and plant growth-promoting agent *Trichoderma asperellum* T203. FEMS Microbiol Lett 305:42–48

Wang FY, Liu RJ, Lin XG, Zhou JM (2004) Arbuscular mycorrhizal status of wild plants in salinealkaline soils of the Yellow River Delta. Mycorrhiza 14:133–137

Welsh DT (2000a) Ecological significance of compatible solute accumulation by micro-organisms: from single cells to global climate. FEMS Microbiol Rev 3(263):290

Welsh DT (2000b) Ecological significance of compatible solute accumulation by micro organisms: from single cells to global climate. FEMS Microbiol Rev 24:263–290

Wu QS, Zou YN, He XH (2010) Contributions of arbuscular mycorrhizal fungi to growth, photosynthesis, root morphology and ionic balance of citrus seedlings under salt stress. Acta Physiol Planta 32:297–304

Wu ZS, Peng Y, Guo L, Li C (2014) Root colonization of encapsulated Klebsiella oxytoca Rs-5 on cotton plants and its promoting growth performance under salinity stress. Eur J Soil Biol 60:81–87

Xia XJ, Wang YJ, Zhou YH, Tao Y, Mao WH, Shi K, Asami T, Chen ZX, Yu JQ (2009) Reactive oxygen species are involved in brassinosteroid-induced stress tolerance in cucumber. Plant Physiol 150:801–814

Xie XN, Yoneyama K, Yoneyama K (2010) The strigolactone story. Annu Rev Phytopathol 48:93–117

Xueming Z, Zhenping H, Yu Z, Huanshi Z, Pei Q (2014) Arbuscular mycorrhizal fungi (AMF) and phosphate-solubilizing fungus (PSF) on tolerance of beach plum (*Prunus maritima*) under salt stress. Aust J Crop Sci 8:945–950

Xun F, Xie B, Liu S, Guo C (2015) Effect of plant growth-promoting bacteria (PGPR) and arbuscular mycorrhizal fungi (AMF) inoculation on oats in saline-alkali soil contaminated by petroleum to enhance phytoremediation. Environ Sci Pollut Res 22:598–608

Yang SJ, Zhang ZL, Xue YX, Zhang ZF, Shi SY (2014) Arbuscular mycorrhizal fungi increase salt tolerance of apple seedlings. Bot Stud 55:70

Younesi O, Moradi A (2014) Effects of plant growth-promoting rhizobacterium (PGPR) and arbuscular mycorrhizal fungus (AMF) on antioxidant enzyme activities in salt-stressed bean (Phaseolus vulgaris L.). Agriculture. 2014 Mar 1 60(1):10–21

Yu SH, Ke L, Wong YS, Tam NFY (2005) Degradation of polycyclic aromatic hydrocarbons by a bacterial consortium enriched from mangrove sediments. Environ Int 31:149–154

Yusran Y, Volker R, Torsten M (2009) Effects of plant prowth-promoting rhizobacteria and *Rhizobium* on mycorrhizal development and drowth of *Paraserianthes falcataria* (L.) nielsen seedlings in two types of soils with contrasting levels of pH. In: Proc Int Plant Nutr Colloquium XVI. UC Davis: Dept Plant Sci

Zhang YF, Wang P, Yang YF, Bi Q, Tian SY, Shi XW (2011) Arbuscular mycorrhizal fungi improve reestablishment of *Leymus chinensis* in bare saline–alkaline soil: implication on vegetation restoration of extremely degraded land. J Arid Environ 75:773–778

Zhang HS, Qin CQ, Qin P (2013) Effects of inoculation of arbuscular mycorrhizal fungi and phosphate-solubilizing fungus with different proportion on P-uptake of *Castor Bean* (*Ricinus communis* L.) and rhizosphere soil enzyme activities in coastal saline soil. Chin Agric Sci Bull 29:101–108

Zhang H, Tang J, Wang L, Liu J, Gurav RG, Sun K (2016) A novel bioremediation strategy for petroleum hydrocarbon pollutants using salt tolerant *Corynebacterium variabile* HRJ4 and biochar. J Environ Sci 47:7–13

Zhou N, Zhao S, Tian CY (2017) Effect of halotolerant rhizobacteria isolated from halophytes on the growth of sugar beet (*Beta vulgaris*L.) under salt stress. FEMS Microbiol Lett 364. https://doi.org/https://doi.org/10.1093/femsle/fnx091

Zhu J, FuX KYD, Zhu JK, Jenney JFE, Adams MW, Zhu Y, Shi H, Yun DJ, Hasegawa PM, Bressan RA (2007) An enhancer mutant of Arabidopsis salt overly sensitive 3 mediates both ion homeostasis and the oxidative stress response. Mol Cell Biol 27:5214–5224

Zhu RF, Tang F, Liu J, Liu FQ, Deng XY, Chen JS (2016) Co-inoculation of arbuscular mycorrhizae and nitrogen fixing bacteria enhance alfalfa yield under saline conditions. Pak J Bot 48:763–769

MIX
Papier aus verantwortungsvollen Quellen
Paper from responsible sources
FSC® C105338

If you have any concerns about our products,
you can contact us on
ProductSafety@springernature.com

In case Publisher is established outside the EU,
the EU authorized representative is:
Springer Nature Customer Service Center GmbH
Europaplatz 3, 69115 Heidelberg, Germany

Printed by Libri Plureos GmbH
in Hamburg, Germany